T0336945

FOOD SAFETY AND QUALITY SYSTEMS IN DEVELOPING COUNTRIES

FOOD SAFETY AND QUALITY SYSTEMS IN DEVELOPING COUNTRIES

Volume III: Technical and Market Considerations

Edited by

ANDRÉ GORDON

Chairman & CEO, Technological Solutions Limited, Kingston, Jamaica, West Indies

ACADEMIC PRESS

An imprint of Elsevier

ELSEVIER

Academic Press is an imprint of Elsevier
125 London Wall, London EC2Y 5AS, United Kingdom
525 B Street, Suite 1650, San Diego, CA 92101, United States
50 Hampshire Street, 5th Floor, Cambridge, MA 02139, United States
The Boulevard, Langford Lane, Kidlington, Oxford OX5 1GB, United Kingdom

Notices
Knowledge and best practice in this field are constantly changing. As new research and experience broaden our understanding, changes in research methods, professional practices, or medical treatment may become necessary.

Practitioners and researchers must always rely on their own experience and knowledge in evaluating and using any information, methods, compounds, or experiments described herein. In using such information or methods they should be mindful of their own safety and the safety of others, including parties for whom they have a professional responsibility.

To the fullest extent of the law, neither the Publisher nor the authors, contributors, or editors, assume any liability for any injury and/or damage to persons or property as a matter of products liability, negligence or otherwise, or from any use or operation of any methods, products, instructions, or ideas contained in the material herein.

British Library Cataloguing-in-Publication Data
A catalogue record for this book is available from the British Library

Library of Congress Cataloging-in-Publication Data
A catalog record for this book is available from the Library of Congress

ISBN: 978-0-12-814272-1

For Information on all Academic Press publications
visit our website at https://www.elsevier.com/books-and-journals

Publisher: Charlotte Cockle
Acquisitions Editor: Patricia Osborn
Editorial Project Manager: Devlin Person
Production Project Manager: Surya Narayanan Jayachandran
Cover Designer: Matthew Limbert

Typeset by MPS Limited, Chennai, India

Dedication

This book is dedicated to the memory of John Fray, Erwin Jones and Pauline Gray,
my friends and mentors and U.D. Gordon, my father, all of whom were instrumental in
laying the groundwork for the work we have done over the years that has advised this work.

Contents

8. Food safety and quality considerations for cassava, a major staple containing a natural toxicant   343

Jose Jackson, Linley Chiwona-Karltun and André Gordon

9. Market & technical considerations for spices: Nutmeg & Mace case study   367

André Gordon

Index   415

List of Contributors

Brian Bedard Director, Food Safety and Animal Health, Agriteam Canada Consulting Limited, Ottawa, Canada

Linley Chiwona-Karltun Department of Urban and Rural Development, Swedish University of Agricultural Sciences, Uppsala, Sweden

Debra DeVlieger National Food Expert, Office of Regulatory Affairs, FDA, Bainbridge Island, Washington, United States

André Gordon Chairman & CEO, Technological Solutions Limited, Kingston, Jamaica, West Indies

Dianne Gordon Jamaica Bauxite Institute, Kingston, Jamaica; UWI SALISES Sustainable Rural and Agricultural Development (SRAD) Cluster, Jamaica

Jose Jackson Alliance for African Partnership, Michigan State University, East Lansing, MI, United States

Helen Kennedy Technical & Consulting Services, Southern Caribbean, Technological Solutions Limited, Port of Spain, Trinidad and Tobago

Aubrey Mendonca Department of Food Science and Human Nutrition, Iowa State University, Ames, IA, United States

Carolina Mueses AgroBioTek Dominicana, Santo Domingo de Guzmán, Distrito Nacional, República Dominicana

Rickey Ong-a-Kwie Principalis Consultants N.V., Paramaribo, Suriname

Melvin A. Pascall Department of Food Science and Technology, The Ohio State University, Columbus, OH, United States

Frank Schreurs President, FJS Consulting Group, Guelph, Ontario, Canada

Emalie Thomas-Popo Interdepartmental Microbiology Graduate Program Iowa State University, Ames, IA, United States

Akhila Vasan Manager, Food Safety at Illinois Tech, Institute for Food Safety and Health (IFSH), Illinois, United States

Rochelle Williams Technical & Regulatory Compliance, Technological Solutions Limited, Kingston, Jamaica

Acknowledgments

This book would not have been possible without the contribution of the 14 authors whose work and time has been put into creating what we believe will be an invaluable resource to the food industry, particularly in developing countries, as the world's food supply continues to become more global. The editor owes a debt of gratitude to the Elsevier team, particularly Pat who made this Volume and this series possible, Indhu and Surya whose behind the scenes work significantly improved the work and, most of all, Devlin, who worked tirelessly to guide this project to completion. Our gratitude also is due to Peta-Shea who helped with research and clerical support and an enormous debt of gratitude is owed to Dr. Dianne Gordon who not only contributed from a unique perspective as an author but was instrumental in the editing of the many and detailed chapters herein.

Technical considerations for the implementation of food safety and quality systems in developing countries

André Gordon[1], Debra DeVlieger[2], Akhila Vasan[3] and Brian Bedard[4]

[1]Chairman & CEO, Technological Solutions Limited, Kingston, Jamaica, West Indies [2]National Food Expert, Office of Regulatory Affairs, FDA, Bainbridge Island, Washington, United States [3]Manager, Food Safety at Illinois Tech, Institute for Food Safety and Health (IFSH), Illinois, United States [4]Director, Food Safety and Animal Health, Agriteam Canada Consulting Limited, Ottawa, Canada

OUTLINE

1

Part I Background to technical considerations for food safety & quality systems in developing countries

Introduction

Ensuring food safety and good quality are essential to assuring a safe and wholesome food supply, especially as the international trade in foods and ingredients continues to grow at record levels. The major issue to be addressed therefore is how to accomplish this goal such that consumers around the world are able to enjoy an increasingly diverse range of foods that deliver the nutrition that they need, without worrying about their safety. This will require uniformity in the way the global agri-food system is managed, particularly as regards the production, handling, transportation, storage, processing, packaging and delivery of food to domestic and export markets globally. In this regard, food safety and quality systems are becoming increasingly important and this sometimes raises the question: "Is the future of food safety the future of food?"

This question prompts a contemplation of the future of the world's food supply and the role that food safety will have to play in it. Quality, while not of the primacy or public health significance of food safety, will inevitably also play a critical role in consumer acceptance of foods and therefore be among the main drivers of economic value creation in the agri-food system. Because of this, Food Safety and Quality Systems (FSQS) and their effective implementation and maintenance within the food industry globally are, and will remain, major determinants of the future trajectory of food regardless of the level of sophistication, type of food being produced or the technology deployed in producing it (Fig. 1.1). As will be seen in this chapter and throughout this volume, FSQS, the variations of requirements associated with them, the many additional component systems that are being required by some markets to be allied to, and based on them, and the considerations that determine their effectiveness are increasingly at the base of both domestic and global food supply.

Consideration of food safety in developing country environments necessitates a much deeper understanding of a range of challenges, paradoxes and the need for effective crisis management. Issues of food traceability, intentional contamination, food defense, food safety culture, food authenticity and food fraud are all connected to creating a more resilient food supply chain. This chain may extend across several geographical boundaries and is likely to function more effectively when based on equitably distributed value along the chain, supporting the need for collective security and safety of the food being delivered. Among the other considerations that will impact the future of food is meeting demand for more plant-based diets, the emerging demand from consumers in developed countries and need from people in developing countries for clean foods and alternative proteins and the changing face of food technology innovation: from blockchains to cognitive cuisine and sustainable gastronomy for more sophisticated consumers. In addition to these considerations are a range of challenges facing the global agri-food system which need to be

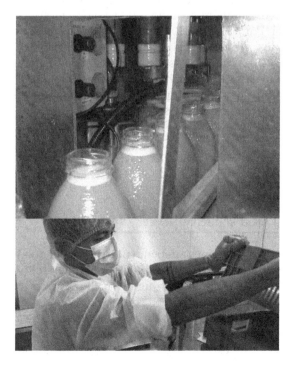

FIGURE 1.1 Automated beverage production and the application of Basic Food Safety Standards in two developing country facilities. *Source: André Gordon (2019).*

effectively addressed if the continued and increasing diversity of food and culinary options in developed country cuisine and sufficiency, quality, nutrition and security of supply are to be guaranteed for developing countries.

The authors have over 100 years of collective experience in facilitating the development of national and regional agri-food systems for domestic food security and trade across both developing and developed countries. This provides a unique perspective on the national, regional and global challenges and considerations that need attention if developed country buyers are to be able to continue to consistently source high quality safe foods from developing countries. Many factors influence the development of the agri-food sector in developing countries and their ability also to trade externally, and firms and participants on both sides of the trade need to have an understanding of these factors and related issues when food safety and quality (FS&Q) are being considered.

This chapter elaborates on the macro-level food safety and quality (FS&Q) issues raised in Volume II of this series (Gordon, 2016) and in Stier (2002a,b), Leake (2016, 2017), Henson and Jaffee (2006), Gordon (2003) and Mendonca and Gordon (2004) as considerations for increasing trade from developing to developed countries. These include the issues of access to the markets targeted which is governed by compliance with various national (Gordon, 2015a,b), and in some cases, state and provincial level regulations for various markets (FSANZ, 2019a), as well as the buyer specific considerations imposed by distributors, retailers, large manufacturers and the major multiples. Even exports to emerging markets will likely require compliance with many of the same or similar regulations as the global food industry has settled on the major requirements for trade in safe foods. This combination of national level market access regulations and those required for success in each market all necessitate a commitment to food safety and quality as defined by Hazard Analysis Critical Control Point (HACCP)- and preventive controls based-food safety and quality systems (Codex Alimentarius Commission, 1995; Food Safety Preventive Controls Alliance (FSPCA, 2016)). The success of the Global Food Safety Initiative (GFSI) and the dominance of the major transnational conglomerates, including giants such as Nestlé, Unilever, ConAgra, Kraft General Foods, as well as the expansion of transnational quick serve restaurants (QSRs) and the hospitality sector is creating opportunities for suppliers from developing countries (Gordon, 2016). This, however, means that these producers have to meet FSQS requirements or they will not only not be able to access export markets that want their products, they will also be excluded from doing business with these firms in their domestic markets.

Food safety and food security

The 2008 food price crisis has demonstrated a countries' dependence on global food production and distribution systems for their food security (Mittal, 2009), the vulnerability of which has been further demonstrated by the unprecedented drought which occurred in Eastern Europe and, more recently in Southern America in 2016, both major sources of supply of different foods. The situation is particularly worrying in developing countries, where challenges of food production, productivity, availability and access prevail. Placed in the context of population increases expected to exceed 9 billion by 2050, food and nutrition security is expected to remain a global issue for years to come (Vasan and Bedard, 2019). Although not explicitly spelled out, food safety and nutrition are both basic expectations in the concept of food security. Foodborne diseases cause significant illness and death worldwide and food safety can contribute significantly to the malnutrition-infection cycle of inadequate dietary intake that causes weight loss, leading increased vulnerability to disease, increased morbidity, and/or severity of disease. Most developed countries have adequate systems and procedures in place to guarantee the production of safe and nutritious foods. Such systems and procedures are yet to be fully established in many developing countries to support increased food production, facilitate market access, and contribute to the security of the global food supply.

As has been noted (Jaffee et al., 2019; Vasan and Bedard, 2019), food safety and food security are now inextricably intertwined. Producing more food without meeting minimal quality and safety requirements undermines the response to the global food and nutrition security challenge and the resilience of vulnerable rural households. The global food safety risk of aflatoxin, for example, is a threat to food security affecting more than 40 primary agricultural products such as grains (e.g. corn, sorghum, rice), legumes (e.g. soybeans), oilseeds (esp. peanuts), some pulses, cassava and even a few critical vegetables. It can also affect livestock and aquatic species resulting in lower conversion ratios, adverse health effects, and significant levels of aflatoxin derivatives in consumer products such as dairy, eggs and meat. Aflatoxins contribute to liver cancer, affect maternal and infant health or cause stunting in young children. The FAO estimates that up to 25% of the global food grain and nut supply is contaminated which can impede trade, reduce marketability, harm consumer brands, result in recall costs, lower sales, and raise production and marketing costs. As a food security issue contamination in staple foods leads to economic losses, increased healthcare costs, reduced physical development, lower IQ and nutritional deficits (Vasan and Bedard, 2019). It also contributes significantly to the overall public health burden (PHB) that countries have to bear and, when this occurs in developing countries, it aggregates to impede the overall economic development and growth prospects for the country.

In their paper *"Rethinking Food and Food Safety"* co-authors Vasan and Bedard (2019), explored the intersection of data science, genomics, and social sciences in food. In what is a unique perspective, they noted that over the last ten years, the concept of food safety and what "food" is had profoundly evolved due to significant shifts in demographics, in food culture and ground-breaking advances in science, technology and how society globally saw its food. Millennials are, in fact, changing the way developed and developing countries are eating and "Generation Xers" will continue and expand this trend, so much so that what was unthinkable eight years ago is now a reality. One such trend is the growing interest in, and the consumption of food derived from insects in developed country markets (Fig. 1.2), for example, which contrast with the now routine inclusion of an explosion of developing country fruits, vegetables, spices, seasoning and flavors as a part of northern diets.

Farmers and manufacturers that process food, the private sector, academia and governmental organizations will continue to forge public private partnerships to support the delivery of tangible results in feeding a growing global population that require more and greater diversity of food choices to meet their specific dietary, health, wellness and nutrition needs in both developed and developing countries. Consumers will continue to demand transparency, catalyzing changes in the food industry requiring the application of a plethora of

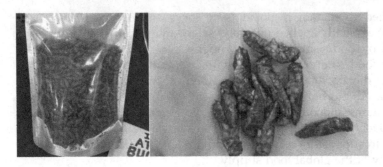

FIGURE 1.2   Insects being sold as food in the American marketplace. *Source: André Gordon (2019).*

technological solutions to be applied. This technology will not be a panacea or silver bullet, but rather an enabler to support the sustained long-term solutions required. Integrating global systems with localized subsistence and urban farming, collecting and analyzing data from consumers, customers, agricultural growth conditions and the environment, and making strategic decisions based on quality data with artificial intelligence (AI) will force a rethink food and the agri-food system in order to meet these demands.

Among the major changes in science have been the application of genomics, epigenetics and rapidly expanding and path-breaking studies on the microbiome. From data science and genomics to mass consumer markets, changes in food and related sciences are facilitating greater precision in developing and delivering consumer-focused products to specific consumers. Developments are also being driven by new findings in xenobiotics[1] and environmental health, as well as the trend by consumers towards wanting personalized nutrition. These improving capabilities and changes are transforming our understanding of food, health and nutrition and creating a new food safety paradigm along with this transformation. As the shifts in the global food supply chain continues driven by consumer preferences and the improving ability of producers in developing countries to meet the demands of regulators and the market in developed countries for safe, high quality food, there has been an inexorable integration of local value chains into global value chains. This highlights the importance of food safety in developing countries to the global food supply, a matter explored and developed in more detail throughout this volume.

Impact of food safety in developing countries on food security and nutrition globally

Food security exists when populations have access at all times to sufficient quantities of safe and nutritious food that meet their dietary needs and food preferences for an active and healthy life. However, for people to achieve nutrition security, factors beyond food come into play. Food safety is an integral part of food and nutrition security. Safe food is defined as food free of hazardous biological (such as virus, bacteria, toxins, parasites, prions), chemical (such as contaminants, pesticides and drugs residues) and physical (foreign bodies) agents. Foodborne diseases cause significant illness and death worldwide through the ingestion or handling of food contaminated by bacteria, viruses, parasites,

---

[1] Xenobiotics are foreign substances or exogenous chemicals that the body does not recognize such as drugs, environmental pollutants, certain food additives and cosmetics.

chemicals and biotoxins. This in turn links to the malnutrition-infection cycle, which is described as inadequate dietary intake that causes weight loss, which leads to growth faltering, and eventually either to increased vulnerability to disease, increased morbidity, and/or severity of disease. On the other hand, disease often leads to problems of altered metabolisms, nutrient loss, and malabsorption, and lack of appetite causing weight loss, faltering growth and under nutrition (Vasan and Bedard, 2019).

Most developed countries and some developing countries in the Americas and Asia, in particular, have adequate systems and procedures in place to guarantee the production of safe food. These countries are therefore able to better feed themselves and, where necessary, are active participants in various global food supply chains required to ensure a consistent supply or a variety of safe, wholesome and nutritious foods to consumers. However, such systems and procedures are yet to be established in many developing countries (Kumar and Kalita, 2017) as would be required to support increased food production, facilitate market access, and contribute to the security of the global food supply (Vasan and Bedard, 2019). This impacts not only the safety and security of domestic food supply in these countries, but also the nutritional status of their populations as the unavailability of appropriately nutritious, safety of food of good quality are matters requiring national attention, become even of greater concern in times of crisis. The seriousness of the situation is evident by the fact that by 2017, the number of undernourished people globally was estimated to have reached 821 million; around one person out of every nine in the world. Undernourishment and severe food insecurity are a concern in almost all subregions of Africa as well as in South America, whereas in most regions of Asia, the undernourishment situation is stable (Food and Agriculture Organization of the United Nations FAO, International Fund of Agricultural Development IFAD, UNICEF, World Food Programme WFP, and World Health Organization WHO, 2018). Consequently, the security of the global food supply in the future, as well as the immediate and future needs of these countries for nutritious, safe and wholesome foods demand that urgent attention is paid to addressing the need for enhanced capacity in affected developing countries for systems that can guarantee safe quality food.

Producing more food without meeting minimal quality and safety requirements would undermine the response to the global food and nutrition security challenge and the resilience required in vulnerable rural households. A major risk for staple crops such as grains and nuts, for example, is associated with the contamination by aflatoxins: these pathogens are produced by fungus strains and known to cause immune-system suppression, growth retardation, liver disease, cancer, and, in high concentrations, death in both humans and domestic animals. Such contamination is very common (FAO estimates that up to 25% of the global food grain and nut supply is so contaminated) and can be reduced significantly. Food-borne disease outbreaks have serious implications for public health, agri-food trade, food security and contribute to post-harvest losses. An unexpected impact of global Avian Influenza outbreaks in 2005—2007 was the food safety risk which created fear amongst consumers. This precipitated a significant, if temporary, decrease in global production and consumption of poultry products. More importantly, it negatively affected incomes, livelihoods and food security of populations in many of the developing countries where the production and sale of poultry products (Fig. 1.3) constitute an essential source of income and of proteins and micronutrients such as iron and niacin, in particular for children and women.

FIGURE 1.3 Intensive production of poultry & poultry products in South America. *Source: André Gordon (2019).*

Assurance of food safety is critical for guaranteeing access to markets. Globalization of the food supply means that food safety risks are shared across borders. In the years to come, global food production will need to increase. According to USDA, the increases in yields alone in developed countries will not be sufficient enough to satisfy the global food demand in 2050. While previously the developed world was the major supplier of food, developing countries are currently becoming key actors in the global food supply chain. The USFDA has reported that over 240,000 suppliers from 150 different countries now provide food to consumers in the US (Vasan and Bedard, 2019). The comparative situation in the EU is likely even more diverse given that it is the largest trading block and has historical relationships with many more and more geographically disperse developing nations. Availability and affordability of food will therefore increasingly depend on how much safe food developing countries are able to produce that will satisfy both their own needs and the needs of importing countries. If developing country producers can access global food supply chains, this will have significant positive consequences for global food supply in the future.

This would also help to ensure more sustainable income and more secure livelihood in these countries. As such, the upgrading of these supply chains to true global food value chains with cooperation from the final buyers back to the initial producers in developing countries, as discussed in Chapter 3 and as is being facilitated by organizations such as the Rainforest Alliance and the Sustainable Agriculture Network (SAN) will bring greater stability, more equity, safer, higher quality food and significantly enhanced sustainability to the global food supply. This is already being borne out by the findings of Impact Reports by both organizations which show that certified farms have less deforestation, more trees and biodiversity, more climate resilience, healthier soil, water and other ecosystems, better-educated children and, critically, higher incomes (Tavares and de Freitas, 2016).

In terms of nutrition, there is a clear, yet untapped, potential of value-chain approaches and analysis, which are linked to the food safety agenda through market access, to address nutrition concerns. Traditionally, efforts to address problems of poor diet quality and resulting micronutrient malnutrition have largely focused on 'quick fix' approaches, such as supplementation, food fortification, and the development of specially formulated and fortified products for different vulnerable groups. These direct nutrition interventions have the potential to improve micronutrient intake in the short term, but their sustainability is questionable if they are implemented without simultaneously addressing the key underlying determinants of under nutrition. Value-chain approaches, with their focus on the functioning of the supply chain, are one way to tackle one of the key underlying determinants of under nutrition—the lack of access to high-quality, safe foods that are essential for balanced diets. This important aspect of securing the global food supply for the future is discussed in the role of FSQS along value chains in Chapter 3 of this volume. As such, food safety is a global public good, and addressing it requires the strongest commitment to global solutions amongst all of the stakeholders along values chains. In particular multi-sectoral cooperation and multi-national alignment of food safety standards, policies and trade agreements are essential pre-requisites Improving food safety globally, would contribute significantly to mitigation of the global food and nutrition security risks.

Global food system resilience

Responding to the challenges of resilience and building a food system that can withstand shocks requires a holistic approach to encompass all aspects of food production, from impacts of climate change and socio-economic changes, to risks along the supply chain, be it manpower, food security, food integrity, smallholder farmers, market access and agricultural trade policies, nutrition and technological innovations in smart agriculture. Despite all of the technological changes and dependability in future agricultural and food manufacturing systems, the critical link in all supply chains is the human being (Vasan and Bedard, 2019). This makes culture and the ability to influence a positive food safety culture one of the major considerations in influencing the pace and trajectory of the security of global food supply chains. Previous work has shown that the failure to address this has been among the major causes of the failures experienced with many well-intentioned, donor-funded food safety programs in developing countries (Jaffee et al., 2019; Stier et al., 2003)

Wheat provides a good example of global food production resilience in the face of constant and varied shocks related to climate variability, social, political and economic factors which impact trade, pricing and availability on a daily basis. The resilience of the supply of wheat grown around the world is primarily influenced by soil conditions, local weather, availability of seed and other inputs, disease resistance, access to labor and commodity pricing. The supply chain from harvesting through milling to ingredients of valued-added manufacturing is further challenged by potential disruptions in transportation (availability of labor and transport), potential conflict, market pricing, food safety and food fraud, customer and consumer demand, dietary preferences (i.e. allergens/gluten-free), exchange rates, pricing, regulatory and trade barriers, and much more. Although typically viewed as a simplistic linear supply chain, the progression of wheat from field to market encompasses a complex multi-dimensional network of transactions and events each with underlying risks which can threaten the supply and availability of wheat or other grains as critical components of global food security. The weaknesses in the national and regional food safety infrastructure helps to create challenges for increased food production, facilitating market access and contributing to the security of the global food supply (Kumar and Kalita, 2017). Since the supply chain at its' core is global with no boundaries between developed and developing countries, enhancing and leveling the FSQS-basis for appropriate production practices in developing countries is critical for a robust global food supply. Selected global value chains are discussed in Chapter 3 with a view to assessing how these address the concerns raised here.

Domestic considerations for food safety in developing countries

Impact of foodborne illnesses on developing countries: a driver in FSQS implementation

When the issue of FSQS implementation in an emerging market/developing country setting is discussed, it is often linked to the need to derive greater economic benefits from exporting food products to external markets (Gordon, 2003,2005,2013,2015a,b,c,d; Mendonca and Gordon, 2004; Henson and Jaffee, 2006; Jaffee et al., 2019). In fact, Stier et al. (2002b) and Jaffee et al. (2019) indicated that much of the focus of the international development partners for developing countries in on helping them to be able to meet the requirements in developed countries, while being able to trade competitively in those markets. However, while compliance with food safety regulations and the requisite standards of international trading partners is important and have had the greatest visibility with export-oriented businesses and policy makers, new commercial opportunities are arising in the domestic environments of emerging economies (Jaffee et al., 2019). Other important considerations are the rising increasing the exposure of developing country populations to food safety hazards and the growing economic costs of unsafe food in these countries.

Leake (2016) indicated that food safety systems implementation and the domestic and regional awareness and approach to these were relatively well developed in the Caribbean as compared to other developing countries. In contrast, interviews and research on the situation in Central America, found that many while the Dominican Republic, Mexico, Costa Rica, Honduras and Panama had governmental support to protect both exports and domestic consumers, the government in most other Central American countries did not (Leake, 2017). Another exception to this is Belize where the Belize Agricultural Health Authority (BAHA) has been among the leading state agencies in the region in championing the cause of HACCP-based food safety systems from in the early 2000s. The regulatory authorities in countries such as Malaysia, the Philippines, Thailand and India, in Asia, as well as Cote d'Ivore, Ethiopia and South Africa in Africa have also implemented various approaches with different levels of sophistication in several food industry subsectors and across the agri-food sector as a whole. Countries such as Brazil and Chile have been leaders in this area in South America, while Columbia, Peru, Guyana and Argentina, are among others who have also increased their efforts to assure food safety for production targeted at both the domestic and export markets. Despite these efforts, significant challenges remain and the impact of foodborne disease (FBD) in developing countries, as in more developed countries is significant.

Foodborne illness outbreaks (FBIOs) in developing countries and the reports of the incidences of microorganisms on various foods is explored in Chapter 5. Reports have come from LAC, Asian and African countries and have included illness caused by the major pathogens *Clostridium botulinum*, *Listeria monocytogenes*, *Salmonella*, *Shigella*, *E. coli*, *Staphylococcus aureus* and *Vibrio* spp., among others (CDC, 2019; Brown, 2018; George et al., 2017; Tassew et al., 2010), as well as parasites which have been the causative agents in outbreaks in India (Peters et al., 2019), Ethiopia (Brown, 2018) and Asia (WHO, 2017), respectively. The increasing availability of data on foodborne illness outbreaks (FBIOs) from developing countries and their associated costs has shed new light on the overall global burden of foodborne disease. In a detailed overview of data gathered and assessed from 14 regions globally covering thirty-one (31) of the most important foodborne hazards that cause illnesses, Jaffee et al. (2019) reported that 2010 figures indicated an estimated **600 million illnesses** and **420,000 premature deaths**. They indicated as well that the data show that the incidence of illnesses caused by chemical hazards was substantially underestimated. Incidents of heavy metal contamination of food in Bangladesh (Hezbullah et al., 2016) and other developing countries (Ali et al., 2013; Bhattacharya et al.; 2010; Laylor, 2008) are examples of this.

So significant is the impact of FBIO's that Jaffee et al. (2019) have reported that the foodborne disease burden (FBD) in developing countries had cost **US$95.2 billion** in productivity lost and **US$15 billion** in treatment costs, imposing an economic cost across developing countries of **US$110 billion/annum**. Further, while the economic costs of unsafe food varies across countries depending on their level of economic development, clearly this cost is exceedingly high and is impacted by other factors including demography, culture, dietary practices, the state of the national environmental health infrastructure and the level of development of the export sector. Consequently, the incidence and risk of exposure of their populations to food safety hazards, as well as the impact on exports, the latter being estimated in 2016 at **US$5−7 billion** on total exports of US$475 billion (Jaffee et al., 2019), suggest that food safety should be both national public health and economic priorities for the governments of developing countries.

Other domestic considerations

The domestic factors influencing the capacity of various developing country sub-regions and countries to fully, consistently and competitively engage in the global food value chains varies by country and region. In Central America, it has been reported that despite efforts and programs to address FSQS implementation, the success required has not been achieved (Leake, 2018). The outcome of interviews with leading regional professionals indicated that the governments in Salvador and other Central American countries did not have a focus on food safety which helped to exacerbate the problem for domestic consumers. In the Caribbean, on the other hand, Leake (2017) reported that, led by Jamaica, there was a strong and growing trend towards seeking national, regional and individual firm-level FSQS compliance. This has been coupled with the development and implementation of regulations and systems to improve the overall FSQS status of the sectors, including food service and distribution, as well as production for domestic consumption and export. Likewise, as discussed in Chapter 3, Cote d'Ivoire (CI) and India (in the Indo-Ivorian Cashew Nut Value Chain) and CI, Ghana and others have evolved effective mechanisms to assist farmers and other value chain participants in meeting the requirements of their target export markets to the benefit of their domestic market and economies. Consequently, while countries in a sub-region face many of the same challenge as their counterparts (such as in the Caribbean/Central American region comparison above), their regional, national and sectorial responses, and the attendant outcomes, have been different. These different outcomes and the slow pace of progress generally in getting FSQS adopted country- and region-wide can be attributable to a range of technical, cultural, economic and other issues which are discussed below. Nevertheless, if robust, safe and sustainably growing agri-food sectors in developing countries are to emerge, the challenges must be successfully addressed. Otherwise, the ability of these countries to feed themselves, nutritional equity across countries and the safety and overall security of global food supply may be at risk.

Technical FSQS considerations related to local and regional regulatory infrastructure, requirements and implementation

A review of formal assessments of national food safety systems of developing countries points to common a range of shortcomings (Jaffee et al., 2019; Stier et al., 2002a,b, 2003). This is supported by the experience of the authors in their involvements in development programs and projects, as well as privately funded systems development and implementation in Africa, the Americas (including the Caribbean) and Asia (Gordon and Mendonca, 2004; Leake, 2018; Vasan and Bedard, 2019). These include:

- The absence of a comprehensive national food safety policy, translating into a lack of prioritization of investments;
- Where such policies exist, there is often a lack of coherence of policy, dispersion of the regulatory authority and oversight and therefore already meagre resources across several governmental ministries and agencies;

- The fragmentation of institutional responsibilities, especially for market surveillance and inspecting food production, processing, and handling facilities and compliance with public health standards;
- Fragmented infrastructure for providing analytical support to the regulators or services to the industry, if at all available;
- The presence of many regulations and standards, yet a lack of clarity on the extent to which these are voluntary or mandatory;
- Where they exist, a lack of coherence and cooperation to enhance overall national and regional capabilities through the creation of a scientifically competent, robust and mutually reinforcing national/regional food safety analytical system comprised of public and private capabilities as is the norm in developed countries;
- A focus on the acquisition of analytical equipment, often without the required attendant scientific and technical competence to use the equipment;
- Where the technical capabilities exist for the use of sophisticated analytical capabilities, it is often plagued by a lack of resources for supplies, servicing and to remunerate and retain the competent scientific professionals required for its effective deployment;
- Inability to apply, or weakness in the application of science-based risk assessment and FBD surveillance to identify and mitigate risk and help to quickly solve foodborne illness outbreaks;
- A focus on hazard rather than risk, often leading to the misallocation of resources;
- Weak, inconsistent or ineffective enforcement that is further exacerbated by undue influence of some stakeholders within the national and regional systems;
- The failure to empower and incentivize the private sector to deliver food safety;
- The lack of effective food safety engagement with consumers, whether in relation to education, risk communication, and other matters; generally underdeveloped consumer advocacy and consumer civil society organizations;
- The lack of consistent and transparent border-in measures to address the risks from growing food imports;
- The structure of management of many firms in developing countries and how decision-making is managed, typically top down without the delegation of authority required for the successful and sustainable implementation of FSQS;
- The weakness in available technical support services such as pest control, sanitation, technical training, repairs and maintenance, if they exist at all;
- Other production related and technical issues among producers include understanding and implementation of basic GMPs, HACCP and/or preventive controls based on hazard analysis, and the availability of suitably qualified and competent technical personnel;
among others.

In addition to these considerations, inconsistencies in the practical implementation of food industry standards and regulations, issues with education and literacy, cultural norms and practices and an overreliance and focus on testing rather than process control also need to be addressed. Other issues include economic pressures on farms and exporters to remain viable and, in some countries, domestic and regional politics (Stier et al., 2002b) all contribute to the difficulties producers and exporters in some emerging economies face with FSQS implementation. The development and support of a strong and knowledgeable, science-based regulatory body or set of effectively coordinated regulatory entities, coupled with their mutually beneficial and reinforcing engagement of the private sector they are regulating, and any exiting private sector expertise are requirements for success as well. These issues need to be recognized and effectively dealt with if effective and sustainable FS&Q systems implementation across developing economies is to become a reality. In addition, there needs to be an understanding of effective drivers of change.

In the Caribbean, a major driver has been the need for local producers to be able to supply transnational hotel chains and, perhaps even more importantly, international quick serve restaurant and other food retail chains. This situation was also noted to be the case in Egypt (Stier et al., 2002a). Other major catalysts are the need to grow revenues by accessing export markets, all of which now require compliance with specific FSQS regulations or market-determined certification standards. In this regard, there is need for caution as some certification bodies whose work is evident in the marketplace could cause the validity of certifications to be questioned if they fail to maintain the same standards which apply in developed country markets, a concern also alluded to by Stier et al. (2002b).

Among the technical considerations are the design and implementation of a plethora of national and regional programs and projects in the Caribbean, Central and South America, several parts of Africa, as well as many Asian countries that are expected to successfully embed FSQS in the agri-food sector in these countries. Unfortunately, the results have largely not met with expectations. Change in the development effectiveness and sustainability trajectory of most development assistance programs in food safety will only come when there is a radical rethink of the approaches being used and an adjustment in the commitment of stakeholders in both developed and developing countries. It will also require an acknowledgment from the major donor agencies and other development partners that the success sought has largely been eluding them or has been too few and far between to build momentum towards making a sustainable difference. This will also mean having to acknowledge that there must be a better way to do things (Stier et al., 2002a). Perhaps most critically, change for the better will only happen when it is driven by people who care about the countries with which they are working or in which they live. Lasting, positive change will not and cannot be made during the kind of "in and out" consulting visits that are the norm in the design of most development programs (Stier et al., 2002a). This is a position that the authors have observed in multiple cases, in many countries throughout the developing world. It underscores the need for those food safety and quality systems (FSQS) practitioners who have better local and cultural understanding to be integrally involved in the design of interventions and programs if they are to be successful and, equally as important, sustainable.

Export considerations for FSQS in developing countries

Most exporters or suppliers of food products or ingredients to transnational business (hotels, QSRs. etc.) face one option. As Stier et al. (2002a) puts it, "The economics are simple: implement HACCP or find another outlet for your products." This has evolved in more recent years to requiring compliance with a range of other HACCP-based food safety systems, including preventive controls, the Global Food Safety Initiative (GFSI)-benchmarked set of standards, as well as other private standards and addenda that specific buyers may demand. The challenge has therefore been for developing country exporters to either themselves alone, together as a group (as shown in the Cote d'Ivore and other value chain examples discussed in Chapter 4) or in partnership with state agencies (see examples of exports to the EU in Chapter 2) to find a way to comply with requirements. The outcome of this evident need is that developing countries themselves, or in partnership with international or regional development assistance organizations have sought to design and implement programs to help their exporter achieve this compliance to the benefit of their economies.

Exports have therefore been a major focus of donor funded programs in FSQS implementation and upgrade of regulatory systems. There have been many programs with the objective of improving the infrastructure in developing countries to enable compliance with the food safety requirements for export to developed country markets. Some of these have been quite successful and transformational. Some have involved development partners working with local interest groups, while others have largely been driven by local imperatives, even if supported with external funding. Examples of the former, include the 1999/2000 US Embassy, FDA, local private sector and government collaboration to lift the import alert on Jamaican ackee fruit, described in detail in Gordon et al. (2015) and Gordon (2015a), the involvement of the US authorities in solving the *Cyclospora* contamination issue with raspberries from Guatemala in 2011 (Gordon, 2015a) and the several targeted programs funded by the EU, the InterAmerican Development Bank (IADB), the Canadians and DFID (DFID) in Africa, the Americas and Oceana. Other examples would be the very effective and far reaching programs implemented by local partners in the Philippines, Malaysia and Thailand that have led to dramatic improvements in quality, safety and competitiveness of traditional food industry sectors such as the coconut, pineapple (fruit) and seafood industries.

The more successful of these projects and programs have been heavily influenced by local project design or led by local food industry practitioners. In general, however, the success of numerous interventions to improve food safety and quality in the food chain in developing countries is, at best, questionable (Henson and Jaffee, 2006; Jaffee et al., 2019). This is because of several shortcomings, the reasons for which have been excellently reviewed by Stein et al. (2002b, 2003) and Jaffee et al. (2019). These issues are the same as or interlinked with those listed above that have also stymied successful implementation of more projects and programs in securing the safety of domestic foods supplies. This has meant that the gaps that existed in national capabilities and among producers has generally remained in Sub-Saharan Africa and other parts of the continent not characterized by large, well-capitalized industrial production systems (e.g. as in South Africa). A similar situation exists in much of Asia and parts of Central and South America, with much of the success being in selected countries within each region where project implementation or design differed from others that were done before. This lack of capabilities and supporting infrastructure often impacts both the technical ability of developing country food production systems to consistently deliver products compliant with import and buyer requirements, as well as to supply their own markets with safe, high quality food. These are discussed in more detail below as this volume explores several aspects of the challenge. It also presents not only details of specific approaches to addressing the shortcomings but also provides a wealth of technical and scientific information to support those interested in changing the current dynamic.

Summary: Technical & market-related FSQS considerations in developing countries

Several developing economies have begun to effectively expand access for their foods in third country markets as their agri-food system becomes more sophisticated. Within this emerging reality, one has to locate the reality of a trading environment in which assuring food safety is being driven by consumer demands which have, in turn, influenced the regulatory requirements for entering markets and the requirement by major buyers for food safety and quality systems (FSQS) certification. This requirement has meant that suppliers along the value chain, including those located in developing countries, have had to attain certification to a range of systems, including FSQS such as the Global Food Safety Initiative (GFSI) benchmarked set of standards, organic and sustainable production practice certifications and buyer specific certifications such as Tesco's Nurture 10 (TN10), the Costco Addendum and Albert Heijn Protocol (AH) for fruit and vegetable suppliers. These also include certifications of ethical production practices such as Fairtrade, Rainforest Alliance, Soil Association (UK) and Safe Quality Food International (SQFI)'s Ethical Sourcing standard certification, as well as others dealing with animal welfare such as the Global Aquaculture Alliance (GAA) and Marine Stewardship Council for sustainable fishing and seafood and Humane Choice for Free Range animal products. The GFSI standards cover from primary production (GlobalG.A.P, SQF, Primus, GAA, Global Red Meat Standard (GRMS), etc.), through manufacturing SQF, British Retail Consortium Global Standard for Food Safety (BRCGS), International Featured Standard (IFS) and Food Safety System Certification (FSSC) 22000, warehousing and distribution (BRC, IFS, FSSC 22000 and SQF) and food service/retail (SQF). The considerations for these market-determined standards and their interrelation to the current and future trajectory of the food industry are discussed in more detail in Chapter 4.

This chapter therefore seeks to provide a background for, and discuss, the issues that affect the efforts of developing countries to improve the food supply for greater food security, better nutrition and health for their populations (domestic consumption) as well as for economic advancement through exports to third country markets or domestic supply to transnational buyers with stringent FSQS requirements. The first part looks at the current status of the food industry in developing countries in Asia/Pacific, Africa, Central and South America and the Caribbean, the major domiciles of developing/emerging nations. It examines the national, regional and international environment and considerations impacting the operation and development of the agri-food sector, as well as the challenges that are influencing important developmental and operational factors and hence the trajectory of the sector. It lays the basis for the examination, in this volume, of the technical and market considerations that drive FSQS development, implementation and application in trade and domestic businesses within developing countries and their associated global value chains. The second part of the chapter discusses in detail selected regulatory and technical requirements for accessing and trading in the major destination markets for developing country exports. Consequently, an overview of European Union (EU) Food Law and its associated regulations and directives are discussed (in this and other chapters), as are the food regulations infrastructure in Australia & New Zealand, the Safe Food for Canadians Act (SFCA) for exports to Canada and key components of the Food Safety Modernization Act (FSMA) of the United States of America (USA).

Part II Selected market-specific considerations for exports from developing countries

Overview of the European Union regulatory structure for foods

Introduction

The European Union (EU) has a system for the management of food production, import, export sale and distribution that is focused on meeting specific objectives for the EU agri-food system in food safety, responsiveness to consumers and equitable trade. Based on the principle of applying an integrated approach to food safety from farm to fork covering all sectors of the food chain, the central goal is to ensure a high level of protection of human health. This is implemented in a manner that is designed to safeguard the right of each European citizen's right to know how the food he or she eats is produced, processed, packaged, labeled and sold. In this context, the food laws in the EU have undergone ongoing revision as the imperatives for food safety, consumer protection and trade have evolved. Any entity or business from any country, EU or Third Country, have therefore to avail themselves of the food laws in order to trade freely on the EU market.

Previously, the EU had a variety of Regulations and Directives that managed the safety of their food supply. These regulated meat products, dairy, fisheries products, fruits and vegetables, as well as all other existing stipulated categories of food separately from each other. In the late 90s, this began to change, with basic Regulations and related Directives being rationalized to give effect to the current system. In the EU system of laws *Regulations* have the effect of law and must be followed as laid down by all EU member states. *Directives* are to be implemented in EU member states but can be done in a manner that is compatible with the local regulatory framework (e.g. in the UK vs. France).

An overview of the existing EU system of governance, control and support of a modern, healthy, safe, responsive and consumer-focused agri-food system can be found on the European Commission (EC)'s user-friendly portal that links the user to all of the relevant information required to navigate the relatively involved set of laws and regulations that comprise the governance of the EU agri-food system (European Commission, 2019a). Among other things, it provides an overview of the EU system for assuring food safety, as well as related areas, including details of the laws governing the EU agri-food system. It allows access to the basic information required through a range of resources that also include key components of the food safety regulatory infrastructure a schematic of which is presented above (Fig. 1.4).

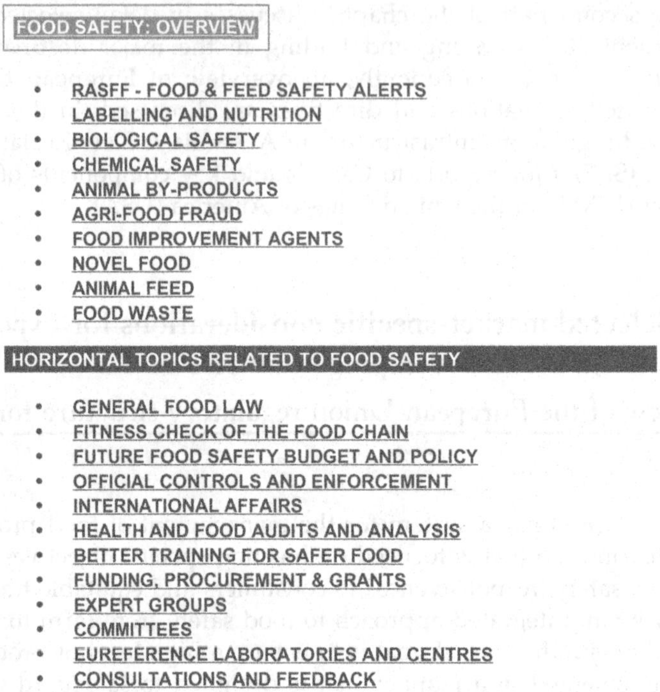

```
┌─────────────────────────┐
│ FOOD SAFETY: OVERVIEW   │
└─────────────────────────┘
```

- RASFF - FOOD & FEED SAFETY ALERTS
- LABELLING AND NUTRITION
- BIOLOGICAL SAFETY
- CHEMICAL SAFETY
- ANIMAL BY-PRODUCTS
- AGRI-FOOD FRAUD
- FOOD IMPROVEMENT AGENTS
- NOVEL FOOD
- ANIMAL FEED
- FOOD WASTE

HORIZONTAL TOPICS RELATED TO FOOD SAFETY

- GENERAL FOOD LAW
- FITNESS CHECK OF THE FOOD CHAIN
- FUTURE FOOD SAFETY BUDGET AND POLICY
- OFFICIAL CONTROLS AND ENFORCEMENT
- INTERNATIONAL AFFAIRS
- HEALTH AND FOOD AUDITS AND ANALYSIS
- BETTER TRAINING FOR SAFER FOOD
- FUNDING, PROCUREMENT & GRANTS
- EXPERT GROUPS
- COMMITTEES
- EUREFERENCE LABORATORIES AND CENTRES
- CONSULTATIONS AND FEEDBACK

FIGURE 1.4 A Schematic: current information available on the EU's food and feed portal. *Source: https://ec. europa.eu/food/safety_en.*

Key EU food laws

While there are many specific and detailed components of EU food law, as for most other developed country jurisdictions, the main food law is the General Food Law, Regulation (EC) No. 178/2002. This laid the basis for the current set of food laws that seek to give effect to the EU's objectives. Selected food laws governing packaged foods are discussed in detail in Chapter 7 and so will not be discussed here. Others are available through links from the portal (above, Fig. 1.3). A summary of other selected, key food laws are presented below.

- Regulation (EC) No. 178/2002: The general requirements regarding food safety, as well as the basic food law, effective from January 2005.
- Regulation (EC) No. 852/2004: The general hygiene requirements concerning food, effective from January 2006.
- Regulation (EC) No. 853/2004: The specific hygiene requirements concerning food of animal origin, effective from January 2006.
- Regulation (EC) No. 854/2004: The organization of official controls on products of animal origin intended for human consumption, effective from January 2006.
- Council Directive 2002/99/EC: The rules regarding animal health for the organization of the production, processing and distribution of products of animal origin applicable form January 2005.

- Regulation (EU) No 1169/2011: Updating the labeling laws of the EU.
- Regulation (EC) No 882/2004: Providing the basis for and an overview of the EU laws governing *Official Controls*.

These, along with a range of other addressing labeling, risk assessment, packaging use and disposal, as well as other food safety issues, provide the legislative underpinning and regulatory backbone for the system of EU food laws that exporters to the EU or import trading partners must be conversant with to avoid challenges in that market.

Selected EU food regulations impacting exports to the EU

For firms seeking to import foods into the EU as part of its highly transglobal food supply system or exporters looking to expand exports to the EU, the main Regulations of interest are:

- Regulation (EC) No. 178/2002 of the European Parliament and of the Council laying down the general principles and requirements of food law, establishing the European Food Safety Authority and laying down procedures in matters of food safety (also referred to as the General Food Law)
- Regulation (EC) No. 852/2004 of the European Parliament and of the Council of 29 April on the hygiene of foodstuffs
- Regulation (EC) No. 853/2004 of the European Parliament and of the Council of 29 April laying down specific hygiene rules for food of animal origin
- Regulation (EC) No. 882/2004 governing Official Controls.

Of these, the two main regulations that affect the majority of developing country exports to the EU are (EC) No. 178/2002 (the General Food Law) and (EC) No. 852/2004 (Hygiene of Foodstuffs). A summary of these is provided below.

Summary of EU regulations (EC) 178/2002

This Regulation laid down the general principles and requirements of EU food law, established the European Food Safety Authority (EFSA) provided the acceptable procedures in matters regarding food safety and presented a common basis for measures governing human food and animal feed. When it was enacted in 2004, having been passed by Council in 2002, it created a cohesive platform for food safety in the EU. It stipulates that companies must ensure that food is safe, establish systems to ensure that they can trace food throughout the food chain, withdraw or recall food from the market where it does not comply with food safety requirements and notify authorities of any action they have taken to secure withdrawal or recall of food or feed. This Regulation also covers the safety of animal feedstuffs to ensure that this does not indirectly cause illness or harm when humans consume animal products.

Regulation (EC) 178/2002 also established the Rapid Alert System for Food and Feed (RASFF), first promulgated in 1979 and, in February 2019, the European Union moved further towards modernization of the EU food safety policy when the European Parliament and the Council reached a provisional agreement on the Commission's proposal for a Regulation on the transparency and sustainability of the EU risk assessment in the food chain. This proposed Regulation, the details of which are available under the General Food Law (European Commission, 2019a), will facilitate more responsiveness to privately undertaken food safety risk assessment, and hence greater flexibility and equivalence to similar efforts among Europe's trading partners.

Summary of EU regulations (EC) 852/2004

Enacted in January 2006 after being passed in 2004, Regulation (EC) 852/2004 sought to harmonize food hygiene legislation across Europe. It laid down the general requirements relating to food hygiene, clarified existing responsibilities of food businesses and made primary producers subject to the hygiene requirements (farm-to-fork). With some exceptions, it applied the law to all food businesses and provided a more flexible and risk-based approach than previous legislation. It set out the minimum requirements for hygienic structure and layout of food premises, personal hygiene, training of food handlers and encouraged the development of guides to good practice. Most importantly, in line with *Codex Alimentarius*, Regulation (EC) 852/2004 mandated food safety management systems based on Hazard Analysis Critical Control Point (HACCP) principles mandatory for all food businesses trading within the European market. While it exempted primary producers from the HACCP approach, by stipulating HACCP-based food safety systems for all other parts of the agri-food system, it laid a sound, science-based platform for the ongoing development the EU food market with food safety and consumer protection at its core. This made trade in the food industry with and within the EU transparent, and provided producers in other markets, importers based in the EU and manufacturers looking to source ingredients and other foods components from Third Country markets the certainty required to expand their food businesses.

Regulation (EC) No. 882/2004 − official controls

Regulation (EC) No. 882/2004 was promulgated to create a uniform and integrated approach to official controls along the agri-food chain in the EU and to allow competent authorities in EU territories to verify compliance with food and animal feed laws. It provides the European Commission with audit and control powers in EU and Third Countries and also the power to take action at the EU level. The regulation allows the Commission to prevent or eliminate risks associated with food, guarantee fair trade practices and protect consumers interests throughout the EU. Known as the Official Controls regulation, it is a key aspect of the EU regulatory governance of the agri-food chain and is among the facilitating facets of the EU's the leading position in global practices in ensuring the safety of its food supply.

EU reference laboratories

The European Food Safety Authority (EFSA) and the EC have established a system of reference laboratories that are regarded as the official laboratories for specific analyzes throughout the EU and receive funding from the EU to fulfill their roles. The legislation outlining the designation and tasks of the European Union Reference Laboratories (EURLs) and the European Union Reference Centres (EURCs) is Regulation (EU) No. 2017/625 which came into force on 29 April 2018. These EURLs are required to submit annual work programs and are the go-to sources in the event of disputes and in support of regulatory action. For example, the Croatian Veterinary Institute, Laboratory for Food Microbiology, the Veterinary and Food Laboratory in Estonia and the Finnish Food Authority are among the approved reference laboratories for Campylobacter while Livsmedelsverket, the Swedish Food Agency, is the EURL for foodborne viruses in the EU. EU Reference Laboratories (EURLs) aim to ensure high-quality, uniform testing in the EU and support Commission activities on risk management and risk assessment in the area of laboratory analysis. Regulation (EC) No. 882/2004 on official controls defines the tasks, duties and requirements for all of the exiting the EURLs, a list of which is provided by the Commission in Annex VII. While this listing indicates the existing EURLs the Commission can change designation of existing EURLs or establish new ones at its discretion, within the established protocols. A current listing of EURLs is available at on the Commission's website (European Commission, 2019b).

Update of labeling legislation

Among the information accessible to users from the food portal (Fig. 1.4) is information on EU labeling laws. As is happening elsewhere worldwide, the EU updated its approach to labeling with a new regulation, Regulation (EU) No. 1169/2011 which came into force on 13 December 2014 and covers the provision of food information to consumers. It also has mandated from 13 December 2016 that all foods sold in the EU must meet its revised requirements for providing nutrition information to consumers in the stipulated formats. This new law combined 2 previous Directives on labeling, discussed in detail in Chapter 7, into one legislation.

Impact of BREXIT[2] and the relationship of EU food laws to the United Kingdom food businesses

The relationship between the United Kingdom (England, Scotland, Wales and Northern Ireland) and the EU as regards requirements for compliance with respective food laws in the trade and cross border movement of food is stipulated by the EU in a Commission notice dated 20 March 2019 (European Commission (EC), 2019c). This will determine the shape of UK/EU food trade in the foreseeable future.

---

[2] BREXIT is the term used to describe the decision by the United Kingdom, by way of a referendum on 23 June 2016, to leave the European Union.

Synopsis of selected important components of the US Food and Drug Administration (FDA)'s Food Safety Modernization Act (FSMA)

Background

Since the passage of the Pure Food and Drug Act of 1906 and subsequent revisions of the Federal Food, Drug and Cosmetic Act, the combined efforts of the government and the food industry have produced a set of standards and practices to improve food safety. These efforts include the adoption of Good Manufacturing Standards which provided the foundation to further food safety programs including current Good Manufacturing Practices, Low-Acid Canned Foods, Hazard Analysis Critical Control Points (HACCP), and more recently Preventive Controls regulations.

HACCP is an important preventive control-management system and can be integrated into any operation. The HACCP concept was first applied to food production during efforts to supply safe food for the U.S. space program in early 1960s, and through recommendations put forth by the National Academy of Sciences and the U.S seafood industry, the U.S Food and Drug Administration (FDA) finalized the Seafood HACCP regulation and USDA published their Pathogen Reduction HACCP regulation in the July 1996. Those regulations were followed by FDA's Juice HACCP regulation in January 2001. By then, HACCP principles, programs and regulations had been endorsed world-wide and provided some assurance of safe food products.

From a public health perspective, however, foodborne illness continued to place a significant burden on the public as the U.S. food supply became more complex, more foods were being imported and new hazards were found in food that were not previously seen. As a result, the FDA Food Safety Modernization Act (FSMA), the most sweeping reform of U.S. food safety laws in more than 70 years, was signed into law in January of 2011. The law provided FDA new enforcement authorities, mandated new risk-based food safety standards and provided new tools to hold imported foods to the same standards as domestic foods.

Among those tools were increased inspections of foreign producers across multiple continents, the experience of Caribbean exporters in the early years of FSMA being discussed by Gordon (2013). This chapter will discuss one of the seven mandated regulations under FSMA, the *Current Good Manufacturing Practice, Hazard Analysis, and Risk based Preventive Controls for Human Food* regulation (hereafter referred to as the cGMP/Preventive Controls for Human Food regulation). Among the other regulations that have impacted exporters are the *Foreign Supplier Verification Program* (FSVP) which is specific to U.S. importers and the *Produce Safety Rule* were both discussed in overview in Saltsman and Gordon (2015). Other important rules that impact exports to the US include the rule on the Sanitary Transportation of Human and Animal Food (Sanitary Transportation) Rule, issued in April 2016 and Mitigation Strategies to Protect Food Against Intentional Adulteration (Food Defense Rule) issued in May 2016. Of the seven (7) rules that are an important part of the FSMA, the most important and most far reaching, impacting many aspects of the operation of the food industry and bringing all areas of the industry not governed by other specific regulations under the FDA's jurisdiction. Being among the more recent of comprehensive international food legislations, it has benefitted from not only domestic considerations, but also being able to be structured to assure equivalence with other legislation in major global economies. It will therefore be used in this volume as the point of reference for food safety regulations, the United States being the major market globally for food imports.

21 CFR 117: cGMP/preventive controls for human food

21 CFR 117 is divided into seven subparts:

Subpart A – *General Provisions* lists definitions and exemptions for certain foods, activities and facilities. It also includes training requirements and records for qualified individuals. A qualified individual is a person who has education, training and experience (or a combination of these) necessary to manufacture, process, pack, or hold clean and safe food as appropriate to their duties and who has received training in at least the principles of food hygiene and food safety. Such a person is designated a *Preventive Controls Qualified Individual* (PCQI) and each facility that packs, handles, manufactures, processes or holds foods intended for sale in the United States of America (US) must have at least one such individual who oversees the process. The PCQI is discussed in more details below.

The training provisions in this subpart also require all staff with a food safety responsibility in the processing facility to be trained at least in cGMP's and also to have training specific to their job functions – especially as it relates to food safety. For example, a warehouseman may need to understand how to store allergenic ingredients, while a metal detector operator would need to understand how to address the hazard of metal inclusion if the metal detector calibration shows that the machine is not working. Employees in a facility should be prepared to be interviewed during an inspection about their training and knowledge as it relates to their duties, and records of their training must be available for review.

Subpart B – *current Good Manufacturing Practices (cGMP)* outlines the minimum sanitary standards that all food processing facilities must meet in order to ensure that food products are safe and processed under sanitary conditions. These updated cGMP's now include the term "allergen" cross-contact in several provisions, as well as a new provision for human food that is diverted to animal food.

The cGMP's apply to all food processors including seafood, juice, dietary supplements and others as they are the basis for determining whether products have been processed under sanitary conditions. However, within the context of food safety and preventive controls, hazards such as allergen cross contact[3] or pathogen cross contamination[4] will be part of a preventive control program (Subpart C and G) and not covered under a GMP program. In those instances, the GMP program focuses strictly on filth[5], not controlling a hazard. The provisions in the GMPs that are specific to hazard control, allergen cross contact, pathogen cross contamination and the process controls were left in the rule as a basis for

[3] Cross contact refers to contamination with a specific allergen by way of coming into contact with personnel, ingredients, packaging material or finished product containing or made or handled in an environment in which the allergen was present (Ref 21 CFR 117).

[4] Cross contamination refers to contamination with a specific pathogen (Ref 21 CFR 117).

[5] "Filth" in the US context means dirt, foul or putrid matter.

FIGURE 1.5 Snack food production in a developing country. *Source: Gordon (2018).*

assessing only qualified facilities[6] that are subject to modified requirements (discussed later in this chapter). Those specific facilities should be aware they will be assessed under the GMPs for both filth and preventive control programs. Qualified facilities subject to modified requirements will be discussed later in Subpart D of the regulation.

Subpart C – *Hazard Analysis and Risk-Based Preventive Controls* and **Subpart G** *Supply Chain Program* apply to most food facilities unless specifically exempted in Subpart A. For example, since seafood and juice products are already under preventive control programs (HACCP) they are exempt from the preventive control requirements in Subparts A and C. Snack foods (Fig. 1.5), however, are not.

Facilities subject to Subparts C and G are required to have and implement a Food Safety Plan (FSP). A Food Safety Plan is a set of written documents that are based on food safety principles and must include:

- A written hazard analysis and if the hazard analysis reveals one or more food hazards that requires a preventive control, the firm must have and implement:
  - A preventive control program. The rule outlines four (4) preventive control programs in Subpart C: process, sanitation, allergen and "other", while Subpart G outlines the supply chain program. All preventive control requirements, including management components for each program will be discussed separately in this section.
  - Recall Plan – if a hazard requiring a preventive control is identified during the hazard analysis, the firm must have a documented recall plan the effectiveness of which has been demonstrated.

[6] A qualified facility is defined below.

Food Safety Plans must be developed and overseen by a *Preventive Control Qualified Individual (PCQI)*. The PCQI can meet the training requirements in the regulation by attending an approved training program delivering the FDA-approved "standardized curriculum" recognized by FDA, such as that developed and administered by the Food Safety Preventive Controls Alliance (FSPCA) or by job experience. If the training requirement is met by attending a course on preventive controls, then training must be documented in records and those records are subject to the record requirements found in Subpart F — General Record Requirements. However, if the training requirements are met by "experience" there is no requirement to document the "experience" in records. Records of classroom training must be available for review.

The PCQI is responsible for preparing and/or overseeing the Food Safety Plan, validation of process controls, review of records and reanalysis of the Food Safety Plan, as required by the regulation. The specific requirements for each of these, inclusive of the frequencies at which these need to be done, where relevant, are discussed below.

Hazard analysis

A written hazard analysis is required for each type of food manufactured, processed, packed or held at the facility. It must list all of the known or reasonably foreseeable hazards, which in HACCP programs are known as "potential hazards" and must determine if any of those hazards require a preventive control or are "significant". The written hazard analysis is required regardless of the outcome (whether or not the hazard is determined to be "significant"), and it should be available upon request for review.

Hazards can be controlled at the facility, by the ingredient supplier, or by the customer. The regulation does not require a facility to specify in their hazard analysis which preventive control program they will use or who or where the hazard will be controlled. Although the rule is flexible in this requirement, the preventive control program and where the hazard will be controlled must be appropriate to the facility and management of food safety.

If the hazard is controlled by the ingredient supplier *a supply chain program* is required. If the hazard is controlled at the facility, *a preventive control program* is required, and if the you rely on your customer to control the hazard, you must identify for your customer, typically on a shipping document, that the food has not been processed to control the hazard.

Preventive control program

Preventive control programs are based on hazard analysis and are expected to be included in a food safety plan, as appropriate. Although the regulation outlines basic preventive control programs, it is flexible in how they can be defined and managed. In other words, *the terms process, sanitation and allergen control programs can be called anything, even a pre-requisite program, as long as the program includes the required management components of monitoring, corrective action, verification, validation (for process controls only) and records*. The supply chain program, however, is a verification procedure and therefore does not require the management components.

Supply chain program controls

Supply chain program requirements are found in Subpart G of the cGMP/PC rule and must be part of the Food Safety Plan if the hazard analysis identifies a hazard requiring a preventive control and you rely on your supplier to control the hazard. For example, the hazard analysis for black pepper identifies *Salmonella* as a hazard requiring a preventive control. If you use the black pepper in a ready-to-eat product with no "kill" step, you would need to rely on your supplier to control the hazard and then would identify the need for a supply chain program. However, if your finished product has a cook or kill step, then you would not need a supply chain program because the hazard will be controlled in your facility.

As a verification program for suppliers, the supply chain program requirements are similar and at times the same as outlined in the *Foreign Supplier Verification Program (FSVP)* which holds importers responsible to verify that their suppliers are controlling hazards, if necessary. The importer supplier verification requirements in FSVP are similar to requirements for importer verification found within the Seafood HACCP and Juice HACCP regulations, which is why seafood and juice products are exempt from the FSVP rules. The FSVP regulation will be discussed later in this chapter.

The general requirements for a supply chain program include the *use of an approved supplier*. Suppliers must be approved by the firm before receiving ingredient or inputs (in the case of packaging) from them or, with proper justification, these can be accepted on a temporary basis. The facility must have written procedure for controls at receipt and receiving records are required where a supply chain program is identified as required.

Once the facility identifies the need for a supply chain program, an *appropriate supplier verification activity must be identified*. Simply put, if the identified hazard can cause serious adverse health consequences or death (SAHCODHA[7]), the appropriate verification activity must be an on-site audit of the supplier's facility. An example of a SAHCODHA hazard is *Salmonella* or undeclared allergens. If not a SAHCODHA hazard, for example aflatoxin, the appropriate verification activity may be sampling and testing by the supplier or on receipt at the receiving facility, or a review of the suppliers' food safety records for the ingredient. In keeping with the flexibility of the rule, other verification activities may be appropriate based on the risk associated with the ingredient and the supplier. Once the appropriate verification activity is identified, the facility must ensure that *the activity is implemented and documented*.

If an onsite audit of the supplier is necessary, it must be conducted by a qualified auditor. Similar requirements to those of the PCQI apply to the qualified auditor, who must be an individual having technical expertise obtained through education, training or experience or a combination thereof. If the education or training requirement is met by attending a course(s), then training must be documented in records and those records are subject to the record requirements found in Subpart F – General Record Requirements. However, if the technical expertise requirements are met by "experience" there is no requirement to document the "experience" in records. Records of classroom training must be available for review. Competent authority inspectors, audit agents of a certification body, and internal auditors could all be considered qualified auditors.

---

[7] SAHCODHA more fully refers to **serious adverse health consequences or death to human or animals**.

The on-site audit itself must at least include a review of the supplier's written plan (HACCP or other Food Safety Plan) if any, and if it is being implemented for the hazard identified in the hazard analysis as requiring a preventive control. The results of the on-site audit must be documented. If sampling and testing are used to verify that the hazard is controlled, typically the test results are captured in a Certificate of Analysis (COA). The COA must be reviewed *prior to receiving the ingredient* to meet the requirement of only receiving ingredients from approved suppliers. If either the on-site audit or the other verification activity shows a non-conformance, a corrective action must be taken and the firm must identify the root cause for the non-conformance as well as indicate the steps taken to correct and to prevent reoccurrence. All corrective actions taken must be documented.

Process controls

Process controls include procedures, practices and methods to ensure control of parameters during operations such as heating, acidifying, irradiating and refrigerating foods. Process controls require documentation of the parameters to be controlled, typically minimum or maximum values (more commonly known as critical limits), as well as monitoring, corrective actions, verification and validation. For the most part, if a facility currently has an effective HACCP plan, it will meet the requirements for a process control program, keeping in mind that most process controls (such as cooking, metal detection, etc.) will require validation.

Process control procedures including setting critical limits, monitoring, verification, validation and record keeping must be documented, with the exception of records of refrigeration temperatures during storage of food that requires refrigeration for safety. These latter monitoring records can either be affirmative records demonstrating control of temperature or exception records[8] as defined in 21 CFR 117 (FDA, 2019; Food Safety Preventive Controls Alliance FSPCA, 2016) demonstrating loss of temperature. However, keep in mind that the temperature of cooler storage must be monitored with some regularity in order to determine if there is a loss of temperature.

Written procedures for process controls should be reviewed for adequacy (do they document what is done, and do they meet the regulatory requirements?) and also to determine if the procedures are being implemented (does the facility do what its documentation says it does?). For example, if the set monitoring frequency is every 4 hours, is monitoring being done at that frequency and is that what the records show? Typically, inspectors will ask for written procedures, review them to see if they meet regulatory requirements, and then interview staff and review records to determine if they are being implemented as described in the written procedures. *This approach is derived from how FDA inspectors are*

---

[8] Exception records are a category of records allowed in 21 CFR 117 and other FDA regulations in which the records need only provide evidence that a limit was missed, when it was breached, 55and not ongoing data showing that the limit was being met. Exception records are therefore those demonstrating loss of control such as the failure to meet or maintain target temperatures or the failure of a product for foreign material with x-ray monitoring. Facilities using exception records must have evidence that the system is working as intended, such as a record that the system has been challenged to a point where an exception record is generated.

trained; they do this to verify that process controls are effective and are being implemented in the manner required to assure their effectiveness.

Food allergen controls

These include practices and procedures to ensure control of allergen cross-contact and undeclared allergens. Food allergen controls also require monitoring, corrective actions, verification and records. Any time a facility receives, stores and uses an allergenic ingredient in any finished product, an allergen control program will be necessary.

Food allergen hazards can be managed in a food allergen control program or in other preventive control programs. For example, undeclared allergens may be included in a process control plan where the critical limit might state *"all products must declare allergens"* and finished product packages are monitored to assure the proper allergen is listed on the package. Likewise, an allergen cross-contact hazard could be managed in a sanitation plan *if controls include equipment cleaning and sanitizing as well as employee practices.*

Written procedures for food allergen control must also be reviewed for adequacy and implemented. Inspectors will assess the written procedures to determine if they meet regulatory requirements and interview staff and review records to determine if they are being implemented.

Sanitation controls

Sanitation controls include practices and procedures to ensure the control of cross-contamination and allergen cross contact. They are always required when a firm is processing a ready-to-eat product that is exposed to environmental pathogens or allergens prior to packaging. For example, if a ready-to-eat (RTE) finished product is handled after a cook step, a sanitation control program will need to be developed and implemented to control pathogen cross-contamination after the cook step, but if that same product was cooked and conveyed in an enclosed system then packaged, it would not require a sanitation control program. Similarly, if a RTE finished product passes through or is exposed or handled in an area where allergens are also handled, a sanitation control program may be required to control the risk of allergen cross contact.

A sanitation control program must include written procedures to prevent cross contamination or allergen cross contact by unclean equipment and employee practices. Sanitation control programs also require monitoring, corrective actions, verification and records. Sanitation verification must include a written environmental monitoring program. The regulation has specific requirements for an environmental monitoring program including identifying the test microorganisms, the locations and number of samples sites, the frequency of sampling, the identity of the test including the analytical method, the identity, nature and competence of the laboratory and corrective action procedures.

Written procedures for sanitation controls must also be reviewed for adequacy and implemented. Inspectors will assess the written procedures to determine if they meet regulatory requirements and interview staff and review records to determine if they are being implemented.

Other control programs

The preventive controls rule also gives processors the flexibility to identify and implement "other" preventive control programs besides those outlined in the rule. However, if an "other" or different control program is identified, it will need to provide an equivalent level of control to the four other stipulated and specifically required controls (process, allergen, sanitation and supply chain). In addition, the written procedures must meet the regulatory requirements and include the management components of monitoring, corrective action, verification and record keeping. Further, if it is a process-type control it will need to have critical limits and process validation, if necessary. "Other" preventive controls are defined in the regulation and clarified in the Food Safety Preventive Controls Alliance (FSPCA) FDA-approved training curriculum and its accompanying manual (Food Safety Preventive Controls Alliance FSPCA, 2016).

Recall plan

The Recall Plan is not really a preventive control program but must be part of the FSP if the hazard analysis determines there is a hazard that requires a preventive control. The recall plan must be written and include procedures that describe the steps in performing the recall. The written plan must include who is assigned responsibility to notify direct consignees and the public and conducting effectiveness checks and disposal of recalled product.

The written recall plan, if needed, will be reviewed during an inspection. However, *the review will only be to assess if the regulatory requirements are met, not how it is being implemented.*

Subpart D - *Modified Requirements* apply to two specific types of facilities, those that are determined to be "qualified" and those that are solely engaged in the storage of refrigerated, unexposed packaged foods that require temperature controls to prevent pathogen growth.

A qualified facility is *a very small business that has self-attested to averaging less the $1 M in annual sales of human food and that the facility has implemented preventive controls to address hazards OR is in compliance with non-federal food safety laws.* Self-attestation can be electronic or by mail. Information on how to submit a self-attestation can be found in the regulation itself, and on FDA's website at:

https://www.fda.gov/Food/GuidanceRegulation/FoodFacilityRegistration/Qualified FacilityAttestation/default.htm

Self-attested qualified facilities are exempt from preventive control Subparts C and G but are still subject to cGMP requirements in Subparts A and B, training, sanitary conditions and practices and process controls. *Inspections of these facilities will be conducted under those provisions.* However, if a very small facility has not self-attested, they will be not be considered a qualified facility and will not be exempt from the preventive control provisions and will be inspected under the full cGMP/PC regulatory requirements.

A qualified facility can also be a facility that is solely engaged in the storage of refrigerated, unexposed packaged foods that require temperature controls to prevent pathogen growth. These facilities are most likely to be warehouses, where either all or a portion of the warehouse stores temperature sensitive food that require refrigeration for safety. The key words here are "solely engaged", because if there are any products being stored that are exposed or unpackaged, they are no longer "qualified". If they are a qualified facility, they are exempt from preventive control provisions in Subparts C and G, but if they store any exposed foods or manufacture, process or pack food, they are not exempt.

Inspections of these qualified facilities will be conducted under the cGMP requirements in Subparts A and B, and also the temperature control requirements found in Subpart D which states that the temperature must be monitored at an adequate frequency, corrective actions must be taken if there is a loss of temperature control, temperature controls must be verified (i.e. through calibration) and records of temperature checks, corrective actions and verification must be maintained and made available for review.

Subpart E – *Withdrawal of a Qualified Facility Exemption* is merely an administrative process. A qualified facility exemption may be withdrawn if there is a foodborne illness directly linked to the qualified facility and/or FDA determines it necessary to protect public health or prevent foodborne illness outbreak. These withdrawal decisions and process are procedural and will not likely be decided as part of a routine inspection.

Subpart F – *Requirements for Records* outlines requirements for all records that must be established as required by the regulation. In general, the Food Safety Plan including the hazard analysis, written preventive control programs, monitoring, corrective action, verification, validation and training records, as well as the recall plan must meet the general requirements outlined in this subpart. The basic requirements that apply to these records include that they:

- must be kept as original records, true copies, or electronically
- contain actual values and observations
- are accurate, indelible and legible
- are created concurrently with the activity being documented
- are detailed enough to provide the history of the work performed

The records themselves must identify the facility, date and, when appropriate, the time of the activity, the signatures and initials of the person performing the activity and where appropriate, the identity of the product.

The owner, operator or agent in charge of the facility must sign and date the FSP upon initial completion and if there are any modifications. All records need to be retained for two years, including general equipment or process adequacy records that have been discontinued.

All records required by the regulation must be available for review and to be copied as necessary.

Records are an integral part of documenting that a food safety system is in place and being implemented. Inspectors will ask for records covering a certain date or timeframe. They review records to determine if the facility is implementing their written procedures consistently over time. Although it is not a requirement, it is suggested to organize and file completed records by date, as is typical for LACF facilities. This helps to identify overlaps and possible gaps in procedures and also, at the end of the day, ensures that all daily records are accounted for and reviewed by management in a timely manner.

Overview of the regulatory environment for exporters to Canada

The Safe Food for Canadians Act (2012) was created to streamline and modernize Canada's food regulatory framework, provide greater congruence with trading partners and clarify the role of Canada's regulatory agency, the Canadian Food Inspection Agency (CFIA) whose role has sometimes not been clear. Traditionally, Health Canada has been responsible oversight of veterinary drugs for animals used for food, as well as for setting the standards that the CFIA implements. Health Canada also has jurisdiction over fortified foods, as shall be discussed later in this section. The CFIA was set up under Canada's Food and Drug Act and is responsible for implementing this Act and its attendant regulations, as well as the standards set by Health Canada. Prior to the Safe Food for Canadians Act (SFCA), the CFIA also had to implement four different Acts, the Fish Inspection Act, the Meat Inspection Act, the Canada Agricultural Products Act, and the Consumer Packaging and Labeling Act, the updates and modifications to which over time had contributed to the lack of clarity of the CFIA's role. All of these have been streamlined under the SFCA under which the CFIA is the sole agency responsible for food safety. The CFIA is now therefore the focal point of Canada's food safety regulations, while Health Canada retains approval and regulatory oversight over aspects of consumer protection, particularly those regarding labeling and the classification of selected food products offered for sale in Canada. These will be discussed in greater detail later on.

Summary of the Safe Food for Canadians Act (2012)

The SFCA has three (3) main objectives which are to improve national food safety oversight to provide better consumer protection, to strengthen and streamline regulatory oversight and authority for the Canadian agri-food sector and enhancing international market opportunities for the Canadian food industry. In accomplishing these objectives, the Act has provided better protection for consumers against deceptive practices, tampering and hoaxes, as well as the authority to address these, inclusive of increased penalties. It also mandated the need for companies to have food traceability which was not addressed in the pre-existing legislation. The SFCA also provides for better control over imports by requiring the registration or licensing of importers and giving the government the authority to hold an importer accountable for the safety of the foods they are importing. By consolidating the provisions of several pre-existing Acts, it has clarified CFIA's inspection and enforcement authority, enabling inspectors to undertake more inspections and increasing compliance and the safety of food across all industry sub-sectors. Finally, as regards it main objectives, the SFC Regulations give the CFIA the authority to certify foods for export, an increasing requirement in jurisdiction with which Canada trades, particularly those in Europe and Oceania. This makes exporting easier for Canadian firms.

The Act also creates a new review mechanism that avoids the slow and costly judicial review system. Objections to decisions taken by the CFIA can be addressed by this new system for recourse. The parties can still ask for a judicial review if they are unsatisfied with the results.

Foreign food safety systems recognition framework

The government of Canada has, under the SFCA, created a Foreign Food Safety Systems Recognition Framework to assist in ensuring that imported foods meet domestic food safety requirements. The SFCA requires importers to be licensed, to develop and maintain a preventive control plan and to have a fixed place of business in Canada, except in cases where the Minister recognizes the other country's food safety system as equal to the Canadian system. This recognition can be achieved where:

- there is history to Canada's trade with the exporting country which shows that both countries' food safety systems are comparable in the results achieved
- trade in the specific food commodity has a history which suggests that the exporting country's controls for that commodity achieve results comparable with Canada's
- the equivalence of a foreign country's food safety system is a prerequisite for importing a commodity as agreed between both countries.

Consequently then, the recognition framework is based on three levels of recognition, foreign food safety systems recognition, commodity-specific recognition and recognition of inspection systems as a pre-requisite for trade (e.g. in mollusks and meat), providing flexibility. The framework does not address anything outside of food safety controls and oversights. It therefore does not cover labeling, grades or compositional standards, for example. Animal and plant health are also excluded and, as such, existing standards remain in place and must be met.

Example of Canada's food laws in action: export of nutritional beverages to Canada

A company looking to export food or beverage products to Canada must engage with an importer who is registered & licensed with Health Canada for the purposes of minimally receiving and taking ownership and responsibility for the products. The importer is responsible for ensuring that the product meets all regulatory requirements which is primarily set out in the Food and Drugs Regulation (FDR). All food and beverage products must also meet the food safety requirements set on in the Safe Food for Canadians Regulation. This preventive controls-based regulation outlines the outcome-based food safety practices that the exporting company must manage. It includes good manufacturing practices, risk-based preventive controls and risk analysis and assessments such as HACCP. All conformance to regulation monitoring is the responsibility of the Canadian Food Inspection Agency, which is a division of Health Canada. If the manufacturer is based in a country that has regulatory equivalency with Canada and is in good standing with the regulatory body in that country then the importer can accept that as being in compliance with Canadian food safety regulations. As of 2019, only the United States and New Zealand have been designated as having regulatory equivalency. Manufacturers located in countries other than those with equivalency must provide the importer with information, audit reports, certifications or other designations that would illustrate compliance with the Safe Food for Canadians Regulation.

The food and beverage classification will have an impact on the nutritional, vitamin, mineral and other factors effected by formulation. The classifications and definitions within the Canadian regulatory framework as per Health Canada's requirements are stipulated in the FDR which specifies which foods may contain added nutrients, which nutrients may be added, and the permitted levels in the food. Reasons for the currently permitted addition to foods of essential nutrients include:

- To restore the levels of vitamins or minerals to the levels that were present in the food before processing or, in the case of amino acids, to provide protein of a nutritional quality that is equivalent to that which was present in the food before processing,
- To make the food that is intended to be sold as a substitute for another food nutritionally equivalent to the food that it is intended to replace in the diet in respect of (a) the levels of added vitamins or minerals, or (b) the quality of protein provided through the addition of amino acids,
- To prevent or correct a deficiency of vitamins or minerals in the population or specific population groups or,
- To modify the levels of vitamins, minerals or amino acids in the food for special dietary use.

An outline of the categories of allowable products follows.

Categories of allowable dietary and health products

Natural Health Products — (NHPs) are defined as:

- Vitamins and minerals
- Herbal remedies
- Homeopathic medicines
- Traditional medicines such as traditional Chinese medicines
- Probiotics
- Other products like amino acids and essential fatty acids

Foods for Special Dietary Use — Food that has been specially processed or formulated to meet the particular requirements of a person in whom a physical or physiological condition exists as a result of a disease, disorder or injury, or for whom a particular effect, including but not limited to weight loss, is to be obtained by a controlled intake of foods.

Infant foods — Food marketed and designated as infant (less than one year of age) and junior foods (having particles of a size to encourage chewing by infants) and strained foods, and human milk substitutes, i.e., infant formulas.

Drugs — Products requiring a prescription from a medical doctor or products containing ingredients that are designated as medicinal in nature. Note that *Cannabis*, its by products and products which use *Cannabis* and its by-products as ingredients are regulated by the ***Cannabis Regulations***.

Supplemented Food - A supplemented food is broadly defined as a pre-packaged product that is manufactured, sold or represented as a food, which contains added vitamins, minerals, amino acids, herbal or bioactive ingredients. These ingredients may perform a physiological role beyond the provision of nutritive requirements.

General Foods and Beverage − All other foods and beverages (Fig. 1.6) would fall within this category provided they meet FDR-permitted levels for the addition of vitamins, minerals and amino acids to food.

Product labeling and composition

Details of the specific labeling requirements for most foods to be offered for sale in Canada are discussed in Chapter 7. However, manufacturers and exporters can use the categorization of products listed above to assist in determining where their products

FIGURE 1.6 General (other) foods as described by the Canadian Regulations. *Source: Gordon (2012, 2019).*

should fit and hence which regulations they need to conform to. Where products or ingredient fall into the Supplemented Food or do not fit into any other category, then approval for sale in Canada must be provided by Health Canada. This is typically provided through the Temporary Market Access (TMA) process which are basically temporary licenses. The temporary nature allows for time to collect additional information that ultimately can be used to slot the product into an existing category or time for Health Canada to amend regulations. These amendments are currently being considered for many products in the Supplemented Food category.

As part of the process of determining categorization and then ultimately regulation, the manufacturer will be working with their importer to understand current labeling and composition of their products and where and how they might need to amend them to meet regulation. Ultimately, they must ensure compliance with the composition, labeling, nutritional labeling and claims provisions of the *Food and Drugs Act and Regulations*, the *Consumer Packaging and Labeling Act and Regulations* and other commodity specific legislation, such as the *Canada Agricultural Products Act and Regulations*.

Rationale

Compliance with Canadian food safety standards and compositional requirements promotes a safe food supply and helps ensure that consumers' expectations are met with respect to food composition and quality. The declared composition, compositional standards, nutrition information and claims must be adequately controlled to prevent misrepresentation and fraud, and to prevent such health hazards as the presence of undeclared food allergens.

Mandatory information[9] on food labels allows consumers to make informed choices by:

* providing basic product information (e.g., the product's common name, its list of ingredients, its net quantity, its durable life date and country of origin, as well as the name and address of the manufacturer, dealer or importer), and
* providing health, safety and nutrition information (e.g., instructions for safe storage and handling, nutrition information in the Nutrition Facts Table, and specific information on products for special dietary use).

Composition

* Control factors that are critical to the safety and integrity of the product are identified. (This includes microbiological, chemical or physical concerns, as well as concerns related to allergens, extraneous matter, etc.) Their specifications and limits are identified (e.g., thermal process, pH and water activity for ready-to-eat fermented meats, salt content for brined mushrooms, sulfur content on dried fruit, freedom from *Salmonella*, etc.)
* Imported food which is sold in Canada must comply with Canadian nutrient composition and nutrition labeling requirements and may concern the following:

[9] **NOTE:** "Mandatory information" is determined by the commodity and its regulatory requirements.

- added nutrients (i.e., vitamins and minerals) permitted in the specific foods are added at appropriate levels as set out in the *Food and Drug Regulations*;
- foods for which there are nutrient composition requirements (e.g., meal replacements, nutritional supplements, flour, etc.) meet these requirements;
- the nutrient content of the product is accurately reflected on the label and in compliance with the *Food and Drug Regulations* (i.e., the list of ingredients and Canadian Nutrition Facts table);
- products for which there are nutrient content claims and/or diet-related health claims meet the Canadian compositional and labeling requirements under the *Food and Drug Regulations*. (See Section 7, Nutrient Content Claims, and Section 8, Diet-Related Health Claims, in the *Guide to Food Labeling and Advertising* .)

Labeling[10]

- Mandatory information is properly declared on food labels in compliance with Canadian food labeling requirements, and that all label claims are accurate and not misleading. Examples of such procedures are listed below.
- Labels are compliant with Canadian legislation (e.g., presence of mandatory labeling), and for accuracy and correctness of information and can include but is not limited to mandatory information, quality claims (e.g., organic, natural, fresh, Kosher, etc.), compositional claims (e.g., no preservatives, etc.), nutrition claims, and standards of identity.
- Labels (e.g. artwork, text and layout) and packaging (e.g. size of the principle display surface, available display surface or the location of the principle display panel) are in compliance with Canadian legislation.

The following guidance documents provide further direction on compositional and labeling issues and are available on the Canadian Food Inspection Agency Website.

Information on Canada's priority allergens: Food Allergens, Gluten and Added Sulfites. Labeling requirements for nutrition labeling: the CFIA document *Evaluation Standard for Nutrition Labeling*.

Importer responsibilities

Manufacturers and importers must work together in order to ensure products are in compliance prior to arrangement to enter the Canadian markets. As stated previously, it is the importers responsibility to ensure this is in alignment. Details of what is required are discussed in Chapter 2 which provides a practical example of how to engage successfully with these regulations.

---

[10] Labeling requirements are set out in the *Guide to Food Labeling and Advertising* . The above is not a complete listing of all labeling requirements.

Overview of the regulatory environment for exporters to Australia & New Zealand

Foodborne illness outbreaks in Australia and New Zealand caused 4.1 million gastroenteritis cases and 76 deaths in 2010 and 11,030 cases and 10 deaths in 2016, respectively, with *Salmonella* and *Campylobacter* being the major causes of illness in Australia, while norovirus, *Giardia*, *Cryptosporidium*, *Campylobacter* and sapovirus were the major etiological agents in New Zealand (Kirk et al., 2014; ESR, 2018). *Listeria* and *Salmonella* were the leading causes of death in Australia, while *Campylobacter* (3) and norovirus (2) caused the most death in New Zealand. Both jurisdictions share common management of their national food safety systems. In Australia and New Zealand, the agri-food system is overseen by a combination of national, state and local organizations, primary among which is Food Standards Australia New Zealand (FSANZ). FSANZ is a statutory authority in the Australian Government Health portfolio. FSANZ develops food standards for Australia and New Zealand, the primary of which are embodied in the Food Standard Code (FSANZ, 2019b). The Code sets out the requirements for the food sector in Australia and New Zealand and indicates those sections that apply to Australia only (Fig. 1.7). This allows persons trading with New Zealand alone or with both jurisdictions to differentiate those products which may not be required to have certain specific attributes depending on their destination. The Australia/New Zealand Food Standards Code and other legislation adhere to the UN Food and Agricultural Organization (FAO) Codex Alimentarius Commission's principles, including the requirement for HACCP-based food safety systems, and are equivalent in their overall application.

The Code is enforced by state and territory departments, agencies and local councils in Australia, the Ministry for Primary Industries in New Zealand and the Australian Department of Agriculture and Water Resources for food imported into Australia. Collectively, these oversee the requirements for the production, import, sale distribution, warehousing, handling, storage and labeling of various kinds of food in Australia and New Zealand and their territories. The various ministries, departments and other food regulation authorities in Australia and New Zealand work together through the Implementation Subcommittee for Food Regulation (ISFR) to ensure seamless implementation and enforcement of food regulation. FSANZ works with other Australian and New Zealand government agencies through IFSR to ensure continuous monitoring of the food supply to compliance with standards for microbiological contaminants, pesticide residue limits and chemical contamination and that food in the jurisdiction is safe (FSANZ, 2019a).

The Australia New Zealand Food Standard Code is available from FSANZ online and has all of the relevant standards governing the production sale and trade in food (FSANZ, 2019a). Published in the Federal Register of Legislation (FRL), the most current information or Act is available for perusal and use in guiding the agri-food industry. An example of this is the latest amendment to the Microbiological Limits in Food, Standard 1.6.1 (Federal Register of Legislation FRL, 2019). Because many of the requirements in the Code, standards and regulations are similar or variations on what has been indicated for the other jurisdictions covered in this section of Chapter 1, it is recommended that the specific

PART 1.1 Preliminary

Standard 1.1.1 Structure of the Code and general provisions
Standard 1.1.2 Definitions used throughout the Code

PART 1.2 Labelling and other information requirements

Standard 1.2.1 Requirements to have labels or otherwise provide information
Standard 1.2.2 Information requirements – food identification
Standard 1.2.3 Information requirements – warning statements, advisory statements and declarations
Standard 1.2.4 Information requirements – statement of ingredients
Standard 1.2.5 Information requirements – date marking of food for sale
Standard 1.2.6 Information requirements – directions for use and storage
Standard 1.2.7 Nutrition, health and related claims
Standard 1.2.8 Nutrition information requirements
Standard 1.2.10 Information requirements – characterizing ingredients and components of food

PART 1.5 Foods requiring pre-market clearance

Standard 1.5.1 Novel foods
Standard 1.5.2 Food produced using gene technology
Standard 1.5.3 Irradiation of food

PART 1.3 Substances added to or present in food

Standard 1.3.3 Processing aids
Standard 1.3.2 Vitamins and minerals
Standard 1.3.1 Food additives

PART 1.6 Microbiological limits and processing requirements

Standard 1.6.1 Microbiological limits in food
Standard 1.6.2 Processing requirements for meat [applies in Australia only]

PART 1.4 Contaminants and residues

Standard 1.4.1 Contaminants and natural toxicants
Standard 1.4.2 Agvet chemicals [applies in Australia only]
Standard 1.4.4 Prohibited and restricted plants and fungi

FIGURE 1.7   Structure of the Australia/New Zealand Food Standards Code.

sections of the standards are consulted, as required. For example, while the Canadian jurisdiction recognizes 11 allergens, the US, eight and the EU fourteen, FSANZ recognizes ten (10). These include the eight (8) recognized by the US Food Allergen Labelling and Consumer Protection Act (FALCPA), with the additional two being lupines and sesame seeds, also recognized in the EU and Canada, respectively.

References

Ali, S., Sardar, K., Hameed, S., Afzal, S., Fatima, S., Shakoor, B.M., et al., 2013. Heavy metals contamination and what are the impacts on living organisms. Greener J. Environ. Manage. Public Saf. 2, 172–179.

Bhattacharya, P., Samal, A.C., Majumdar, J., Santra, S.C., 2010. Arsenic contamination in rice, wheat, pulses, and vegetables: a study in an arsenic affected area of West Bengal, India. Water Air Soil Pollut. 213 (1–4), 3–13.

Codex Alimentarius Commission, 1995. Guidelines for the Application of the Hazard Analysis Critical Control Point (HACCP) system (CAC/GL 18-1993). Codex Alimentarius, Vol. 1 B, General Requirements (Food Hygiene). FAO/WHO, Rome, pp. 21–30.

European Commission, 2019a. Food safety. Available at <https://ec.europa.eu/food/safety_en>.

European Commission, 2019b. General food law. Available at <https://ec.europa.eu/food/safety/general_food_law_en>. Accessed on 29 June 2019.

European Commission, 2019c. European Union reference laboratories. Available at <https://ec.europa.eu/food/ref-labs_en>. Accessed on 29 June 2019.

Federal Register of Legislation (FRL), 2019. Microbiological limits in foods. Available at <https://www.legislation.gov.au/Details/F2018C00939>. Accessed on 30 June 2019.

Food and Agriculture Organization of the United Nations (FAO), International Fund of Agricultural Development (IFAD), UNICEF, World Food Programme (WFP), & World Health Organization (WHO), 2018. The state of food security and nutrition in the world 2018. Building Climate Resilience for Food Security and Nutrition. FAO, Rome, Italy. Available from: www.fao.org/state-of-food-security-nutrition/en.

Food Safety Preventive Controls Alliance (FSPCA), 2016. FSPCA Preventive Controls for Human Food Training Curriculum, first ed. FSPCA, Participants Manual. Version 1.2.

FSANZ, 2019a. Monitoring the safety of the food supply. Available at <http://www.foodstandards.gov.au/science/surveillance/Pages/default.aspx>. Accessed on 30 June 2019.

FSANZ, 2019b. Food Standard Code. Available at http://www.foodstandards.gov.au/code/Pages/default.aspx. Accessed on 30 June 2019.

Gordon, C.L.A., 2003. Technical Issues in the positioning of the Caribbean in the Free Trade Area of the Americas (FTAA). Institute of Food Technologists Annual Meeting, 12–16 July 2003, Chicago, IL (Abstract).

Gordon A., 2005. Food Safety in the Caribbean. Challenges in the Global Food Market. Invited Lecture to the US Food and Drug Administration, Bethesda, MD, USA.

Gordon, A., 2013. Island Regulations: The Impact of the Food Safety Modernization Act on Caribbean Inspections, Audits, Imports and Exports. In: 2013 IAFP Annual Meeting. 28–31 July 2013. International Association for Food Protection (Abstract).

Gordon, A. (Ed.), 2015a. Food Safety and Quality Systems in Developing Countries: Volume One: Export Challenges and Implementation Strategies. Academic Press, London, UK.

Gordon, A., 2015b. Dealing with trade challenges: science-based solutions to market-access interruption. In: Gordon, A. (Ed.), Food Safety and Quality Systems in Developing Countries: Volume One: Export Challenges and Implementation Strategies. Academic Press, London, UK, pp. 115–128.

Gordon, A., 2015c. Exporting Traditional fruits and vegetables to the United States: trade, food science, and sanitary and phytosanitary/technical barriers to trade considerations. In: Gordon, A. (Ed.), Food Safety and Quality Systems in Developing Countries: Volume One: Export Challenges and Implementation Strategies. Academic Press, London, UK, p. 4.

Gordon, A., 2015d. Influence of Maturity, Source, Handling and Processing on the Safety of Canned Ackees (Blighia sapida). In: 2015 IAFP Annual Meeting. 25–28 July 2015. International Association for Food Protection (Abstract).

Gordon, A., Saltsman, J., Ware, G., Kerr, J., 2015. Re-entering the US Market with Jamaican Ackees: a case study. In: Gordon, A. (Ed.), Food Safety and Quality Systems in Developing Countries: Volume One: Export Challenges and Implementation Strategies. Academic Press, London, UK, pp. 91–114.

Henson, S., Jaffee, S., 2006. Food safety standards and trade: enhancing competitiveness and avoiding exclusion of developing countries. Eur. J. Dev. Res. 18 (4), 593–621.

Institute of Environmental Science and Research Ltd (ESR), 2018. ESRAccessed on 28 June 2019. Annual Summary of Outbreaks in New Zealand 2016. Available from: https://surv.esr.cri.nz/PDF_surveillance/AnnualRpt/AnnualOutbreak/2016/2016OutbreakRpt.pdf.

Jaffee, S., Henson, S., Unnevehr, L., Grace, D., Cassou, E., 2019. The Safe Food Imperative: Accelerating Progress in Low- and Middle-Income Countries. Agriculture and Food Series. World Bank, Washington, DC. Available from: https://doi.org.10.1596/978-1-4648-1345-0, License: Creative Commons Attribution CC BY 3.0 IGO.

Kirk, M., Ford, L., Glass, K., Hall, G., 2014. Foodborne illness, Australia, Circa 2000 and Circa 2010. Emerg. Infect. Dis. 20 (11), 1857–1864. Available from: https://doi.org/10.3201/eid2011.131315.

Kumar, D., Kalita, P., 2017. Reducing postharvest losses during storage of grain crops to strengthen food security in developing countries. Foods 6, 8.

Laylor, G.C., 2008. Review of cadmium transfers from soil to humans and its health effects in the Jamaican environment. Sci. Total Environ. 400 (1–3), 162–172.

Leake, L., 2016. Handling Food Safety in Paradise, Part 1: The Caribbean. Food Safety & Quality28 December 2016. Available from: http://www.foodqualityandsafety.com/article/handling-food-safety-issues-paradise-part-1-caribbean/?singlepage = 1&theme = print-friendly.

Leake, L., 2017. Handling Food Safety in Paradise, Part 2: Central America. Food Safety & Quality26 January 2017. Available from: http://www.foodqualityandsafety.com/article/handling-food-safety-issues-paradise-part-1-caribbean/?singlepage = 1&theme = print-friendly.

Mendonca, A., Gordon, C.L.A., 2004. Control of food safety and quality in Caribbean countries: Implications for international trade. Institute of Food Technologists Annual Meeting, 12–16 July 2004, Las Vegas, NV (Abstract).

Mittal, A., 2009. The 2008 food price crisis: rethinking food security policies. G24 Discussion Paper #26, UNCTAD.

Stier, R.F., Ahmed, M.S., Weinstein, H., 2002a. Constraints to HACCP implementation in developing countries. Food Safety Mag. April 2002/May 2002 Issue. Available from: http://www.foodsafetymagazine.com/magazine-archive1/aprilmay-2002/beyond-haccp-constraints-to-haccp-implementation-in-developing-countries/.

Stier, R.F., Ahmed, M.S., Weinstein, H., 2002b. Constraints to HACCP implementation in developing countries II. Food Safety Mag. October 2002/November 2002 Issue. Available from: https://www.foodsafetymagazine.com/magazine-archive1/octobernovember-2002/constraints-to-haccp-implementation-in-developing-countries-part-ii/.

Stier, R.F., Lauren, M., Posnick, Sc.D., Kim, H., 2003. Constraints to HACCP implementation in developing countries III. Food Safety Magazine December 2002/January 2003 Issue. Available from: http://www.foodsafetymagazine.com/magazine-archive1/december-2002january-2003/constraints-to-haccp-implementation-in-developing-countries-part-iii/.

Tavares, P.A., de Freitas, A., 2016. Are sustainable farm certifications making a difference? Greenbiz Accessed on 28 June 2019. Available from: https://www.greenbiz.com/article/are-sustainable-farming-certifications-making-difference.

Vasan, A., Bedard, B.G., 2019. Global food security in the 21st century—resilience of the food supply. Cereal Foods World 64 (2).

Further reading

Council of the European Union, 2003. Council Directive 2002/99/EC of 16 December 2002 laying down the animal health rules governing the production, processing, distribution and introduction of products of animal origin for human consumption. Available at: <https://eur-lex.europa.eu/LexUriServ/LexUriServ.do?uri = OJ:L:2003:018:0011:0020:EN:PDF>. Accessed on 29 June 2019.

European Commission, 2019d. Notice to Stakeholders. Withdrawal of The United Kingdom and EU Food Law and EU Rules on Quality Schemes. EC Directorate-General for Health and Food Safety; Directorate-General for Agriculture and Rural Development, BrusselsAccessed on 29 June 2019. Available from: https://ec.europa.eu/info/sites/info/files/eu_food_law_en.pdf.

Havelaar, A., Kirk, M.D., Torgerson, P.R., Gibb, H.J., Hald, T., Lake, R.J., et al., 2015. World Health Organization Global estimates and regional comparisons of the burden of foodborne disease in 2010. PLoS Med. 12 (2).

Morgan, V., Gordon, C.L.A., N'dife, M.K., 2003. The case for a Caribbean ethnic foods market: Sensory profiling of Jamaican jerk seasoning and its explosion in the U.S. market. Institute of Food Technologists Annual Meeting, 12–16 July 2003, Chicago, IL (Abstract).

Addressing trade and market access issues based on food safety strategies

André Gordon[1] and Frank Schreurs[2]

[1]Chairman & CEO, Technological Solutions Limited, Kingston, Jamaica, West Indies
[2]President, FJS Consulting Group, Guelph, Ontario, Canada

Introduction

Background

In Chapter 1, the global food supply chain and the issues that are important to securing an adequate, safe, wholesome and nutritious food supply that is likely to be based on transnational global value chains was discussed. It would have become evident that for this to become a reality, food safety and quality systems practitioners in developing countries will have to work with their developed country counterparts and other local and international partners to overcome many significant challenges that exist and will arise as FSQS standards continue to change to meet market needs. Because the stability of the global food system will continue to rely on this complex of transnational, private sector-driven food industry networks, many originating in developing countries, the major buyers at the end of the value chain will have to be mindful of these realities. While some countries will continue to lead in the upgrading of their food industry and its capacity to supply compliant products, significant parts of the food industry in developing countries will continue to be characterized by vulnerabilities unique to their realities. These include those related to shelf life issues, variability of raw materials, climate change-influenced phenomena and economic, political and social shifts with the potential to weaken resilience of the food supply (Stone, 2015).

Significant local or global disruption in the food supply networks can create considerable public health risks and affect the provision of safe, nutritious, traceable, culturally acceptable food, which is available on demand, in ample quantity and at an affordable price. These food supply network vulnerabilities are being mitigated through industry investments in a combination of organizational capabilities including geographic dispersion of facilities, training of staff in the appropriate FSQS required for compliance and flexibility in sourcing and input/product fulfillment. They can also be addressed through developing operational redundancy and efficiency, awareness raising and anticipation to be able to respond to disruptions. At the primary production level, they have to be mitigated through the adoption of climate-smart agricultural practices using many of the approaches discussed in more detail in Chapter 4. Regardless of the circumstance, the supply of food in developed countries where the majority of the supply base is located in developing countries will always face interruption as long as the technical support available is unable to successfully navigate the market access issues that will invariably arise from time to time. It is therefore critical to the long term development of a robust global food system that food industry practitioners in developed and developing countries understand what is required to successfully mitigate the risk of market access interruptions and, critically, the strategies and approaches to employ when they do arise.

This chapter therefore seeks to explore these market and technical considerations that may occasion market access interruptions and further to contextualize them using specific examples of how key regulations in selected markets have been addressed, thereby attaining compliance with market requirements. It gives an overview of considerations for selected exports from developing countries to the US, Canadian and EU markets, as well as examines in detail the approaches that are required to overcome market access prohibitions and comply with regulatory requirements in these markets using case studies with seafood, mangoes, a major spice (curry powder) and a range of powdered beverages. It also explores the issue of the importance of attaining certification to a recognized food safety and quality system, using the case of an iconic developing country brand, Pickapeppa.

Strategies to deal with cultural issues

An examination of the resilience of food production systems that are compliant with developed importing country requirements necessitates a holistic review of approaches to attain compliance that encompasses all aspects of food production. This includes the impact of climate change and socio-economic changes on risks along the supply chain affecting food security, food integrity or food safety. These are dependent on many factors that impact food production in developing countries, including smallholder farmer knowledge, the nutritional or medicinal properties of the food, detailed product knowledge, and understanding of the process behind the application of traditional production systems (Fig. 2.1) or technological innovations to producing the food. Regardless of the technological approaches that may be used in existing and future agri-food production systems, tradition and cultural practices, as well as understanding of the current imperatives, dictates that the fundamental driving force in all supply chains will remain the persons involved (Vassan and Beddard, 2019). Consequently, how such a person is trained (Fig. 2.2), the facilities provided for them (Fig. 2.3) and their understanding of and orientation towards personal hygiene, sanitation, food safety and quality requirements for the food they are producing or handling will be critical to the integrity of the supply chain and its capacity to deliver what is being demanded of it.

FIGURE 2.1 Noni (*Morinda citrifolia*): an example of a product attributed with culturally significant health-giving properties manufactured using traditional methods. *Source: André Gordon (2017).*

FIGURE 2.2   Hands-on training of farmers and production staff in HACCP-based systems in two developing country pack houses. *Source: André Gordon (2012, 2014).*

FIGURE 2.3   Provision of adequate staff facilities and implementation of record-keeping in a West African facility. *Source: André Gordon (2012).*

This is particularly more important where the transformation of the production and delivery of traditional agricultural products to an international market, as has been discussed at length in Gordon (2015), in Chapter 3 (following this chapter) and when the production of cassava (Chapter 8) and nutmeg and spices (Chapter 9) are explored. These all bring into focus the importance of *food safety culture*, a concept championed by Yiannas (2008). This takes on even greater significance when this transformation involves long held cultural beliefs and traditional practice, as against operations within a modern food production environment. Consequently, as we explore the concept of supply chains and the imperative for equity in trans-border food supply systems to ensure sustainability of the global food chain (Chapter 3), the need for transitioning many of these to cooperative, transnational value chains and the central role of its human participants will become much more evident. Without going into the details presented in the excellent treatments of matter of food safety culture presented by Yiannas (2008, 2015) and Jespersen et al. (2017), this chapter will discuss practical examples of how cultural changes were enacted, including through the upgrading of facilities made available to staff and record-keeping (Fig. 2.3), as well as transformation of how staff saw their roles in the firm.

Case studies of system change at the national (seafood) and industry/sectoral level (mangoes), as well as the firm-level strategies applied to address market access and expansion issues are discussed below. Also discussed are the approaches required to meet market access requirements and market expansion by the application of food safety and quality systems through the certification of a unique developing country brand.

Case study 1: Getting seafood from developing countries back into the EU: sea change in food safety systems

Introduction

The seafood and aquaculture industries are an important part of the global food supply system and are of particular importance to developing countries as they are not only a major source of food supply and of growing importance to food security, but also major contributors to many economies. The industry, which includes from the harvesting of a range of fish, seafood and seaweed from the wild or from managed culture,[1] to pond-grown fish and other aquatic animals is of major significance and an important source of food globally. Seaweed alone is a US$6 billion global industry in which China and Indonesia produce over 23 million tons each year that is consumed in many dishes across Southeast Asia, while countries such as Chile, Canada, Norway and the US also are notable players (Nayar and Bott, 2014). More traditional products that are a part of the culinary traditions in developed countries such as shrimp (Fig. 2.4), lobsters and pond and sea fish are major exports from developing countries. Delicacies such as sea cucumber (from the class *Holothuroidea*), echinoderms that are harvested in the wild or grown on aquaculture farms in Asia, and queen conch (*Strombus giga*) are even more prized in specific markets and command high prices in China and France, respectively. Whatever its origin, seafood comprises a major part of the diets of most developed countries, are

FIGURE 2.4 Shrimp in a South American seafood plant. *Source: André Gordon (2010).*

[1] That is, fresh water or saline pond-raised, or grown near shore in managed areas sequestered from the marine environment.

therefore in demand and represent a market segment, the demand for which continues to grow globally.

Against this background, the introduction of changes in the European regulatory regime for seafood entering the European Union common market as a result of the "COMMISSION DECISION of 20 May 1994" and "COUNCIL DIRECTIVE of 22 July 1991", (Directive 91/67/EEC) created major dislocations across developing countries where fisheries industries dependent on the EU market were located. As reported by the US Food and Drug Administration, in a communication to all US exporters in August 1995:

> *"The EU will require that by January 1, 1996, all seafood companies in EU member states and those making product to ship to the EU have in place an effective food processing control program. This program must be based on the principles of HACCP (Hazard Analysis Critical Control Point). This means that processing plants shipping to the EU will need to institute preventive controls, such as HACCP as soon as possible. Consequently, firms that ship to the EU are being advised to implement HACCP before HACCP is required in the US. The January 1, 1996 effective date of the EU requirement is before the anticipated effective date of FDA's Seafood HACCP regulation".*

This meant developed country seafood exporters such as those in the US now had to implement Hazard Analysis Critical Control Point (HACCP) systems if they were to be able to continue to export to the EU, a major shift even in those food production systems. These were the same requirements that impacted developing countries with exports to the EU which were curtailed between early to mid-1998 and early 1999 if the specific country was not on the *"Approved Third Country List"* and if the specific firm was itself also not on the specific *"Approved Third Country Establishments"* List which the EU maintains for exporters. The Caribbean was no different and countries and companies had to dramatically change the way they operated (i.e. a *sea change*) if they were to continue to export to the EU. The impact of this prohibition on the Caribbean seafood export sector was traumatic: countries and companies that had been exporting for many years now found themselves having to qualify to be put on the respective lists. This could only be achieved by undertaking the kinds of industry-wide reforms that were thought to be impossible in the region at the time (in the 1990s). This was largely because of the systemic, nation-wide transformations of culture, particularly the implementation of an unfamiliar food safety culture, supported by appropriate legislation and a regulatory infrastructure which, in many cases, was weak if at all extant. The author was privileged to have been a part of the transformation in several countries[2] and describes below the approach to transforming the seafood industry in Jamaica to regain access to the EU market in 1998–2000.

Approach to the market access challenge

The EU's implementation of Directive 91/67/EEC resulted in the urgent need for affected seafood export sectors to respond as quickly as possible or face long term exclusion from their market. As the pioneers and leading proponents of food safety systems

---

[2] Through TSL, Dr Gordon worked with the seafood industry in Suriname, St. Lucia, and particularly Grenada and Jamaica, as well as the various regional and sub-regional mechanism set up to assist the process in regaining compliance with the new HACCP-based food safety rules and national regulatory compliance required.

implementation at the time in Jamaica, Technological Solutions Limited (TSL)[3] was asked by the Government of Jamaica (GOJ) and the lead players in the seafood export sector to lead a national programme to regain access to the EU market after Jamaican exports, including the lucrative Queen Conch (*Strombus giga*), were prohibited in 1998. The approach taken mirrors the one used subsequently for regaining access for Jamaican ackees to the US market which is described in detail in Gordon (2015) and was as follows:

- Formed a National Collaboration with the critical GOJ entities and private sector including the Veterinary Services Division (VSD) of the Ministry of Agriculture (MOA), the Ministry of Foreign Affairs and Foreign Trade (MFAFT), the Jamaica Exporters Association (JEA), the Fisheries Industry and the MOA Fisheries Department.
- Developed an overall strategy to re-opening the EU Market including firm-level, sectoral-level, national-level and bilateral (Jamaica-EU) strategic objectives and approaches.
- Developed a strategy to engage the EU in actively supporting the GOJ and the sector in attaining compliance.
- Did the detailed research in understanding the EU SPS and TBT requirements and to identify what would be required in any local regulations that may need to be developed.
- Developed and implemented the national level programmes, systems and procedures required to comply with the EU directives.
- Required firms to implement approved HACCP-based food safety programmes, supported by appropriate analytical data verifying their effectiveness.

The subsequent implementation of the overall strategic programme with the full support and collaboration of all arms of the GOJ and support of the industry made the success attained possible. The outcome was regaining access to the market and transformation of the sector such that it is now the norm for seafood, when not frozen or in a refrigerated cooler, to be kept on ice and its temperature monitored (Fig. 2.5).

Industry-wide transformation: the implementation of HACCP-based food safety systems

A major part of the programme of achieving compliance involved the upgrading of physical facilities used for processing seafood products to make them compliant with EU standards and requiring firms to implement approved HACCP-based food safety programmes, supported by appropriate analytical data verifying their effectiveness. This involved working with each of the lead firms in the sector, including Miles Franklin, DYC Fishing, B&D Trawling and Ton-Rick Ltd. to first identify the areas of their physical facilities and practices that needed improvement and then implementing all of the required Prerequisite programs (PRPs), precedent to implementing HACCP for *Strombus giga* and each of the other products being processed and packaged for export to the EU (and US). The approach used is the same as outlined previously in Gordon (2015) for ackees and Gordon (2016) with a variety of firms, including comprehensive training of staff at all levels of the organization, development and implementation of the documentation and records for the PRPs and HACCP plans and guiding the physical improvements, plant layout and design modifications and those in personnel and product flows required. The development of a detailed and rigorous

---

[3] TSL is the lead author's firm.

FIGURE 2.5   Seafood kept on ice. Temperature being monitored (kept at or below 4 °C). *Source: Rickey Ong-a-Kwie (2018).*

analytical programme that met the requirements of Directive 91/67/EEC for pathogen, phytoplankton, heavy metal, pesticide residue and other analyzes was developed and implemented, the data originally being largely generated by TSL's laboratory as the only facility at the time that could both do the analyzes and handle the volumes required. Fig. 2.6 gives an indication of what one of the major conch processing companies transformed a part of their processing facility to look like after implementation, while Fig. 2.7 shows the transformation that occurred in staff attire, equipment (e.g. stainless-steel tables) and facilities.

TSL, the firm leading the implementation had to contend with different requirements for exports to the US (a major market), the EU (the main market) and local regulations. Although the information was not available at the time, a review of the literature showed that other developing countries were also dealing with the same challenges. Higuera-Ciapara and Noriega-Orozco (2000) compared the requirements for HACCP systems implementation in their native Mexico and the US and EU at the time of the implementation of the new requirements for HACCP implementation in the seafood industry and their impact on the industry. As was the case in the Jamaican situation, they noted that the major differences were with the content and scope of what was required to be in the HACCP plan and the documentation, implementation and verification of the prerequisite programmes (PRPs) for the different jurisdictions. This was a challenge at the time and this conflict of requirements remains a challenge today as practitioners seek to implement food safety systems to comply with multiple requirements from regulatory (and certification) bodies. In the Jamaican situation (later extended to other countries in the region with which the authors worked), the approach taken was to target the most comprehensive set of requirements and use these as the base for the documentation, implementation and verification of the HACCP programme. This approach was also taken to the PRPs and became a standard approach in future implementations. While seeming more difficult to achieve in the initial instances, it became apparent that this was the right approach as it significantly reduced the need for modification of documentation, systems and practices later on when new requirements came into force or new markets were being targeted for access.

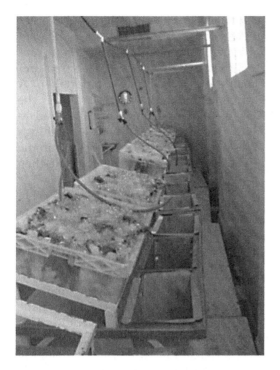

FIGURE 2.6 Processing of Queen Conch (*Strombus giga*) for export to Europe. *Source: from Gordon (2015), with permission.*

FIGURE 2.7 Processing of sliced frozen fish at a seafood facility in 1999. *Source: André Gordon (1999).*

Attaining equivalence to the EU regulatory infrastructure and requirements

While working on implementing changes in the processing firms, changes were also required at the sectoral level (Fig. 2.7) and in the regulatory environment. In seeking to attain equivalence to the EU regulatory requirements and their monitoring and enforcement infrastructure, far-reaching changes had to be made in the local regulatory environment and backed by the appropriate legislation. To advise this, a detailed review of pre-existing regulations governing food safety, aquaculture and aquatic animal health, seafood harvesting and exports, facilities and fisheries vessels requirements and food handling, including personal health and hygiene was undertaken. A task force comprised of public and private sector stakeholders was formed under the auspices of the VSD, advised by TSL, and began to undertake the work required to transform the sector and make it compliant with the EU Directives. Several things were required to be done as indicated above. When broken down into programmatic elements for implementation, these included developing a framework for and then implementing:

- inspections of the processing facilities, fishing areas and fisheries vessels;
- conformity assessment of the products being exported to certify their compliance with EU requirements;
- monitoring and being able to demonstrate control over the harvesting areas, specifically for Queen Conch (*Strombus giga*);

The Competent Authority (the VSD of the MOA) had to demonstrate the ability to undertake these inspections to meet the EU requirements and produce or have made available credible analytical results that the EU would be comfortable in accepting. This required extensive training of the regulatory personnel by a collaborative effort of the EU, the InterAmerican Institute for Cooperation on Agriculture (IICA), Technological Solutions Limited (TSL) and the VSD themselves. In addition, the local public and private sector laboratories had to collaborate to get the analytical work required done and the VSD and Fisheries Division had to have the harvesting areas mapped for a range of analytes, including heavy metals. The fishing and factory vessels had to be transformed to include better frozen storage facilities on board and the implementation of HACCP plans. And finally, the development of legislation that was equivalent to the EU legislative framework had to be completed. After much research, discussion and deliberations, the task force was able to have prepared, sent to Parliament and passed in 1999 an equivalent piece of legislation entitled the *"Meat and Meat Products and Meat By-Products (Inspection and Export Act)"*.

In summary, the industry and the regulatory framework was transformed in Jamaica and the regulatory institutions with primary oversight responsibilities for the export of seafood (including Queen conch), the Fisheries Division of the MOA and the VSD, inclusive of their laboratory and analytical capabilities, were significantly strengthened. The national analytical capabilities were able to attain compliance with EU requirements by also incorporating the capabilities of private laboratories (TSL's laboratory), as well as those of other entities such as the University of the West Indies (UWI) and Jamaica Bureau of Standards (JBS). The staff of the VSD were retrained to be able to incorporate HACCP-based inspections into their monitoring and approval process for the industry. Finally, the revised Meat Export Act was promulgated on March 12, 1999, the gazetting of which met a key criterion in according the Jamaican legislative framework and system of regulations equivalence with those of the EU.

At the time of the coming into force of the EU regulations regarding seafood imports, McDorman (1997) undertook a comprehensive review of the impact of these regulations on seafood exports from the South Pacific to both the United States of America (USA) and the European Union (EU). A comprehensive legislative review of the existing legislation (food, health and fisheries) in the Cook Islands, the Federated States of Micronesia, Fiji, Kiribati, Marshall Islands, Nauru, New Caledonia, Niue, Palau, Papua New Guinea, Samoa, the Solomon Islands, Tonga, Tuvalu, and Vanuatu was also done. McDorman (1997) found that only three (3) countries had legislative export standards for fish and none of these required HACCP systems to be implemented. As part of the process, draft legislation was also provided to help each jurisdiction develop and implement their own legislation to attain equivalency to the EU legislation. The findings of this review were similar to the situation in Jamaica at the time and would likely have mirrored the situation in most seafood exporting countries as HACCP-based food safety systems were uncommon in the global seafood industry outside of the EU in the 1990s to early 2000s. The implementation of the US Seafood HACCP regulation in 1997/98, in conjunction with previously imposed EU requirements subsequently led to the transformation of the sector globally.

Editors Note:

Since this time, the industry has changed significantly and the application of HACCP-based food safety systems is now the norm, largely catalyzed by the seafood HACCP requirements in both the US and EU, as well as the EU's lead role in mandating HACCP-based food safety system in keeping with Codex Alimentarius. As early as 2006, the EU were forced to modify their requirements because of their subsequent ratification of the World Trade Organization (WTO) Agreement on SPS Measures which recognized the World Organization for Animal Health (OIE) as responsible for setting the rules on animal health and the trade in animals for food, including aquatic animals. They therefore had to modify Directive 91/67/EEC as the requirements set out in it were more stringent than those required by the OIE, particularly as regards the analytical framework (EUR-Lex, 2019).

Case study 2: retaining access to the USA market for mango exports

Background

The United States Department of Agriculture (USDA) and the United States Agency for International Development (USAID) partnered to establish and support agricultural development and exports in Haiti during the period 2000–2007 through a project known as the Hillside Agricultural Program (HAP). One objective of one of the sub-projects under HAP was the development of the capabilities of small farmer mango (*Mangifera indica*) exporters to be able to significantly increase the volume of exports they sent to the US market. Project personnel worked with farmers to adapt market-based approaches and to treat the farming and export operations as a business and introducing improvements in resource management, agricultural production, marketing systems and post-harvest technologies. Crops targeted included coffee, cocao and mango, as well as several other destined for the local and diaspora markets.

The project was run on behalf of the US Government entities by Development Alternatives International (DAI)[4] who, through a partnership with Fintrac,[5] brought the lead author in to work with the small farmers and other value chain stakeholders over the period 2004–2006. The objective of this series of interventions was to improve the practices of farmers and their packing and exporting operations to allow them to be better able to meet the requirements of the US market. Among the outcomes of the project were that small farmers were able to increase their market share of total mango exports from 7% to 20% and, by 2007, approximately 2.6 million boxes of mangoes were being exported annually at a value of approximately $15 million, becoming the main export earner for Haiti, surpassing coffee. This case study describes the process of working with what ranged from organized operations employing traditional practices and a range of technologies, to basic, rudimentary operation and transforming them to a modern, food safety-sensitive export sector.

Description of the mango export industry in Haiti (2004–2006)

Structure of the sector

Mangoes were the second largest agricultural export to the US from Haiti in the early 2000s. The industry was characterized by farmers who supplied mangoes to several small producers and exporter of mangoes or larger, more organized exporters who employed better technologies in their packhouses. Several of the exporters were themselves farmers and ran a vertically integrated operation where they packed and exported from the farm or supplied from their farms to a related company that did the packing and exporting. In all cases, pack houses/exporters also relied on buying from *fournisseurs* (suppliers) who gathered from multiple farms and delivered to the buyer the quantities they had available for sale. These *fournisseurs* were not typically linked to any one operation (although some had their preferred suppliers) and were therefore not always amenable to following the dictates of the exporters, particularly when mango prices were high, or the supply was less than the demand. This added a layer of complexity to the supply chain in terms of assuring compliance with export standards. This was one of the challenges that had to be successfully addressed on this project during the process of seeking to transform the industry.

The supply chain and production process

Mangoes grown on commercial orchards, on farming operations that had arisen because of the natural growth of stands of mango in an area, or in the wild were harvested and delivered to pack houses where the were off-loaded prior to examination and sorting. In some cases, the mangoes were packed in field crates and delivered by commercial trucks (Fig. 2.8). In others, the transportation and delivery conditions were less than ideal, with product being transported in passenger vehicles (Fig. 2.9). In these

---

[4] DAI was the main contractor who sub-contracted major aspects of the project to Fintrac.

[5] Fintrac was the entity that brought the lead author in to address specific aspects of the post-harvest technology component, particularly those involving food safety.

FIGURE 2.8   Delivery of mangoes to a hot water treatment facility in Haiti. *Source: André Gordon (2004).*

FIGURE 2.9   (A) Delivery of mangoes to a Pack House in 2004; (B) worker standing in field crate sorting mangoes on receipt at hot water treatment facility. *Source: André Gordon (2004, 2006).*

cases, the handling of the mangoes could vary them being in field crates (black arrow) to being in direct contact with the passenger(s) or persons traveling along with the vehicle to help with offloading (yellow arrow, Fig. 2.9A). In these instances, the persons delivering the product were likely to be *fournisseurs* and were not employed to the packing house, making the enforcement of acceptable standards for transport and delivery a challenge.

Once received at the facility, the mangoes were either offloaded in the field crates in which they came or transferred to crates, after which they would be examined (Fig. 2.9B), sorted and unacceptable mangoes discarded. The mangoes would then undergo washing to clean them, after which they would either be kept in crates (Fig. 2.8) or in flumes or troughs Fig. 2.10A, prior to going for hot water treatment. Mangoes destined for the US had to be immersed in/under hot water at 46 °C/117 F for 60–110 minutes depending on the size and variety of fruit in a USDA-APHIS certified hot water treatment system. Various types of systems were approved and were being used in the industry in Haiti. Some were batch-type, full immersion systems (arrow, Fig. 2.8) while other were continuous spray or water immersion systems (Fig. 2.10B), depending on the size and sophistication of the operation. While at the time of this case study, the US was only allowing water immersion or water spray heated-mangoes from USDA-APHIS approved systems, more recently, the USDA-APHIS has approved the use of irradiation as an alternative treatment and mangoes from Mexico, India, Jamaica and other countries are now being allowed in after an approved irradiation treatment.

After hot water treatment, the mangoes undergo drying, final checks and packaging in carton boxes, prior to shipment to the US. During this process, the shipment could be inspected and pre-cleared by USDA-APHIS staff domiciled in Haiti and placed there to assists with programmes such as these, and certified prior to export. Alternatively, they could be shipped and have the inspection take place at port of entry. In either case, full traceability back to the farm was required and many of the personnel involved were not yet trained on these systems, ensuring that challenges would occasionally arise.

The post-harvest intervention component of the programme

Outline of the intervention

The component of the HAP that dealt with post-harvest handling of the product, introduction of food safety-based standards and the improvement of practices was comprised of a series of interventions. The first intervention involved visits to selected food processing facilities to get an understanding of the practices and norms in the industry at the time, followed by visits to selected mango packing houses and exporters. This was

FIGURE 2.10   (A) Receival and washing area for mangoes; (B) conveyor-based hot water treatment system. *Source: André Gordon (2006).*

followed by the preparation and delivery in Creole and English of an introduction to Food Safety and food safety principles to owners and managers operating in the mango industry. Also included were one-on-one sessions with owners to address specific issues and queries they had about making sure their operations could meet the requirements and setting up arrangements for future interventions.

On the second series of interventions, visits were made to, and work was done on a firm by firm basis. Visit were made to each of the major processing plants, pack houses and exporters. This included detailed audits of each operation with the elaboration of corrective action and development plans and in-house tailored training of management and workers in different session, again in Creole and English (French was not popular among the Haitian workers who spoke only Creole). Also involved was support in development and design of layouts and personnel, product and materials flows for facilities to support remodeling or, in the case of JMB S.A., the development of a new facility (Fig. 2.11B).

Further support and interventions involved a study tour visit to Jamaica to see how the agro-processing sector had implemented HACCP programmes there, supported by detailed, in-house training in GMPs, HACCP and, for their technical staff, in selected basic laboratory analyzes. Société Haïtienne Agro-Industrielle (SHAISA), the largest agro-processors in Haiti at the time, was the firm that was selected and benefitted from the intensive extensive support to their management and key technical staff, while JMB, S.A. benefitted from assistance with their new facility.

Observations and recommendations on practices

As a result of the visits to the packing operations, audits on the facilities and operations and general observations, several issues with the facilities being used by some operations, practices in the industry and issues with handling, including the lack of recognition and application of GMPs and other food safety principles and practices were identified. Among these were the way in which mangoes, once harvested were being transported (Fig. 2.9A), the way the mangoes were being handled during inspection and sorting in

(A)   (B)

FIGURE 2.11 (A) Outdoor packing of mangoes (arrow) in 2004 (B) conveyor/sorting table being installed at new plant in Haiti in 2006. *Source: André Gordon (2004, 2006).*

some facilities (Fig. 2.9B), the way mangoes were being packed in others (Fig. 2.11A) and the sanitary facilities provided for staff in some plants (Fig. 2.12A). Other deficiencies included the effectiveness and frequency of cleaning and sanitation, staff training and the absence of a structured way of dealing with and improving the sanitary practices of *fournisseurs*.

As a result of these observations, the following were recommended, and assistance provided to the exporters in implementation:

1. Standards need to be developed and implemented for all suppliers (*fournisseurs*), outlining acceptable transportation and handling practices.
2. The practice of packing mangoes in an unsuitable and unsanitary environment, such as an outdoor garage (Fig. 2.11A) must be discontinued and, where possible, the sourcing and installation of good handling facilities (Fig. 2.11B) should be encouraged.
3. Good Agricultural Practices (GAPs) and Good Manufacturing Practices (GMPs) supported by good personal hygiene practices needed to be implemented at selected facilities where issues existed.
4. All staff needed to be trained in GAP (where relevant), GMPs and be given an introduction to HACCP. Training had to be done in a language they could understand (Creole).
5. All management, technical staff and key support personnel needed to be trained in HACCP and its implementation.
6. HACCP-based post-harvest handling food safety programmes needed to be implemented in exporting facilities.
7. Physical facilities, where deficient (such as sanitary facilities, Fig. 2.12A) needed urgent improvement to at least match the best practices already available at some Haitian facilities (Fig. 2.12B).
8. Analytical support for microbiological testing (Fig. 2.13), where possible, needed to be developed and engaged. Where required, technical staff needed to be provided with the training required to effectively discharged their functions in this regard.

(A)   (B)

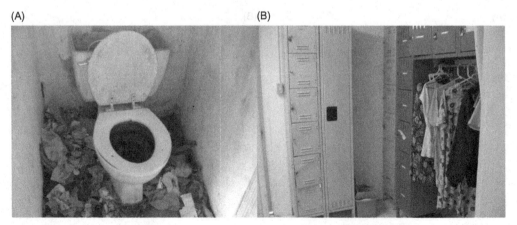

FIGURE 2.12   (A) Unacceptable sanitary facilities; (B) good facilities for staff. *Source: André Gordon (2004, 2006).*

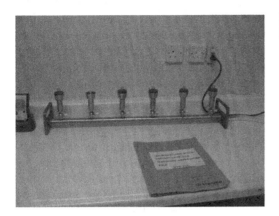

FIGURE 2.13 Laboratory facilities for undertaking basic analyzes at a plant. *Source: André Gordon, 2006.*

Outcome of the interventions

All of the training programmes, at various levels and to varying audiences as described previously were delivered, supported by in-house, hands on work with each pack house on the programme. These were delivered in both Creole and English, significantly enhancing the understanding and subsequent application of the best practices discussed. One-on-one sessions with owners to address queries and specific issues with the implementation were also held with significant success in getting improvements to staff facilities, staff participation in the process and significantly increased compliance rates with US import standards among participants. Several operations were transformed to meet the requirements, thereby assuring them of unfettered access to the US market. The layout of facilities was improved, efficiencies and increased losses or rejection due to non-compliance were significantly reduced. Overall, the intervention has the intended effect of improving the returns to the sector, the exporters and the farmers themselves.

Case study 3: Removal of curry powder from FDA's red list

Background

The United States Food and Drug Administration (US FDA) has imposed an import alert on curry powder for potential *Salmonella* contamination, requiring "Detention Without Physical Examination" (DPWE) based on its assessment of product entering US port of entry over time. Many spices, curry being no exception, have been shown to be contaminated with this pathogen if not properly processed and handled prior to shipment and so to protect US consumers from this potentially uncontrolled risk, the FDA has imposed the import alert. This covers cumin, curry powder and curry products from most individual suppliers in multiple countries, including the developing countries of India, China, Sri Lanka, Bangladesh, Jamaica and Trinidad and Tobago, as well as developed countries such as Belgium and the Netherlands (US FDA, 2019a). Consequently, all curry products from all companies that have not been granted exemption to the alert by the FDA, regardless of origin, are placed on the FDA's "Red List" and subject to regulatory action on arrival. This requires all product that is not exempt to be held at port of entry on arrival, the importer and the firm exporting the product is notified and the product has to be sampled by a FDA-approved laboratory using a pre-determined sampling protocol and tested for the presence of *Salmonella*. The impact of this has been a significant reduction in the availability of curry powder and curry spice blends from some of the preferred sources in the Far East, Europe and the Caribbean, creating a vacuum which has yet to be filled.

It is against this background that the case of Chief Brand Products Limited (CBPL), a 62-year-old company based in Trinidad and Tobago in the Caribbean and specializing in the production of dry spices and spice blends with ingredients imported mainly from the Far East should be considered. CBPL had been manufacturing and exporting a range of spices to the US for many years without having had an issue at the port of entry. On

SAMPLING PLAN

The following are to be sampled and tested and the relevant equipment and personnel swabbed at Chief Brand Products Limited in Trinidad. Samples can be sent to CARIRI with duplicate samples for analyses to TSL's Testing Laboratory:

Raw Materials
- Crushed chillies
- Fennel
- Mangriel
- Aniseed
- Mustard
- Pimento
- Ajwan
- Turmeric powder

Machine/Equipment
- Blender 1
- 3th (Grinder)
- Blender 2
- Surge bin
- 2DH 1 (Grinder)
- 2DH 2 (Grinder)

Production Staff at these locations (swabbing of hands)
- Batching
- Packaging into drums
- Packaging into pouches

Packaging Material & Equipment (according to what is being shipped to the US)
- GP2
- GP3
- Twin auger 2
- Twin auger 1
- Prodo pack
- Packaging Film[1]

---
[1] Send samples of each packaging film used.

FIGURE 2.14 Areas sampled for microbiological assessment at CPBL.

28 July 2016, a container destined for Florida in the United States of America was held at port of entry, sampled and tested for the presence of *Salmonella* sp. The organism was found to be present in some bags of curry powder consigned to CBPL's customer in the state. CBPL's curry powders were refused entry and all future shipment put on import alert and subject to DPWE under Import Alert 99-19, "Detention Without Physical Examination Of Food Products Due To The Presence Of Salmonella".

Approach to the problem

In order to be removed from the import alert, CBPL had to identify the cause of the contamination and institute effective mitigation measure such that the FDA would have confidence that they could consistently manufacture compliant, safe curry products. CBPL requested technical support from Technological Solutions Limited (TSL) to develop a strategy for implementation that would result in their exemption from the import alert. This support also included the preparation of document and files to support a petition to the FDA for CBPL to be included on the Green List for curry products to enter into the United States of America (USA) without any restrictions once they followed the agreed protocols for the exemption. CBPL, guided by TSL, undertook detailed investigations of the events that may have led to contamination of the product. The details of this investigation, the findings, the mitigating programmes that were developed and implemented, the approach to gaining exemption from Import Alert 99-19 and the outcome of the implementation of the strategic plan for exemption are presented in this case study.

Investigation and remediation of the cause of salmonella contamination of curry powder

In developing an approach to identifying the root cause of the problem and developing and implementing effective mitigating measures, TSL had to be mindful of the need to present a sustainable solution to the FDA, along with sufficient evidence of an effective system to prevent recurrence. A review of CPBL's inputs and their overall process for making the curry powder, as well as their systems and practices in general was identified as necessary. An assessment was therefore done of the process, inclusive of a comprehensive review of the raw materials involved, the process in detail and an identification of risk through a detailed risk analysis. The investigation included the preparation of a sampling plan (Fig. 2.14), undertaking analyzes of swabs taken at various points throughout the production process and microbiological assessment of each of the inputs into the process and product. It concluded with a determination of the cause of the *Salmonella* contamination and the advised the way forward in preventing the issues identified from recurring.

The cause of the contamination

Below (Fig. 2.15) are selected results of the microbiological analyzes of swabs undertaken for selected areas identified by the sampling plan. The results showed that the equipment, inclusive of the roaster (Fig. 2.16), and the staff responsible for the production

PARTICULARS	E. coli	Salmonella
Samples: Swabs		
Production Worker		
Before	Absent	Absent
After	Absent	Absent
Roasting Equipment		
Before	Absent	Absent
After	Absent	Absent
Roasting Worker		
Before	Absent	Absent
After	Absent	Absent

FIGURE 2.15 Example of some results from the microbiological assessment on the production process for curry powder at CPBL.

[a] *Note that the values for "before" are from swabs that were taken during production and the "after" swabs were taken after sanitation with alcohol.*

FIGURE 2.16 High quality, precision roasters used by Chief Brand Products Limited (CBPL) to process curry powder. *Source: Gordon (2018), Courtesy of CBPL.*

process were consistently free of *Salmonella* or *E. coli* contamination. Analyzes of the raw materials and finished product from the batches held indicated that the source of the contamination was one ingredient sourced from a supplier overseas. Subsequent analyzes of other batches of the ingredient from other suppliers were negative for *Salmonella* or other pathogens of potential concern, including *Clostridium perfringens* and *E. coli* O157H7. While it is possible that contaminated raw materials can result in persistent infections being established in processing lines, no such problem was found as was evident by the swab checks done on the entire processing line of which the results in Fig. 2.15 are a snapshot. Since CBPL employed sanitation practices consistent with GMPs, it was inferred based on the results that the contamination of the curry product (for which detention was effected) was likely due to a faulty batch of one of the ingredients used in the manufacture of CPBL's curry powders.

Although this was the first violative shipment recorded in the company's history of exports, data review indicated that while the records for roasting were mostly complete, the roasting time and temperature on the problem batch were not available. Consequently, it was determined that very high levels of the pathogen *Salmonella* in an ingredient possibly exacerbated by a short cook during roasting may have resulted in the survival of low, but detectable number of the pathogen in the finished product. It was also noted that at the time of the problem, the firm had not yet fully implemented a HACCP plan for this product.

Implementing sustainable preventative programmes

Having determined that an effective HACCP-based food safety programme had not yet been implemented by CBPL, work began to address this immediately on the reporting of the problem at port of entry. A HACCP Plan for all curry products was developed and validated as effective prior to its implementation. This process involved training all staff in Good Manufacturing Practices (GMPs), training the newly formed food safety team in HACCP, undertaking a rigorous hazard analysis tailored to the realities of the production process and facility, and preparing HACCP plans and other documentation and records for use in the production of curry powder and other products. An ongoing sampling program, as well as other systems aligned with the manufacture of safe food (see Chapter 4 for a full discussion of the FSQS requirements for manufacturing) were also implemented. There has not been a repeat of the positive results for *Salmonella* or *E. coli* since the batch that was detained as a result of the preventive measures implemented by CBPL to ensure that only safe products are distributed by them. With the implementation, monitoring and ongoing review of the food safety program, a significantly reduced risk of CBPL's products being contaminated and refused further entry into the United States of America (USA) was assured.

Getting the import alert on CPBL's curry products removed

Having implemented the necessary controls, including ensuring the robustness of the roasting process CCP and the control of incoming raw materials, the process of getting Import Alert 99-19 on CBPL's curry products lifted began. The approach to addressing the problem was documented, inclusive of the implementation of mitigating programmes and procedures. Various critical pieces of information and documentation that could demonstrate the identification of the root cause of the problem and effectiveness of the preventive controls implemented to prevent a recurrence, were collated. Evidence of the competence of the personnel involved (inclusive of training records), as well as all relevant information that had been gathered during the process of problem solving were also prepared. All of these were compiled to support a petition to the FDA to have the import alert lifted. The petition requested the removal of CBPL and its curry products from the FDA's Red List and placement on the Green List, which allows free access of the items so listed into the US market. An excerpt of the letter sent to the FDA requesting removal of CBPL from Import Alert 99-19 and placement on the Green List is presented in Fig. 2.17 (below).

Re: Request for Chief Brand Products Limited's Petition to Remove Curry Products from Import Alert 99-19

This is further to my correspondence with you last year on preparing a petition on behalf of Chief Brand Products Limited to have their curry powders removed from the Import Alert 99-19 / Red List. This is to facilitate the resumption of exports of said products to the United States of America.

The following products were detained on August 29, 2016 and placed on DWPE/Import Alert 99-19/Salmonella with entry number 102-0001463-6:

- Cariherb Madras Curry (25CT)
- Cariherb Madras Curry (10CT)
- Curry Powder (200CT)
- Curry Powder (220CT)
- Curry Powder (205CT)
- Curry Powder Duck and Goat (70CT)
- Kala Brand Curry Powder (20CT)

The firm has put in place all of the required processes and procedures to meet the requirements of the FDA and have implemented procedures to specifically analyse and approve products regarding safety prior to exporting to the United States of America (USA). As a specialist who has worked closely with the Agroprocessing Sectors across the Caribbean for over 25 years, and after assessing the products, equipment, staff and practices at Chief Brand Products Limited (CBP), and making recommendations on how to secure the safety of the products manufactured at the facility, I am comfortable to submit this cover letter on behalf of CBP ......

Import Alert 99-19

Import Alert Name: *Detention Without Physical Examination Of Food Products Due To The Presence Of Salmonella*

FIGURE 2.17   Excerpt of letter to FDA requesting removal of CBPL curry from red list.

Having gotten a positive response from the FDA to the request submitted, a formal petition was prepared, including all of the relevant information required to support it, and submitted. Among the information submitted were:

- The cover letter requesting the lifting of the import alert
- The HACCP Plan
- Validation information to support the HACCP plan
- A description of the preventive controls that were developed and applied
- Proof of the effectiveness of the preventive measures
- Results for five (5) compliant shipments

November 30, 2017

André Gordon, Managing Director

TECHNOLOGICAL SOLUTIONS LIMITED
Unit 31, The Trade Centre
30-32 Red Hills Road, Kingston 10
JAMAICA, W.I.
TEL: 876-632-3245-6, FAX: 876-632-3249

Email: andre.gordon2@tsltech.com

CASE #540541

Dear André Gordon:

This letter is in response to your September 2017 request to remove Curry Powder produced by Chief Brand Products Limited, Chaguanas, Trinidad & Tobago from detention without physical examination under Import Alert #99-19 "Detention Without Physical Examination Of Food Products Due To The Presence Of Salmonella"

The information you provided, as well as FDA's national entry data, was reviewed. The data indicates that Curry Powder produced by Chief Brand Products Limited, Chaguanas, Trinidad & Tobago has met the criteria for removal from detention without physical examination.

Routine coverage of entries will resume. Should detentions occur for the same or related reasons, detention without physical examination may be reinstated.

Enclosed is a copy of the advisory to our FDA field offices.

Further correspondence in this matter may be addressed to U.S. Food & Drug Administration, Division of Import Operations, 12420 Parklawn Drive, ELEM-3109, Rockville, MD 20857, attention: Sam Rudnitsky, 562-256-9214, email: Samuel.rudnitsky@fda.hhs.gov.

Sincerely,

Elvia J. Cervantes

John E. Verbeten
Director, Division of Import Operations

JEV:sr

Enclosure

cc: Subject File IA 99-19

U.S. Food & Drug Administration
12420 Parklawn Dr. Elem 3109
Rockville, Maryland 20857
www.fda.gov

FIGURE 2.18   FDA letter to technological solutions limited (TSL) regarding Chief Brand Products Limited's curry powder's removal from DPWE. *Source: Courtesy of Chief Brand Products Limited, Trinidad & Tobago.*

Subject:
Revision to Import Alert: 99-19, "Detention Without Physical Examination Of Food Products Due To The Presence Of Salmonella".

Revision Type:
Remove from RED List

Firm Name(s)/ FEI:
Chief Brand Products
144-150 Ackbar Road
Charlieville, Chaguanas
Trinidad & Tobago
FEI: 3010116510

Product Description /Product Code(s):
Curry Powder
28E[][]04
28F[][]04

Problem(s):

Note:

Remarks:
Firm listed twice on Red List with same product codes. Remove both listings.

CMS Case Number:
540541

FOI:
No purging required

Published By:
CDR Sam Rudnitsky

Date Published:
11/24/2017

Under the Authority of: John E. Verbeten

FIGURE 2.19 Published FDA notice regarding the removal of Chief Brand Products Limited's curry powder's from detention without physical examination. *Source: Courtesy of Chief Brand Products Limited, Trinidad & Tobago.*

Conclusion

The approach to getting the import alert on CPBL's curry products lifted was successful as indicated by the letter to TSL from the FDA and instructions issued for the removal sent to CBPL (Figs. 2.18 and 2.19) below. The determination of CBPL to comply with the requirements and the overall standards and high-quality, effective processes, equipment and systems employed and implemented played a major role in this success.

Case study 4: Fortified beverages exported to Canada

Background

Technological Solutions Limited (TSL) was asked to provide technical assistance through a review of the information related to the case in which beverage products shipped to Canada were prohibited entry into the country. This involved a review of the information relevant to the case, the categorization under Canadian food laws and the compliance of the products with these laws. A review of the regulations was done in order to determine exactly what caused the products to be non-compliant. Recommendations of the course of action to take as regards future shipments of the products to Canada was then provided. This case study provides a synopsis of the findings and recommendations for the beverage products involved with an example of a Flavoured Drink Mix being used to illustrate the complexity of the case, the findings, the application and interpretation of related Canadian legislation and regulations and recommendations for action. This builds on the information presented in Chapter 1 on Canadian regulations for foods.

Description of the case

In 2016, Technological Solutions Limited (TSL) was requested to undertake an assessment of the compliance of beverage products with the Canadian market requirements. This arose out of a situation in which a shipment of fortified beverage mixes and dehydrated porridge mixes to Canada was initially held and then rejected and had to be returned to the exporter. The products were shipped to Canada from a developing country in which they were made and were imported by a Canadian counterpart. The beverage mixes were packaged in sachets and were of different flavors and consistency (one being a porridge mix). A total of 11 different stock keeping units (SKUs) was affected. They were detained under section 101 of the Customs Act and inspected by the Canadian Food Inspection Agency (CFIA) on January 15, 2016. A letter received from the Department of Justice of Canada further indicated that the products were found to be fortified with vitamins and minerals not permitted in those types of products, which made them non-compliant to D.03.002(1) of the Food and Drugs Regulations. The letter further explained that the importation of these fortified products is prohibited in accordance with A.01.040 of the Food and Drug Regulations.

A review of the relevant sections of the Canadian Food and Drugs Act and Regulations follows, along with the recommended course of action for the product that was prohibited, as well as what should be done to get future shipment to be compliant.

Regulatory review, findings and recommended course of action

The relevant sections of Canada's food laws were reviewed. For clarity, these are presented, where relevant. In the case of the rejected products, the applicable statutes are below:

Canada's Food and Drugs Regulation A.01.040 - Importation subsection A.01.044(1) is copied below.

> **A.01.044 (1)** *Where a person seeks to import a food or drug into Canada for sale and the sale would constitute a violation of the Act or these Regulations, that person may, if the sale of the food or drug would be in conformity with the Act and these Regulations after its relabeling or modification, import it into Canada on condition that*
> *(a) the person gives to an inspector notice of the proposed importation; and*
> *(b) the food or drug will be relabeled or modified as may be necessary to enable its sale to be lawful in Canada.*
> *(2) No person shall sell a food or drug that has been imported into Canada under subsection (1) unless the food or drug has been relabeled or modified within three months after the importation or within such longer period as may be specified by*
> *(a) in the case of a drug, the Director; or*
> *(b) in the case of food, the Director or the President of the Canadian Food Inspection Agency.*
> *SOR/92-626, s. 3; SOR/95-548, s. 5; SOR/2000-184, s. 61; SOR/2000-317, s. 18.*

Action

For the products that were rejected, since the relabeling or modification options were no longer available since the goods were returned, there was no further action to be taken to bring the products into compliance based on this subsection in the Food and Drugs Regulations.

Food and Drugs Regulation D.03.001 – Addition of Vitamins, Mineral Nutrients or Amino Acids to Foods: Subsections D.03.002 (1) and (2) are copied below.

> **D.03.002 (1)** *Subject to section D.03.003, no person shall sell a food to which a vitamin, mineral nutrient or amino acid has been added unless the food is listed in <u>Column I</u> of the Table to this section and the vitamin, mineral nutrient or amino acid, as the case may be, is listed opposite that food in <u>Column II</u> of the Table.*
> *(2) No milk or milk product or derivative listed in Column I of the Table to this section applies to the lacteal secretion obtained from the mammary gland of any animal other than a cow, genus Bos, or a product or derivative of such secretion unless that animal is identified therein.*

Our understanding of the products (nature, contents, and intended use) indicated that the category that most suitably described the products was "meal replacements and nutritional supplements". From the list of twenty-seven (27) different categories, is the section of the table that relevant to these types of products was extracted. Further information is also presented below.

For further assessment of the compliance, samples of the labels for the affected products were requested so that the ingredients listing of each could be compared to the extracted requirements as seen above. The results of the comparison indicated that while all of the Flavoured Drink Mixes and Flavoured Drinks were compliant, the Porridge Mixes were

TABLE 2.1   Extracted information from the food and drugs regulation that applies to meal replacement and nutritional supplements.

Column I	Column II
Food	Vitamin, mineral nutrient or amino acid
Meal replacements and nutritional supplements	Vitamins — alpha-tocopherol, biotin, D-pantothenic acid, folic acid, niacin, riboflavin, thiamine, vitamin A, vitamin B6, vitamin B12, vitamin C, vitamin D
	Minerals — calcium, chloride, chromium, copper, iodine, iron, magnesium, manganese, molybdenum, phosphorus, potassium, selenium, sodium, zinc

not. This was because the latter contained *Thiamine Mononitrate (Vitamin B1)* and *Niacinamide (Vitamin B3),* neither of which are allowed to be added to products in this category in Canada. This was based on Canada's Food and Drugs Regulation D.03.002(1) and (2) and the fact that the Porridge Mixes were categorized as "meal replacements" or "nutritional supplements". On the other hand, vitamins and minerals listed in the ingredients of the six (6) Flavoured Drink Mixes and four (4) Flavoured Drinks products were compliant with the regulations, as these products also classified as "meal replacement" and "nutritional supplements" did not contained disallowed ingredients.

Recommendations

In light of the findings, TSL suggested that a letter be prepared by the company's lawyer, to seek clarification as to what category the food was rejected as, and to also indicate that the intended purpose and nature of the beverage products were as meal replacements. This response was expected to highlight the ingredients in each product as compared to the vitamins and minerals that were allowed so that it would be clear that the products were compliant. This letter was sent in response to the letter received from Department of Justice of Canada. It was also suggested that a formal application should be made to have the products categorized as meal replacement or nutritional supplements to allow compliance with Canada's import laws. For the products which did not comply, TSL recommended that the formulations were to be changed (if feasible) to use vitamin premixes which contain the vitamins and minerals that are allowed in Canada, as shown in Table 2.1 (below).

TSL was subsequently requested to help them to develop an approach to get the drink mixes and related products into the Canadian market. This included a determination of the appropriate category in which the products should be placed and an assessment of the products for compliance with the Canadian market requirements for those categories. TSL collaborated with its associate, FJS Consulting on this assignment. Out of this, an approach to market entry was crafted that would ensure the compliance of the products and provide technical assistance to the manufacturer for future shipments of the products to Canada. The outcome of the assessments is presented below with specific recommendations as to what will need to be done in respective areas covered. An overall summary of the recommendations is also presented at the end of the section.

Label assessment & findings

Label assessment for compliance to Canada Regulations was done for the Porridge Mix and a Flavoured Drink Mix representative of all the provided drink mixes being considered for export to Canada. The analysis presented below is an example of the details provided on both compliance and non-compliance to the regulations. The analysis provided the logic and recommendations as to under which regulations would these products best be documented for acceptance in Canada as well as any further work that may be required to accomplish this. For the purposes of this example and in consideration of the granular details involved, only an example of the flavoured drink mix will be provided below.

Flavoured drink mix

Scope

In Canada, nutritional supplements fall under the broader category of "Foods for Special Dietary Use." These foods are specially formulated to meet particular requirements of people with certain physical or physiological conditions resulting from disease or injury, or those for which a particular effect is required through a controlled intake of food. The drink mixes in question were not intended for this subset of the population; the advice was that they were intended to be a more a type of food with added vitamins and minerals that were intended for the general public. These additions, however, are not permitted by the Canadian regulations – in other words, once fortified, these foods would fall into the category of *Supplemented Foods*. The addition of vitamins and minerals is highly regulated and as Supplemented Foods, these products would have to be approved by Health Canada through the Temporary Marketing Authorization (TMA) application process, which could take 3–6 months or longer. The assessment, findings and recommendation presented below therefore report the guidance provided to the exporter on mandatory food labeling information for the products according to the requirements of the Canadian Food and Drug Regulations, with an emphasis on **Supplemented Foods.**

General information

- All information must appear in both English and French with equal prominence.
- All information must appear in type height of at least 1/16" (1.6 mm), except where noted. This is based on the height of an upper-case letter where words appear in upper case. When words appear in lower case or in a mixture of upper and lower case, it is based on the height of the lower-case letter "o".
- While no specific font type is stated, the information on a food label must be clearly and prominently displayed on the label, and readily discernible to the purchaser or consumer under the customary conditions of purchase and use.

For greater clarity for the discussions on categorizations and options that follow, further excerpts of the Canadian Food and Drug Act are presented below.

Excerpts from Canada's Food and Drugs Act

5. (1) No person shall label, package, treat, process, sell or advertise any food in a manner that is false, misleading or deceptive or is likely to create an erroneous impression regarding its character, value, quantity, composition, merit or safety.

Common name The common name of a food is the name prescribed in the Food and Drug Regulations or other federal regulation, or where that is not applicable, when not prescribed by regulation, the name by which the food is commonly known. It must accurately represent the product and not be misleading.

Location: The common name must appear on the principal display panel (PDP), which is the part of the package that is displayed to the consumer under normal conditions of sale.

Findings

Our assessment found that the name of the product was *not an acceptable common name by virtue of the decriptions used* and recommendations were provided in how to address this.

Net quantity The Net Quantity is the amount of the food being sold, declared in metric units. The statement "Net Weight/Poids net" is not required but permitted if present in both English and French.

Location: The Net Quantity must appear on the principal display panel (PDP), which is the part of the package that is displayed to the consumer under normal conditions of sale.

Nutrition facts table Unless otherwise exempted, all pre-packaged products must carry a Nutrition Facts Table (NFT) in order to convey information to the consumer about the nutrient content of the food.

Location: The table may appear anywhere on the package but should be printed on a continuous surface (i.e. not around any corners). It may be oriented vertically or rotated on its side where vertical space is limited.

Format: The Canadian food regulations dictate the sizes and presentation formats that can be used. Regulations require that the NFT be printed in a dark color (i.e. black) against a white or neutral background using "san serif" fonts such as Arial or Helvetica. Serif or other decorative fonts are not permitted in the table.

Findings

The nutrition table on the label was not compliant. There were several noncompliant elements some of which are indicated below:

> Based on the dimensions of the label artwork pdf, the label needed reformatting.
>
> Salt cannot be declared in the NFT; it should be replaced with sodium.
>
> For Supplemented Foods, the absolute amounts (mg, mcg, etc.) and the % DV of both the core (Vitamin A, Vitamin C, calcium and iron) and added vitamins and minerals (assuming they are permitted) must be declared in the NFT.

Ingredient list and allergens Ingredients must be listed by common name in descending order by weight, including the components of compound ingredients when applicable. There are priority allergens and sensitizing agents that must be named specifically on a food label within the ingredient list or immediately following it in a "Contains" statement when they are included as an ingredient or sub-ingredient: eggs, milk, peanuts, tree nuts, fish, crustaceans, shellfish, mustard, sesame, soy, sulfites, wheat and gluten from sources such as oats, rye and barley.

Location: The ingredient list and allergen declaration must be grouped together and may appear anywhere on the package except for the bottom. Ingredients should be listed in descending order by weight and all components of compound ingredients must be declared. Ingredient specifications were not provided, but the following comments on ingredient list was included in the technical report:

- It is suggested not to use upper case letters or bold type to identify allergens. This isn't specifically stated in the regulations, but it is not common practice.
- *Contains sulfites* can be shortened to *sulfites* (in parentheses after soy protein isolate).
- Change *flavorings* to *flavor* or *natural flavor* (if applicable). If the flavors are artificial, *artificial flavor* must be declared. Nature identical flavors are considered to be "artificial" in Canada.
- It is not necessary to declare the function of a food additive (e.g. emulsifier) or its e-number.
- The proper form of calcium phosphate must be declared (*mono, di* or *tricalcium phosphate*). All three forms are permitted in this type of food as a pH adjuster. Only the tri- form is permitted as an emulsifier. If it is used for another purpose or if it a different form, it may not be permitted. The maximum legal level was not verified.
- The proper form of potassium phosphate must be declared (*mono, di* or *tripotassium phosphate*). The di- and tri- forms are permitted in this type of food as a pH adjuster. Only the tri form is permitted as an emulsifier. If it is used for another purpose or if it a different form, it may not be permitted. The maximum legal level was not verified.
- Change *diacetyltartaric ester of mono- and diglycerides of fatty acids* to *diacetyl tartaric acid esters of mono- and diglycerides*.
- *Magnesium phosphate* may not be permitted as an additive.
- The quantity of acesulfame-potassium and sucralose must be grouped with the ingredient list, expressed as *Contains x mg acesulfame-potassium and x mg sucralose per 30 g serving*

Dealer name and address (Domicile)   The name and address of the responsible party must be declared in a way that is sufficient for mailing purposes, i.e. − name, city, province, country, postal code, etc.

Location: This information may appear anywhere on the package except for the bottom. There were no issues with the English domicile statement as it appears on the label PDP.

Regarding the web address   Internet advertising is covered by the Food and Drugs Act and the Consumer Packaging and Labelling Act advertising definitions, and as such, it is subject to the same criteria as other advertising (e.g. labels). Non-compliant internet advertisements may be subject to enforcement action by the Canadian Food Inspection Agency (CFIA). The CFIA has jurisdiction over Canadian-based websites that advertise foods. If multiple countries have sections on a Canadian-based website, any non-compliances on the Canadian section of the site may be subject to enforcement action. Advertising on websites that are not based in Canada does not fall under CFIA jurisdiction.

Date codes If the shelf life of the product is 90 days or less, the "best before date" must appear on the label. If the shelf life is greater than 90 days, the best before date may still be displayed, as long as it appears in the proper format (see below).

The words BEST BEFORE/MEILLEUR AVANT must be grouped together with the date. Use this wording to replace BEST IF USED BEFORE.

The format of the durable life date is a two-letter bilingual abbreviation for the month and the date, e.g. JN 28. The year (4 digits or the last 2 digits) may be added for clarity, e.g. 15 JN 28 or 2015 JN 28. The bilingual symbols for the months in the durable life date are as follows:

JA for JANUARY	**JL** for JULY
FE for FEBRUARY	**AU** for AUGUST
MR for MARCH	**SE** for SEPTEMBER
AL for APRIL	**OC** for OCTOBER

Claims

Lactose-free Although there is no criteria in the regulations, this means that there is no detectable lactose in the food using an acceptable analytical method. The CFIA *may ask manufacturers to provide documentation that substantiates that the claim.*

Rich in protein
— *The product must have a minimum protein rating of 40 (manufacturer must confirm). If the protein rating is less than 40, then the claim must be removed from all packages bearing this claim.*
— The protein rating (PR) calculation is # grams of soy protein per 30 g, multiplied by 2*
— The protein efficiency ratio of soy protein is 2. *It may be higher for soy protein isolate and PR calculation is required to confirmaccurcy of the information provided.*

Rich in calcium
i. The product must contain at least 25% of the DV *before rounding* to comply. This claim was compliant based on the information on the artwork.

Fortified with iron
i. Iron is not permitted to be added to Supplemented Foods, therefore this is not an acceptable claim.

Recommendation

It was recommended that this claim be removed. Additional recommendations were made based on the observed PDP including the removal of the national flag of the manufacturer or the inclusion of the statement "Product of XXX" adjacent to the flag.

Sweeteners

— *The PDP must show a statement such as:*
   (Contains or Sweetened with) acesulfame-potassium, sucralose, corn syrup solids and sugar.
— *The above must be declared in letters that are at least the same size and prominence as the numerical portion of the net quantity declaration.*

Artificial flavors/picture on PDP

If artificial flavors are used, then "artificial flavor" or "simulated flavor" must appear adjacent to the picture indicating the flavor e.g. chocolate on the PDP, in the same type height as that used for the numerals of the net weight declaration.

Information specific to supplemented foods

In order to be considered eligible for a TMA, the food should provide no more than the maximum levels, including both naturally occurring and added sources, for the vitamins and minerals established for either Path 1 or Path 2.

The following minerals listed on the artwork provided for the products were not permitted in Supplemented Foods:

— Iron
— Iodine

The following vitamins listed on the artwork did not have maximum levels established for Supplemented Foods. *Any levels above those found in currently marketed supplemented foods will be assessed by Health Canada during the TMA approval process.*

— Riboflavin
— Thiamine
— Vitamin B_{12}

Findings

The amount of Vitamin D appears to be compliant (below the Path 1 maximum level). The label artwork did not state the amount of magnesium and therefore its compliance could not be verified.

The amounts of the following ingredients are above the Path 1 maximum levels: zinc, phosphorus, potassium and calcium. As such, additional cautionary statements are required on the label, including:

— *Not intended for children OR For adults only*
— *If you take a daily supplement, you may be getting too much calcium and zinc by consuming this product*
— *Do not exceed 1 serving per day*
— *Do not consume this product with other supplemented foods*

Recommendation

The product formulation could be adjusted to reduce these elements to be within the current existing limits — which was undertaken by the maufacturer

When Supplemented Foods require cautionary labeling, Health Canada recommends that all forms of advertising of these products also contain the following information (or similar wording):

— *This product may not be suitable for everyone.*
— *Read the label and follow the directions for use.*

Our recommendation would be to avoid this kind of cautionary labeling.

Product labeling and composition

Manufacturers can use the categorization of products listed above to assist in determining where their products should fit and hence which regulations they need to conform to. Where products or ingredients fall into the Supplemented Food category or do not fit into any other category then approval for sale in Canada must be provided by Health Canada. This is typically provided through the Temporary Market Access (TMA) process which is basically the issuing of temporary licenses. The temporary nature allows for time to collect additional information that ultimately can be used to slot the product into an existing category or time for Health Canada to amend regulations. These amendments are currently being considered for many products in the Supplemented Food category.

As part of the process of determining categorization and then ultimately regulation, the manufacturer will be working with their importer to understand current labeling and composition of their products and where and how they might need to amend them to meet regulation. Ultimately, they must ensure compliance with the composition, labeling, nutritional labeling and claims provisions of the *Food and Drugs Act and Regulations*, the *Consumer Packaging and Labelling Act and Regulations* and other commodity specific legislation, such as the *Canada Agricultural Products Act and Regulations*.

Rationale

Compliance with Canadian food safety standards and compositional requirements promotes a safe food supply and helps ensure that consumers' expectations are met with respect to food composition and quality. The declared composition, compositional standards, nutrition information and claims must be adequately controlled to prevent misrepresentation and fraud, and to prevent such health hazards as the presence of undeclared food allergens.

Mandatory information on food labels allows consumers to make informed choices by:

- providing basic product information (e.g., the product's common name, its list of ingredients, its net quantity, its durable life date and country of origin, as well as the name and address of the manufacturer, dealer or importer), and
- providing health, safety and nutrition information (e.g., instructions for safe storage and handling, nutrition information in the Nutrition Facts table, and specific information on products for special dietary use).

NOTE: "Mandatory information" is determined by the commodity and its regulatory requirements.

Composition

- Control factors that are critical to the safety and integrity of the product are identified (This includes microbiological, chemical or physical concerns, as well as concerns related to allergens, extraneous matter, etc.). Their specifications and limits are identified (e.g., thermal process, pH and water activity for ready-to-eat fermented meats, salt content for brined mushrooms, sulfur content of dried fruit, freedom from *Salmonella*, etc.)
- Imported food which is sold in Canada must comply with Canadian nutrient composition and nutrition labeling requirements and considerations may include the following:

- added nutrients (i.e., vitamins and minerals) permitted in the specific foods are added at appropriate levels as set out in the *Food and Drug Regulations*;
- foods for which there are nutrient composition requirements (e.g., meal replacements, nutritional supplements, flour, etc.) meet these requirements;
- the nutrient content of the product is accurately reflected on the label and in compliance with the *Food and Drug Regulations* (i.e., the list of ingredients and Canadian Nutrition Facts table);
- products for which there are nutrient content claims and/or diet-related health claims meet the Canadian compositional and labeling requirements under the *Food and Drug Regulations*. (See Section 7, Nutrient Content Claims, and Section 8, Diet-Related Health Claims, in the *Guide to Food Labelling and Advertising*.)

Importer responsibilities

Manufacturers and importers must work together in order to ensure products are in compliance prior to arrangement to enter the Canadian markets. As stated previously, it is the importers' responsibility to ensure this is in alignment. The consequences if they are not could be delayed entry, but more common place is refused entry simply because re-labeling cannot change the composition and formulation of products if they are incorrect to begin with. This can lead to costly storage and transportation costs or in the case of short shelf life product, loss of product as well. Here are some specific practices that should be in place with the importer for both initial and on-going shipments:

- Procedures are in place to enable the verification of a product's formulation, its list of ingredients, its nutrient content and nutrient values as stated on the label. These procedures may include, but are not limited to:
  - a documented communication system with suppliers which permits access to product specific information (e.g., names and amounts of ingredients, components, additives, and if applicable, added nutrients);
  - an overview of the method(s) used to determine nutrient values (e.g., laboratory analysis, the use of a database for finished foods, the use of an ingredient/recipe data base, published values, the use of a consultant or other technical expertise, etc.);
  - the name and address of the person responsible for the values;
  - possession of, or ability to obtain, complete and current written specification sheets detailing product ingredient and nutrient information;
  - product testing and/or information linkages with suppliers of products that are sole sources of nutrients (such as baby formula) and foods for special dietary use (such as meal replacements and nutrition supplements).
- Procedures are in place to verify the supplier's (manufacturer's) food safety system and its compliance to Safe Food for Canadians regulations.
- In the case of corrective labels applied in Canada, incoming labels from labeling and/or printing companies are reviewed against signed-off proofs.

- All pamphlets, posters, handouts and other Canadian advertising materials developed and/or distributed by the importer are reviewed and verified for accuracy and compliance with Canadian legislation. (See Section 3 of the *Guide to Food Labelling and Advertising*.)
- The importers have obtained allergy-related information from their suppliers and manufacturers. This information can be used to conduct hazard analysis and assess the need for label changes, including the need for any precautionary statements on the label (e.g., "may contain peanuts").
- Labels are reviewed both for compliance with Canadian legislation (e.g., presence of mandatory labeling), and for accuracy and correctness of information. This assessment includes, but is not limited to:
  - mandatory information, quality claims (e.g., organic, natural, fresh, Kosher, etc.), compositional claims (e.g., no preservatives, etc.), nutrition claims, and standards of identity.
- Products are periodically reviewed against the Nutrition Labelling Compliance Test to assess the accuracy of nutrient values.

Recommendations

It is highly recommended that the importer work with a labeling and regulatory expert to ensure all labels are up to date. The expert could be on staff, could be a resource with the supplier/manufacturer or could be under contract, as required. Appropriate data, history and analytical results are required to enable appropriate product and label evaluation so the supplier/manufacturer should be prepared to produce this for the importer and their technical experts.

Summary and recommendations

The dry blended drink and porridge mixes examined in this case fall under the "Supplemented Foods" category in the Canadian market. The addition of vitamins and minerals is highly regulated and as supplemented foods, they must be approved by Health Canada through the Temporary Marketing Authorization (TMA) application process, which could take 3−6 months or longer.

The labels of all products intended for export to Canada by the exporter needed to be revised in order to be brought into compliance with Canada's Food and Drug Act. Revisions were needed to the Nutrition Facts table. Specific details about what was required for the declarations for the drink and porridge mixes were provided in a technical report to the exporter for their action. Based on the current formulations being manufactured, reformulation of the products would be required to make them compliant with Canadian regulations as some of the vitamins and minerals declared on the current label at the existing levels are non-compliant.

Other areas that were found to need specific attention included certain aspects of the cautionary labeling on the product. Assistance was provided to the exporter to prepare for, and when ready start the process of presenting TMA submissions to Health Canada for the products for which entry to the market was being sought. It was also recommended that assistance with review of the final label artwork be sought and that guidance

FIGURE 2.20 Pickapeppa sauce (the original at left) and related products. Source: Courtesy of The Pickapeppa Company Ltd., 2019.

with the nutrition facts table also be sought. An approach which would see the local technical support provider, TSL, working in collaboration with a Canadian associate to ensure the completion of the technical process in Canada, as well as the accuracy of the English/French translations required was recommended as well.

Case study 5: Expanding market success through food safety system certification: The Pickapeppa Company

Background

The Pickapeppa Company Limited (TPCL) had a humble beginning in the kitchen of Norman Nash who formulated their flagship product in June 1921. Twenty-four years later the rights and business were sold to the Lyn Kee Chow family, the present owners, with operations at Shooters Hill, Manchester, Jamaica. The company prides itself in maintaining the distinct quality, taste and flavor of Pickapeppa Sauce (Fig. 2.20) today the same as it was originally. The process of crafting the flagship Pickapeppa Sauce is a meticulous operation, which requires patience and takes time. The details of the formula are a trade secret, known only to family members, passed down from generation to generation and kept securely in a bank vault. This sauce, like a fine wine, is still being aged in casks to achieve its unique flavor today as it was on the first day it was made. Pickapeppa manufactures a range of premium quality, artisanal, all-natural sauce products. The firm uses only 100% natural ingredients, inclusive of the manufacture of their own cane vinegar. No artificial colorings, thickeners, preservatives or flavorings are used, the products are gluten free and are suitable for vegetarians, having a shelf life of three years from date of manufacture. The company employs more than forty persons from the local communities, several of whom have been with the firm for their entire working life.

As is done for fine wine, the flagship original Pickapeppa Sauce is aged in oak barrels in the company's warehouses (Fig. 2.21). The present Pickapeppa product range in 148 mL (5 oz) glass bottles and industrial ingredients sizes includes Pickapeppa Original Sauce, Hot Pepper Sauce, Spicy Jerk Marinade (Fig. 2.20) and Original Jerk Marinade. More than

FIGURE 2.21 Pickapeppa sauce being aged in oak barrels. Source: Courtesy of The Pickapeppa Company Ltd., 2019.

90% of Pickapeppa products are exported. Pickapeppa products can be found in all major supermarkets in the Caribbean, USA, Canada, United Kingdom, Cayman Islands, New Zealand, and in many other countries.

While the iconic Pickapeppa brand has a worldwide presence, it is particularly successful in the USA where its sauce products are widely distributed and are available in major retail chains such as Safeway/Albertsons, Walmart, Publix, Whole Foods and Target, as well as many small corner shops in a range of communities. For a small privately-owned family business from a developing country to have achieved such global recognition and, in many cases, to have crossed from the ethnic shelves to become mainstream alongside brands such as Heinz, Tabasco and Worcester Sauce is testimony to its quality and the ability The Pickapeppa Co. Ltd. to adapt to continually changing demands from the market. Among these changing demands has been that for the firm, despite its long standing and renown, to demonstrate the safety of its products through certification.

The need for Food Safety & Quality Systems (FSQS) Certification

The company began getting increasing requests for documentation and other evidence regarding the compliance of its products with food safety and quality standards from 2015, with the requests become more frequent and more detailed as time passed. Up until that time, like many other companies in Jamaica, a country with a reasonably well developed regulatory apparatus, Pickapeppa had relied on approvals and certification from the local regulator, Bureau of Standards Jamaica. It readily became apparent to the Pickapeppa management team that the USA market, through fear of litigation in the event of food safety issues arising, was not satisfied with this national level of assurance. Indeed, major customers were unequivocally demanding that the company had to attain a Global Food Safety Initiative (GFSI) certification if they were to continue in business together. Some also made it quite clear that they would not be taking any new customers from 2016 unless they were GFSI certificated. The choice was stark — get a GFSI certification or forfeit a key segment the USA export market that had been carefully built over the last sixty years.

A global food safety initiative

In 2016, the International Finance Committee (IFC) of the World Bank embarked on a program in Jamaica to persuade a cluster of condiment manufacturers to achieve a Global Food Safety Initiative—benchmarked standard certification to make them more competitive in the international market. Pickapeppa joined this program. After some discussion within the cluster, it was decided that Food Safety System Certification 22000 (FSSC 22000) would be the most appropriate form of certification as most members had already been operating in accordance with HACCP or ISO standards. The FSSC 22000 Scheme had been given full recognition since 2010 and GFSI recognition demonstrates that the Scheme meets the highest standards globally leading to international food industry acceptance. Furthermore, the scheme is widely accepted by Accreditation Bodies worldwide and supported by important stakeholders like Food Drink Europe (FDE) and the American Grocery Manufacturers Association (GMA), now the Consumer Brands Association (CBA).

A key concern of the management at the outset was that Pickapeppa's factory had been constructed in the 1940s and was therefore far from being a modern facility. It was therefore anticipated that substantial structural and layout change would be required to meet the certification standard. Fortunately, one of the first activities scheduled under the IFC program was for an experienced international consultant to visit the plant at Shooters Hill, Manchester, Jamaica for a two-day intensive audit of existing management systems, processes and the facility. The positive feedback from this audit was instrumental in the decision to embark on the FSSC 22000 certification program.

At this point because of the increasing pressure from overseas markets, it was decided to fast track the certification program and set the challenging target to achieve it in just six months. This was felt to be the least disruptive approach for the business and its customers. To support this intensive program, Pickapeppa worked with Quality Circle International Limited, a Jamaican owned consulting company based in Texas. This was to prove an excellent strategy as it provided a one-on-one service in all aspects of the certification requirements. As a small business with wide ranging calls on available funds close control on finances was essential as the process of attaining compliance required repairs and upgrading of the factory floor to facilitate rearrangements of the workflow plan; building new raw materials and finished goods stores and hygienic metal cladding of internal ceiling areas and walls.

The factory was just one aspect of the overall work towards compliance: internal audit systems were updated, focus on staff training was intensified, and record keeping which had been largely paper-based was to be remodeled and iteratively loaded to a specially designed database with Cloud backup.

By the end of the six-month schedule of work that had had been agreed, the company was ready for the Stage 1 audit by Intertek. The firm indicated that the audit was even more wide ranging and detailed than they had expected, but they had been well prepared and there were no issues that would have precluded the scheduling of the Stage 2 audit two months after the stage 1 audit was completed. The final audit covered much of the same ground as Stage 1, though in greater depth. Again, the firm was successful, with some minor non-conformances which were soon cleared. The rest is history as Pickapeppa continues on the annual cycle of FSSC 22000 audits.

The benefit of a GFSI certificated FSQS

The benefits of attaining certification to a GFSI-benchmarked program and what it means for the firm are substantial and are discussed in more detail in Chapter 4. This case study, however, facilitates a view of the benefits from the perspective of an iconic developing country brand that has achieved a measure of global success, such as Pickapeppa. They describe the benefits, in their own words, as follows:

"During the process of preparing for FSSC we soon began to see the benefits of our substantial investment in our team and the factory. Not only did the factory workflow make management of our processes more efficient but the top down bottom up approach to intensive training meant that everyone in the team identified with the task and wanted to be part of the success. Job descriptions rather than being largely generic now included ownership of specified documented processes in a way that had not been possible before and thus a shift in culture was achieved. Most importantly, the demand by our customers for large numbers of documents relating to manufacturing and quality processes dwindled and were to a large extent in many cases superseded with a simple presentation of our FSSC 22000 certificate".

The changes to Pickapeppa's operations and business have given them more confidence in going for greater global reach and expansion of their presence, brand and range. As such, the firm has introduced new products such as its *Spicy Mango Sauce, Hot Mango Sauce* and *Gingery Mango Sauce.* As they have indicated, certification *"has also impacted how our customers see us as a company committed to deliver excellence"*.

Food Safety System Certification

Pickapeppa Company Limited is certified under Food Safety System Certification FSSC 22000 v 4.1.

References

EUR-Lex, 2019. Council Directive 2006/88/EC of 24 October 2006 on animal health requirements for aquaculture animals and products thereof, and on the prevention and control of certain diseases in aquatic animals. Accessed on 5 August 2019 from https://eur-lex.europa.eu/legal-content/EN/ALL/?uri = CELEX:32006L0088# ntr16-L_2006328EN.01001401-E0016.

Gordon, A. (Ed.). 2015. Food Safety and Quality Systems in Developing Countries: Volume One: Export Challenges and Implementation Strategies. Academic Press.

Gordon, A. (Ed.). 2016. Food Safety and Quality Systems in Developing Countries: Volume II: Case Studies of Effective Implementation. Academic Press.

Higuera-Ciapara, I., Noriega-Orozco, L.O., 2000. Mandatory aspects of the seafood HACCP system for the USA, Mexico and Europe. Food Control 11 (3), 225–229.

Jespersen, L., Griffiths, M., Wallace, C.A., 2017. Comparative analysis of existing food safety culture evaluation systems. Food Control 79, 371–379.

McDorman, T.L., 1997. Seafood Safety Standards (With Special Reference to HACCP): Review of the Import Regulations of the US and EU and the Relevant Laws of the South Pacific Region. Secretariat of the Pacific Community.

Nayar, S., Bott, K., 2014. Current status of global cultivated seaweed production and markets. World Aquacult. 45 (2), 32–37.

Stone, J., Rahimifard, S., Woolley, E., 2015. An overview of resilience factors in food supply chains. Presented at the 11th Biennial Conference of the European Society for Ecological Economics, Leeds, 30th June-3rd July, European Society for Ecological Economics.

Yiannas, F., 2008. Food Safety Culture: Creating a Behavior-based Food Safety Management System. Springer Science & Business Media.

Yiannas, F., 2015. Food Safety = Behavior: 30 Proven Techniques to Enhance Employee Compliance. Springer.

Further reading

European Commission, 2019a. EU import conditions for seafood and other fishery products. Accessed on 28 August 2019 from: https://ec.europa.eu/food/sites/food/files/safety/docs/ia_trade_import-cond-fish_en.pdf.

European Commission, 2019b. Third country establishments. List by section. Section VIII: Fisheries products. Accessed on 28 August 2019 from: https://webgate.ec.europa.eu/sanco/traces/output/non_eu_listsPerActivity_en.htm.

Jespersen, L., Griffiths, M., Maclaurin, T., Chapman, B., Wallace, C.A., 2016. Measurement of food safety culture using survey and maturity profiling tools. Food Control 66, 174–182.

Powell, D.A., Erdozain, S., Dodd, C., Costa, R., Morley, K., Chapman, B.J., 2013. Audits and inspections are never enough: A critique to enhance food safety. Food Control 30 (2), 686–691.

Seafood HACCP Alliance, 2008. Seafood exporters to the EU. Accessed on 28 August 2019 from: http://www.haccpalliance.org/sub/news/seafood.pdf.

3

Food safety and quality systems implementation along value chains

André Gordon[1] and Dianne Gordon[2]

[1]Chairman & CEO, Technological Solutions Limited, Kingston, Jamaica, West Indies [2]Jamaica Bauxite Institute, Kingston, Jamaica; UWI SALISES Sustainable Rural and Agricultural Development (SRAD) Cluster, Jamaica

Introduction

The supply of food to developed country markets from emerging economies continues to increase, leading buyers, processors, exporters and international manufacturers to place greater focus on the systems that deliver these foods to the market. However, in most food production and supply chains located in developing countries, there are a range of challenges that often mitigate against long term, sustainable and profitable engagement of the primary producers, as well those entities involved in the transportation, distribution, sale and marketing of the agri-food products all the way through to the final consumer. These challenges include how to ensure consistency of supply, overall competitiveness and fair distribution of the economic benefits. One of the more efficient and sustainable approaches to ensuring this supply is increasingly to have it organized within global value chains where the activities required to deliver a product to the final consumer involve different countries and close collaboration across entities. As a result, global food value chains are becoming increasing complex and transnational in scope, with different nodes (links) in the chain being located in different countries, all required to cooperate to deliver higher value, safe, wholesome quality foods that meet consumer's expectations.

One major challenge is how to ensure that all participants understand the requirements of the target market, how to meet the demand in the regional and global marketplace for specialized, differentiated and branded food products, including requirements for food safety and quality. With the increased demands for safe, high quality food by consumers in the global market, the assurance of food safety and quality is among the more important non-economic considerations in food supply chains. Food value chains are now required, therefore, to deliver safe, quality food, giving considerations to the environmental and social dimensions of how food is produced, while meeting requirements for cost and reliability of supply. Building food safety and quality through systems implementation along the value chain not only enhances value at each step but also reduces losses and often reduces costs (Gordon, 2016). Developing countries have become, and will remain for the foreseeable future, important participants in these globalized value chains, meeting the challenges of delivering safe, high quality agri-food products through their production, processing and distribution systems.

This chapter explores the concept of value chains in the global food industry, addressing the role and importance of food safety and quality systems (FSQS) in building and strengthening value chains, providing an overview of how value is created through food quality and safety protocols in the agri-food sector. Some of the challenges and opportunities for stakeholders along food value chains in developing countries and globally are discussed. These include achieving quality and food safety standards, accountability for sourcing, handling, and traceability and quality assurance of food along the chain. Specific case studies of different food value chains are employed to facilitate a discussion of the complexity of the efforts required to implement food safety and quality systems that add value and improve FSQ outcomes. These cases include value chains involving *Colocasia esculenta* or dasheen (taro), sweet (bell) peppers (*Capsicum annum*) as an example of a local vegetable value chain, papaya (*Carica papaya*) for fruits, cashew (*Anacardium occidentale*) for tree nuts and cocoa (*Theobroma cacao*). In each case, the focus will be on highlighting the activities and practices that add value and contribute to maintaining or enhancing the safety of the food, while maintaining quality.

The value chain concept

The value chain concept was developed by Michael Porter (1985) in his analysis of the development of competitive advantage of businesses, particularly in the manufacturing sector. Porter defined a value chain as the set of activities that an organization carries out to create value for its customers (Porter, 1985). The main focus of Porter's value chain analysis was on how the organization's in-house activities interacted and on the identification of potential sources of competitive advantage (University of Cambridge Institute for Manufacturing IfM, 2019). Porter's value chain was initially represented by a systematic mapping of the organization's primary and support activities that create value.[1] The concept has since evolved to incorporate the entire system of upstream suppliers and downstream buyers, with the ultimate focus on the final customer (Kaplinsky and Morris, 2001).

A definition of value chains now commonly used is "the full range of activities which are required to bring a product or service from conception, through the intermediary phases of production, delivery to final consumers, and final disposal after use." (Kaplinsky, 2000). According to Ensign (2001), value is determined by the unique combination of attributes in a product or service that are important to the customer. Value is added or created as specific functions or activities are executed to provide a product or service. Companies conduct value chain analysis by examining how every activity and linkage within the chain that is required to create a product can be fine-tuned to enhance customer value. The ultimate goal of this analysis is to create and strengthen competitive advantage, which can result from various configurations of activities, delivering maximum value at the least possible cost. The value chain concept has been widely adopted as a tool for understanding the structure and components of value creation in various types of operations, with food production and supply systems being no different (Gereffi and Fernandez-Stark, 2011; Gereffi and Kaplinsky, 2001). In this context, the well-known concepts of a supply chain (Taylor, 1919) and supply chain management (Mentzer et al., 2001) have been expanded to facilitate a broader assessment of not only the production, processing, handling and delivery of food through the series of steps leading from primary production through to the final consumer, but also an assessment of how value can be best delivered at each step along the chain.

Food value chains: activities and participants

The food industry provides interesting opportunities for the study of value chains, particularly from the perspective of how food safety and quality systems interact with and enhance value to participants. Food value chains comprise all value-adding activities that are required to bring food products to the final consumers (FAO, 2014). Agri-food value chains inherently involve several stages of discreet activities involving many participants. These activities include agricultural production, trading and buying the raw products, storage, processing

---

[1] Primary activities are those undertaken to design, produce, market, deliver a product. These include inbound logistics, operations, outbound logistics, sales, marketing and customer service. Support activities include those such as procurement, research and development, finance, legal, corporate management, process enhancement that support a product in the marketplace.

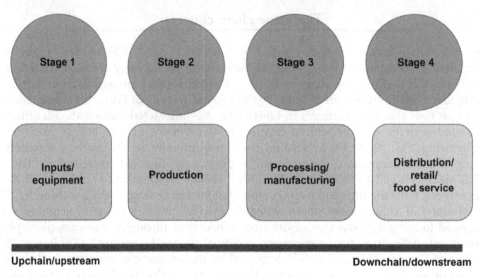

FIGURE 3.1   Stages in a simple food and agribusiness value chain.

and manufacturing of the primary and value-added products, marketing, distribution and consumption. The participants include input supply companies, farmers, agents, traders and retailers, processors, distributors, food companies, regulators and final consumers. In agri-food value chains, activities are coordinated around primary and supporting activities that add value to products. All activities are geared towards satisfying the varying demands of the final consumer. Agri-food value chains also display diversity in that they may involve small-scale subsistence farmers and/or large high-tech commercial farms, small and medium-sized enterprises and/or large transnational corporations. Even the smallest food service outlet displays elements of a value chain from sourcing and purchasing of ingredients, equipment and utensils, through hiring of staff, advertising and preparation of food all the way to serving or delivering the meal to the customer.

According to Humphrey and Memedovic (2006), the agri-food value chain can be divided into four main stages i.e., input supply, production, processing, and delivery to consumers. Fig. 3.1, illustrates a simple agri-food value chain. In the food value chain, input supply is described as upstream activities providing agricultural products and services to farmers as primary customers. Input supplies in the agri-food value chain include chemicals, nutrients, seeds, seedlings, and farm equipment. Production comprises all activities involved in the creation of raw food materials, including, for example, crops, milk and livestock. The raw food materials are passed on to the next link of the value chain to processors or manufacturers that convert them into branded or unbranded packaged or unpackaged products. Food processing companies include producers of packaged foods, dairy, meat and other food products, including beverages (juices, soft drinks, sparkling water, energy drinks, beers, wines, spirits, teas, etc.). Downstream[2], these products are

---

[2] Downstream activities are activities related to handling the products from the core activities, i.e. the output of the core activities, such as harvesting, live transport, processing, exporting and distribution.

then marketed or distributed to wholesalers, retailers and food service companies, and are then sold to the final consumers. This stage, the distribution, sale and marketing of food to the end user, is the final stage in the agri-food value chain.

When managing a value chain, the objective is to maximize value at each stage of the chain while minimizing costs. In food value chains, this means optimizing production and handling, employing industry best practices of the type, scope and scale appropriate to the activities at that node of the chain, while ensuing the delivery of a fully compliant, safe, high quality food to the next node in the chain. Effectively managed food value chains identify and link the processes or activities that take each food product from raw material production in the farm or field through the actualization of the product design concept (for a manufactured item) to the specific target market. This process also involves the supplier of inputs, including all of the basic agricultural items (seeds/seedlings/plantlets, fertilizer, soil amendments, etc. for plant-based foods or feed, vaccines, supplements, etc. for foods of animal origin), as well as full traceability of inputs back to source. By its very nature then, the value addition process involves several participants and many stages of discreet activities from input supply, production, processing, sales and marketing, to distribution and delivery to the final consumers (Fig. 3.1). Each stage of the value chain adds incremental value to the product by enhancing the utility of the food item and maintaining or enhancing the safety of the food while maintaining quality.

Although the value chain is conveniently divided into separate stages for illustrative purposes, an important feature of agri-food value chains is consumer-focussed collaboration and business relationships among the value chain actors. Food industry value chains can be viewed as empowering for all stakeholders at all points along the chain, as stakeholders recognize innovative opportunities to collaborate as strategic partners to increase the value of their products (FAO, 2014). It is the collaboration, strategic alliances and partnerships linking the stakeholders at each stage in the chain (i.e. input suppliers, producers, processors, marketers, distributors, food service companies, exporters) that adds incremental value that accrues along the chain and is fully realized when the final value chain participant delivers the product to the final customer (e.g. a hotel, restaurant or an individual).

An important component of successful food value chains is the technical support provided to the producers by various entities. In developing countries, these are often public sector-based because the governments and/or universities are the ones with the requisite trained personnel, resources and equipment to deliver the technical support. Support is also provided by regional or international organizations with specific mandates or programs for the food sector. However, technical support provision in these countries is increasingly involving consulting and technical services providers such as the authors firm[3] or, as in the case of the St. Vincent and the Grenadines dasheen (taro—*Colocasia esculenta*) and produce value chain depicted below, National Properties Limited/WIBDECO (Fig. 3.2). This model is a more complete depiction of an agri-food value chain than that provided in Fig. 3.1. It captures critical inputs, including food safety and quality systems requirements, as well as other influences and considerations, charting the flow starting with sourcing raw materials through to sale to various customers. Critically, it also captures the important role

---

[3] The author's firm, Technological Solutions Limited (TSL) is one such private sector entity working in developing countries.

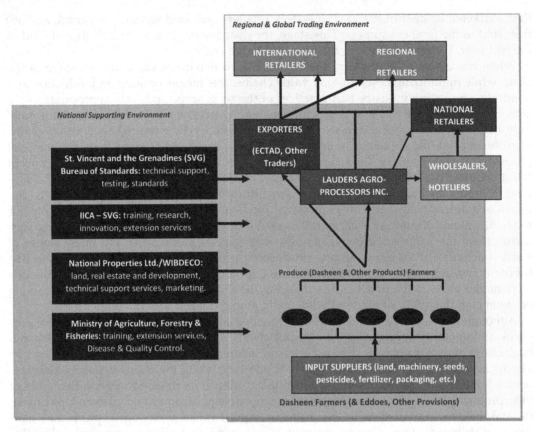

FIGURE 3.2 Example of a simple transnational dasheen-based produce value chain in the country of St. Vincent & the Grenadines, 2010.

played by the different supporting entities, without whose input the value chain may not be effective, competitive or sustainable. Competitive advantage is achieved though this collaboration, both upstream (among producers and input suppliers) and downstream (among distributors and retailers), which progressively creates additional value which, in sustainable, equitable value chains, is shared by all value chain participants.

The model shown in Fig. 3.2 demonstrates how produce in St. Vincent and the Grenadines in 2010 was moved from the small dasheen farmers through to the domestic retail chain and to the international market. It shows the key entities in the national supporting environment which played a central role in enabling the farmers and other members of the value chain to deliver outputs that met the needs of the next link in the chain. These entities are often critical in developing country environments where, as has been discussed in Chapter 1, technical know-how and resources are often limited. At the level of the farm, these entities included the input suppliers who were expected to supply viable seeds, fertilizer, soil amendments, pesticides and other inputs, as well as the Ministry of Agriculture that provided extension support, training for the farmers and disease and quality control.

Technical support, training, marketing and, importantly land for farming were provided by private sector interests who themselves were important customers of the primary production derived from the farms. These firms therefore epitomized the critical role that downstream participants play in a vibrant value chain by reaching back to be actively involved with supporting and enabling the production of the foods that they have an interest in marketing to their local and international customers. Other critical technical support was provided by consultancy firms, as well as local organizations such as the Bureau of Standards and a regional agriculture/agribusiness technical organization, the Inter-American Institute for Cooperation on Agriculture (IICA). These entities provided analytical services and well-needed guidance with international standards and market requirements, as well as guidance on packaging, labeling and post-harvest handling, for the sections of the value chain that produced market-ready, packaged product in various formats (including diced, sliced, cubed and frozen) to meet consumer demands. All of this provided the supporting infrastructure that allowed local, regional and transnational traders to move compliant, high quality product to markets, as required.

Fig. 3.2 gives a good overview of the value chain and its participants as well as the market and various routes to market. However, invariably where multiple markets (local and export), product variants, customers and market intermediaries are involved, such a model is a simplification. While it allows an understanding of the general relationships and where various aspects of food science, food safety and quality systems (FSQS) may be applied, it does not represent the totality of the interactions. It also does not allow for a full understanding of the exposures to risk and where steps and actions may be needed to mitigate food safety and quality challenges, or improve efficiencies, customer satisfaction and therefore value further down the value chain. A more complete, if more complex, but more representative global value chain model illustrating all of the various aspects and considerations that contribute to defining the nature and operation of such a value chain is shown in Fig. 3.3. Using the same Dasheen (Taro)−based produce value chain in St. Vincent and the Grenadines, this model demonstrates a fuller, more detailed description of all of the considerations, interactions and contributions that typify value chains where the markets are both local and global. While both types of models can be used, the model in Fig. 3.3 allows the specific inclusion of FSQS considerations at each stage of the chain, thereby facilitating an assessment of their role in the overall value chain. This is combined with specific inputs, market locations, value chain participants, support services and technical constraints to enable a better assessment of the issues that determine success (Fig. 3.3).

The approach taken to documenting the nature and structure of the value chain represented in Fig. 3.3 outlines in detail the location, participants and their interrelationships, the value addition along the chain (as a percentage of the final selling price), the nature and relative size of the markets and the support service providers and where along the chain they provide technical support. It also details the nature of the support provided, the institutional and other considerations, and operational inputs required at each step, as well as the specific technical requirements and food safety and quality systems components, requirements and considerations. These are associated with a particular segment of the value chain and related to specific value chain participants, technical constraints and value chain supporters. Examples would be the need for HACCP compliance, sanitation

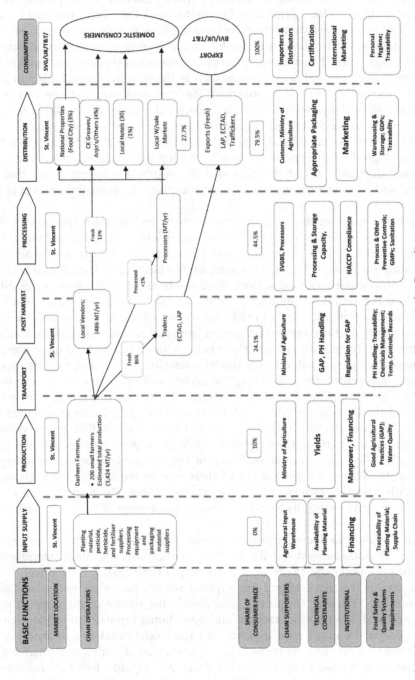

FIGURE 3.3 A detailed agro-processing (dasheen) value chain, St. Vincent & the Grenadines.

and good manufacturing practice (GMP) programmes and process and other preventive controls at the "processing" step of the value chain (Fig. 3.3).

Value addition and value creation along agri-food value chains

Value addition and value creation within the food industry occurs by different means and at different stages of the value chain. Coltrain, Barton, and Boland (2000) define adding value as "the process to economically add value to a product by changing its current place, time and form characteristics to characteristics more preferred in the marketplace." Value addition is achieved by converting a raw food product to a finished or semi-finished product that has more value in the marketplace. Value addition in the agri-food industry is achieved when food value chain participants convert an unrefined food product to a finished or semi-finished product for which final consumers are willing to pay more than they would pay for unrefined generic products. This can occur at different points along the chain, as different forms of the product emerge through various stages of handling and processing.

Value creation, on the other hand, is occasioned by delivering incremental value through measures such as product differentiation i.e. producing a food product that is different from similar products in the marketplace. Although value can be created at different stages of the food value chain, the main focus is usually around the factors of production (to produce a higher quality product) and around marketing (to achieve a better branded product). This reflects an important but growing feature of consumer preferences, i.e., the recognition that consumer demand may not necessarily be driven by price but by the level of value-addition and product differentiation. Product differentiation of course implies higher production and processing costs. As such, the final product which is differentiated and geared toward satisfying a specific niche in the marketplace should have a competitive advantage over generic products, thereby attracting higher prices. However, the key to successful differentiation is to obtain a price premium that is greater than the additional costs associated with differentiation (Porter, 1985).

Products can be differentiated by various attributes such as geographical location (e.g. Fiji bottled water, Jamaican Blue Mountain coffee, Belgian chocolates, French champagne), environmental stewardship (ecolabelling, fair trade), nutritional content, quality, safety, functionality, as well as marketing approaches and delivery systems (De Silva, 2011). Another approach to delivering value and differentiating agri-food products is through the creation and use of *certification marks*. These are collectively or privately-owned distinctive marks granted to products that meet the specific requirements of the mark owners, which requirements seek to assure compliance of the product with stipulated standards which are known by the market. The mark therefore signifies to the market that products that carry the certification mark meet the specific characteristics embodied by the standards. These may include food safety, quality, origin and other characteristics which are specified in the standard. An example of this is the set of "Jamaica's Finest" certification marks created in support of selected exports products from the Caribbean which require compliance with specific FSQS requirements, as well as authenticity, origin and other specific characteristics (Fig. 3.4). To be able to carry the mark, firms making the specific product have to undergo audits by an approved audit entity at

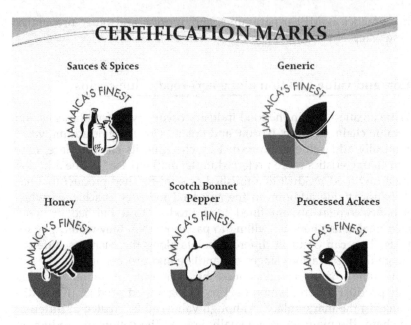

CERTIFICATION MARKS

Sauces & Spices

Generic

Honey

Scotch Bonnet
Pepper

Processed Ackees

(Source: The Competitiveness Company, 2015)

FIGURE 3.4 Examples of certification marks developed by the competitiveness Company[4] for use by Jamaican exporters. *Source: Used with permission from The Competitiveness Company (TCC).*

pre-determined intervals to assure conformance to the requirements. Using approaches such as these, products can differentiate[4] themselves in the marketplace and create additional value for the value chain delivering them to market.

In recent years, there has been a shift by major buyers toward products delivered from production systems that have attained certification of their food safety and quality status to standards such as Hazard Analysis Critical Control Points (HACCP) and globally benchmarked food safety systems such as GlobalG.A.P., Safe Quality Food (SQF), International Featured Standards (IFC) and the British Retail Consortium (BRC) set of standards. These and other certification systems have therefore become a major component of the process of value creation and value addition in the global agri-food system, Developing country producers and exporters must therefore embrace these certification systems if they are to not only access and expand lucrative markets, but also to benefit from greater returns through the higher prices paid for products from certified farms or facilities.

Synergies in value vs. supply chains

Value creation and value chain management has emerged as critical business survival strategies in the food and agricultural sector (Kampen, 2011). Meeting the needs of new,

[4] The Competitiveness Company is a Jamaican-based Caribbean firm that works to improve competitiveness in Caribbean firms, sectors and clusters through industry-best practices, cluster-based collaboration and enhancing value addition along product and service value chains. The editor is Chairman of this company.

more sophisticated and more demanding consumers and importers means that the food industry is challenged to create a sustainable supply of differentiated foods products, using more innovative crop and animal production methods, post-harvest practices, as well as new production, transportation and marketing technology and protocols. Along with technological advances and increased competition, the evolution of consumer demand for safe, high quality, and convenient products is changing the agribusiness sector. Food safety and traceability have become major concerns with new food safety protocols and more involved market entry requirements and regulations, as well as buyer requirements for certification along the distribution chain being put in place and constantly updated. The survival of companies therefore relies on their ability to satisfy the market entry requirements, as well as consumer demands with differentiated, safer, and higher quality value added products. The new agribusiness markets are therefore characterized by companies that are able to focus on value-added activities that allow them to meet these consumer expectations while maximizing profits from the value creation process (Coltrain, Barton, and Boland, 2000).

A novel feature of successful food industry value chains is the importance placed on the business relationships among all the stakeholders within the chain. Accordingly, all stakeholders, including small and large scale farmers should be treated as strategic partners and not just as interchangeable input suppliers. According to the FAO (2014), food value chains can be empowering for all stakeholders at the various nodes in the chain, as stakeholders recognize innovative opportunities to contribute and increase the value of their products. This is somewhat in contrast to supply chains, in which activities are coordinated around supplying raw materials to manufacturing and distribution to ensure products reach consumers (De Silva, 2011). In supply chains, each entity focuses on its own business and on presenting products for sale to the next link of the chain using efficient logistics. The main objective is to maximize profits by minimizing the number of links and bottlenecks in the chain and minimizing the time and costs to present products for sale. By contrast, value chains go beyond just keeping costs to a minimum, focusing on value addition and creation with every actor in the chain focussed on working together to produce what the final consumer wants and providing a more mutually beneficial business environment for all stakeholders. Thus, whereas both supply chains and value chains seek to maximize net revenue, the method by which value chains maximize net revenue is different. While it is important to highlight the differences in focus between supply and value chains, it must be noted however, that an efficient supply chain is essential to developing a viable value chain.

The challenges of delivering highly differentiated food products from farm to table in accordance with complex social, environmental, quality and safety food standards has spurred even greater consumer-focussed collaboration and mutual support among value chain participants. These food value chain participants recognize that value creation in a highly differentiated, competitive food industry and successfully delivering that differentiated final product that conforms with the demands of the discerning consumer requires careful planning and coordination. This includes planning and coordination of planting and production schedules, manufacturing schedules and the building of important partnerships with buyers and distributors. Food value chain participants also recognize the need to agree and adhere to a set of protocols that ensure product consistency and a set of operational principles that guide their interactions with each other.

As participants in food supply chains begin to collaborate more closely, they begin to align their operations in a manner that results in seamless transitions from one element or node in the chain to the other. This requires coordination across a range of activities and areas in a process that sometimes leads to vertical coordination, a process which involves establishing linkages and finding synergies that traverse the entire chain. This coordination improves the performance of the chain as a whole, and has often led to participants increasingly engaging in greater levels of vertical collaboration, such as formal contractual arrangements or vertical integration along the chain in order to improve efficiencies and create more value in the final product being marketed (Hendrikse and Bijman, 2002; Sporleder, 2006). Properly structured, this product will deliver better returns to all chain participants, enhancing value overall. Invariably, higher returns for product exported to more demanding markets, sold in higher value retail markets (e.g. Carrefour, Walmart or Sainsbury)[5] or sold to transnational buyers such as hotel chains (e.g. Marriott, Hyatt) or quick serve restaurants (KFC, Pizza Hut, Burger King, Nandos, etc.) are derived and can redound to the benefit of all participants within the value chain.

Food safety and quality assurance along food industry value chains

A critical aspect of the management of the safety, quality and added value along the chain is the ability to control, manage, track and trace each input into the delivery of the final product and ensure that both inputs and the final product are fully compliant with market and regulatory requirements. Food safety and traceability have become major concerns with new food safety requirements and more rigid international protocols and regulations being constantly put in place. Companies' survival relies on their ability to satisfy the end consumer demand including meeting the required food safety standards, at minimal costs. Food safety and quality (FSQ) management within agri-food value chains begin with an effective supply chain management system. FSQ management must be implemented at every node of the value chain. The importance of ensuring that the food value chains are safe and compliant with relevant regulations cannot be overestimated. Compliant food supply chain management systems help to assure and maintain safe input into the value chain, improving competitiveness and increasing the desirability of the final product, while also helping to protect consumers. Unfortunately, there are several factors which can negatively affect the management of food and other inputs within the value chains in which producers in developing countries participate. Foremost among these are the costs of implementing supply chain risk management strategies (including FSQS[6]), unfamiliarity with, or poor understanding of local and international regulatory compliance, poor communication, asymmetric information and inadequate mapping of supplies and traceability (SGS, 2014). Any of these factors could lead to upstream

[5] These are all transnational multiples (large distributors who retail through their own branded retail supermarkets or stores) who have their own demanding food safety and quality standards which have to be met to do business with them

[6] Food safety and quality systems.

disruptions that threaten safety and quality assurance and consequently negatively impact value addition upstream.

Emphasis on food safety and quality has heightened the need for agri-food value chain stakeholders to be focussed on control over the food safety and quality improvement aspects of the production process, including traceability, as well as delivering specific product characteristic. With this focus on differentiated food products and on the conditions under which they are produced, one of the most important factors in food value chain management is that linkages are established between input suppliers, producers, buyers, retailers and distributers operating within the chain, as is evident in the dasheen (taro) and produce export value chain example already discussed (Figs. 3.2 and 3.3). According to Dyer (2000), no longer can there be arm's length informal interactions between buyers and suppliers. Meeting the conditions for FSQ necessitates buyers, producers and suppliers being able to exert control over the inputs used in primary production, the processes, systems and services employed in delivering the final product to market, and being able to track and trace all inputs and assure compliance of all aspects of the process and with the product requirements. This entails ensuring that the entire value chain, from buyers and input suppliers to manufacturers and distributers, agrees on acceptable standards, appropriate sourcing and production methods, compliant postharvest storage and handling practices, appropriate processing and manufacturing technologies, and compliant transport and delivery to the market. It involves open sharing of knowledge and information through multiple levels of focussed training in compliance with FSQ and the other certification requirements that determine remunerative participation in regional and international agri-food value chains. It also involves both vertical and horizontal coordination, providing opportunities to add value to the food product at appropriate points all along the chain.

Summary

Applying the typical approach to value chain analysis which evaluates value addition along the chain and assesses the greatest opportunities for maximizing profitability and competitive advantage in the context of enhanced FSQ effectively combines market and regulatory imperatives with harvesting the best value for producers along the chain. This approach can provide new opportunities for food industry participants to harvest better value from their efforts to embed food safety and quality systems (FSQS) best practices into their production and handling. When applied in a developing country context, the impact on the global food supply is significant, given the increasingly inter-connected and sophisticated processes required to deliver the more diverse culinary options being demanded by global consumers.

In the next section we present case studies from developing countries that address value addition and value creation by stakeholders across the food value chain: producers, primary and value-added processors, retailers and distributors, consumers, and government. The cases focus on food quality and safety, providing examples of how value is created through FSQS and FSQS protocols in the agri-food sector. These case studies of different food value chains are employed as a useful means of demonstrating the complexity of the efforts to implement FSQS to add value and improve food safety and quality outcomes. They identify some of the challenges and opportunities for stakeholders linked along the food value chain, locally and globally, to achieve quality and safety standards and accountability for the sourcing, handling, and quality control of food.

Case studies

Case study 1: enabling safe and sustainable fresh vegetable value chains

While there is a demand for safe, high quality vegetables, especially from the local retail and hotel/restaurant sector in Jamaica, recent estimates indicate that much less than 1% of vegetables consumed, valued at US$18.1 million, are imported (GlobalEDGE, 2019). Much of this small but significant quantity of imports goes directly to the hotel and the higher-end retail sectors with the rest of their demand being supplied locally. Because these markets demand higher quality products that are safe, and delivered consistently to meet their needs, many local producers are unable to routinely comply with the standards to become approved suppliers to these more lucrative markets. Consequently, meeting the local demand for safe, high-quality vegetables in a manner that is sustainable and also profitable to all stakeholders in the value chain has been a challenge. This is especially true for small-scale farmers (Gordon, 2018). Satisfying current concerns regarding the use of chemicals, field and pack house operations, traceability, field to fork linkages within a fresh vegetable value chain requires the involvement of all key stakeholders at different nodes of the value chain. These stakeholders include input suppliers, farmers, collectors and food distributors, value-added processing facilities and marketers that sell to the final consumers.

Organized small-scale producers of high-quality vegetables capable of monitoring and maintaining their own food safety and quality standards through collective action are particularly attractive to buyers seeking to access high value markets. The existing literature is, however, replete with evidence of the difficulties that smallholders face trying to meet market requirements for food safety in regional and global markets (Reardon and Berdegué, 2002). Notwithstanding, there are some credible examples where given adequate organizational support, technological support and training, small-scale farmer groups have been successfully integrated into modern vegetable value chains. The case study outlined here describes a high-value vegetable value chain based on the production by small scale farmers upstream, successfully pioneered by the Jamaica Bauxite Institute (JBI)[7] in partnership with the Jamaica Social Investment Fund (JSIF)[8] through the establishment of *water harvesting and greenhouse production clusters* in several bauxite communities (Gordon, 2019).

---

[7] The JBI is a state-owned institution that regulates the bauxite Mining Sector in Jamaica and is also responsible for the rehabilitation of mined out lands and supporting viable community development programmes in mining areas.

[8] JSIF is a Government of Jamaica (GOJ) owned and run limited liability company that was established as part of the government's poverty alleviation program. It primarily channels its internationally sourced financial assistance to funding small-scale community-based projects.

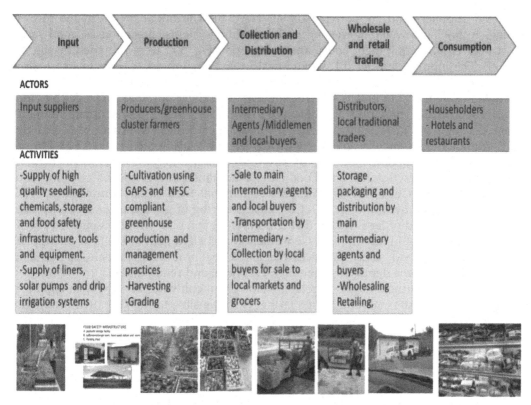

FIGURE 3.5   Greenhouse cluster fresh vegetable value chain for bell peppers and tomatoes.

The *JBI/JSIF Water Harvesting and Greenhouse Production Cluster Project* is a technology driven intervention that focusses on building the capacities of small-scale farmers, uses innovative, climate-smart technologies, low cost greenhouse construction, modern food safety infrastructure and practices, and draws on collective partnerships/organizational arrangements and reliable market linkages (Fig. 3.5). All these components work together to produce reliable greenhouse farmers capable of meeting the demands of the local retail and hotel sector for safe, high-quality vegetables, while increasing their earning potential. This case illustrates the important links in a sustainable vegetable production value chain driven by a modern dynamic domestic market and involving small scale greenhouse farmers. The case study highlights the critical food safety considerations along the value chain. These include those required from selection, preparation and deployment of planting material through to the storage and delivery of high-quality produce to mainly hotels serving an international clientele, retail outlets (supermarkets) and restaurants (Fig. 3.5).

Important linkages and activities in the greenhouse vegetable value chain

The process of value creation in this greenhouse vegetable value chain begins upstream with modern, innovative inputs provided by the intervention through a network of contracted suppliers and support personnel (Fig. 3.5). The initial inputs in the form of training, technology, water harvesting and irrigation systems, greenhouses, food safety infrastructure, farm equipment and planting material (Fig. 3.7) were provided by the JBI[9] and JSIF. The inputs from the JBI were critical in enabling access to climate-smart water storage and irrigation systems to provide year-round access to low-cost water for improving resilience against the effects of drought and periods of low-rain fall. Modern food safety infrastructure provided through JSIFs Rural Economic Development Initiative Project included pesticide storage, vegetable packing facilities, bathroom and changing rooms facilities to ensure compliance with food safety requirements (Fig. 3.7). Human capacity building, especially training in safe agricultural and post-harvest practices i.e. Good Agricultural Practices (GAPs), Food Safety, Sanitation and Hygiene and Good Food Handling Practices (GFPs), all forming part of national food safety compliance (NFSC) was also undertaken by JSIF (Fig. 3.5). Collective sourcing of good quality seedlings from one main grower, and of fertilizers and chemicals from common suppliers enabled consistency in the quality of input supplies. Coordination and alignment of the flow of supplies, use of inputs and food safety practices facilitated the production of safe, high-quality vegetables that meet the demands of the high-end hotel and retail market.

The greenhouse farmers in this vegetable value chain cultivate collectively in eight separate greenhouse clusters, although each farmer is responsible for his/her own production and record keeping. The main crops produced are colored and green (bell) peppers (*Capsicum annuum*), tomatoes (*Solanum lycopersicum*), lettuce (*Lactuca sativa*), cabbage (*Brassica oleracea*) and broccoli (*Brassica oleracea* var. *italica*). Collective farming and harvesting allow for better management of production and for consistency in the application of safe and effective use of fertilizers and chemicals. Collective production also facilitates better adherence to protocols for safe handling of produce and for minimizing post-harvest losses through knowledge sharing within the group and between the farmers and supporting agencies. Climate-smart water collection and storage (Fig. 3.7) has also guaranteed the availability of compliant irrigation water to facilitate year-round production, even under drought conditions. Irrigation water used in the production of vegetables, the water used during produce preparation and packing, and water used for sanitary purposes (hand washing and rest rooms) must meet the minimum requirements, including potability.

After reaping, the vegetables are taken to the pack houses in crates. Here they are sorted and weighed and readied for collection (Figs. 3.5 and 3.7). The collection of vegetables for entry into the marketing channels is mainly performed via intermediary traders/purveyors (Fig. 3.6). It is through linkages and contractual relationships with these intermediaries, that the cluster farmers can access the higher value markets i.e. the large hotels and supermarkets since they lack the compliant storage facilities (including chill storage), market knowledge and organized logistics to directly access these markets.

The system relies heavily on careful vertical coordination (as described earlier in the Chapter) i.e. scheduling and coordination of production and reaping, and the development of trust and information flows between the farmers and the distributor/purveyors. Approximately 90% of the vegetables that are produced by these clusters are collected,

---

[9] Dr. Dianne Gordon led the team that spearheaded the design and implementation of this project and continues to manage it.

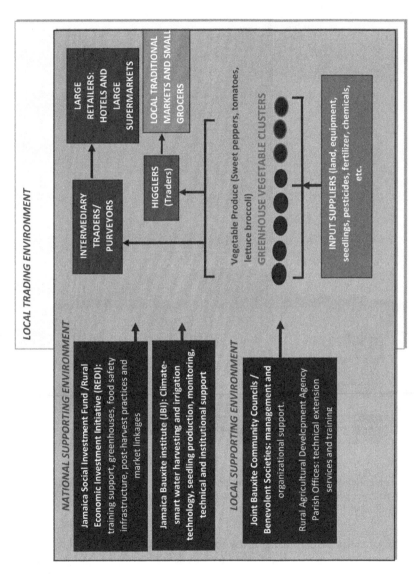

FIGURE 3.6 Summarized national and local supporting and local trading environment for greenhouse cluster vegetable value chain.

(Source: A Gordon, 2018; D. Gordon, 2018)

FIGURE 3.7 The greenhouse cluster set up and supporting infrastructure. (A) Greenhouse exterior; (B) greenhouse interior; (C) bell peppers growing; (D) harvested vegetables; (E) water source; (F) sanitary facilities; (G) storage. *Source: A Gordon (2018) and D Gordon (2018).*

stored and distributed by two main purveyors directly supplying the large supermarkets and hotels. Efficient storage and transportation in keeping with food safety and quality requirements, and timely delivery by the intermediaries are key factor in realizing value and ensuring that the final consumers' requirements for a high-quality (and high value) vegetables are met. As a result, the larger share of the value is captured at this stage by the intermediaries. The remaining 10% of the produce, is taken by higglers (local itinerant traders) who sell directly to local public markets and smaller groceries.

Food safety & quality considerations along the greenhouse vegetable value chain

It would be evident that the only sustainable way for this vegetable cluster to successfully participate in this value chain is to consistently supply high quality vegetables to the hotels serving an international clientele and to larger, up-market supermarkets and retailers. This means that the safety and quality of the vegetables being produced are among the key considerations for the farmers and the supporting partners who provide the guidance and technical inputs (Fig. 3.6). Consequently, a program had to be developed to address FSQ concerns for each aspect of the production and delivery of the vegetables to the market, particularly the upstream component of the value chain since the traders and purveyors already had systems (including storage and transportation) that were compliant with the demands of their market. Among these upstream considerations were:

- the adequacy and quality of the water used,
- pest control
- the type, storage, use and application of chemicals,
- personal hygiene and the adequacy of sanitary facility on the farms
- pre- and post-harvest handling,
- post-harvest storage
- product, process and systems knowledge of value chain participants — training and capacity building requirements

Addressing these considerations therefore required that all farmers were trained in a wide variety of areas including:

  i) the use, handling and storage of fertilizers, soil and grow bag amendments, nutrient applications and pesticides
 ii) contamination prevention during growing and harvesting
iii) the use of appropriate harvesting containers (including keeping produce off the ground)
 iv) pest identification and control
  v) Good Agricultural Practices (GAPs) and Good Food Handling Practices (GFPs)
 vi) food safety (general), personal hygiene and food handling
vii) traceability

Stage of production, appropriate data gathering and record keeping was also a part of the capacity building for the farmers, in particular, with basic GAP requirements such as record keeping on the mixing, dispensing and application of fertilizers and pesticides and the accurate capturing of post-harvest interval (PHI) data for pesticide application being among the focus areas. This was greatly assisted by the centralized storage, issuing and handling of chemicals and other supplies, as well as central provision of data gathering and record keeping support, although each farmer/greenhouse operator was responsible for his/her own record-keeping. In summary, this uncomplicated, but focused targeting of production stage-specific practices required to deliver wholesome, high quality vegetables facilitated the implementation of effective FSQ systems in each of the clusters and along the value chain which they supported. This included full traceability of each lot of product back to the specific cluster from which it came, the specific farmer and the greenhouse, inclusive of the details of chemical use and application to the specific crop during its growing cycle.

Summary

Since start-up in 2014, the JBI/JSIF Water Harvesting and Greenhouse Clusters have collectively and consistently produced high yields of quality produce, adhering to basic Good Agricultural Practices (GAPs) and food safety standards for entry into the local and international hotel chains (Jamaica Bauxite Institute JBI, 2016; Jamaica Social Investment Fund JSIF, 2016). Production by the greenhouse cluster farmers is a critical node in this food value chain. As important stakeholders, the farmers have seen positive income gains from their participation, with annual returns exceeding 300% more than their previous earnings as open-field subsistence farmers. This project adds to the existing cases of successful participation of small farmers in high value vegetable markets. However, in order for these greenhouse cluster farmers to remain competitive and increase their participation in this important value chain, it is critical that they not only continue to satisfy the demand for high quality produce, but that as important stakeholders in the value chain, their earnings are sufficiently remunerative to offset transaction and production costs and ensure reinvestment in the clusters.

Case study 2: value chains producing safe, high quality papaya for the US and United Kingdom markets

Introduction

Fruits, along with vegetables, have gained greater importance in the diets of consumers in metropolitan countries as the focus has shifted to healthier eating and balanced diets. This has made these markets demand much more tasty, flavorful and diverse fruits from developing countries. Along with mangoes, one such fruit enjoying high demand in the European, North American, Australasian and Japanese markets, among others, is papaya (*Carica papaya*), also known as pawpaw in some countries in the Americas. Among papaya exported to these countries, those producers that have a variety more naturally suited to the lifestyles of millennials and baby boomers find their product in even higher demand. Specific market considerations are driving demand, with the safety of the product being offered for sale, its quality and the consistency of quality and its convenience, largely in terms of size (can a single person consume manageable portions without significant waste or the need to store the remainder?) being major considerations. As such, safe, high quality papaya that can be easily consumed by a single individual either at one sitting or easily stored for completion later, are in great demand, even exceeding the general demand for the fruit. In this context, the Solo variety of papaya (Solo papaya), a smaller, deliciously sweet, orange fleshed fruit (Fig. 3.8) produced in Hawaii and a number of farms in the Americas, as well as other small varieties, are in great demand, particularly in North America and the EU. The United Kingdom is a major importer of the product from the Caribbean and the US is the major importer of product from Mexico and Central America.

FIGURE 3.8 A ripe solo papaya.

While the product is in high demand, recent repetitions of foodborne illness outbreaks caused by *Salmonella* have created significant concern for importers and retailers of the fruit, particularly in North America where the outbreaks have occurred. In 2011 and again in 2017 there were outbreaks caused by papaya imported into the US from Mexico (Centres for Disease Control and Prevention CDC, 2017; Larsen, 2019; Mba-Jonas et al., 2018). The most recent *Salmonella* foodborne illness outbreak in 2017 resulted in 220 cases of infections being recorded in 23 states across the US, with 68 hospitalizations and one death. The outbreak resulted in a nationwide recall. The outbreak started in mid-May 2017 and was declared complete in November 2017. It involved Maradol papaya from four farms located in four states in Mexico (see Chapter 8 for more details). This resulted in papaya from Mexico being put on import alert by the US FDA, thereby being subject to automatic detention at all US ports of entry (US Food and Drug Administration FDA, 2017; Centres for Disease Control and Prevention CDC, 2017), the product now only able to be imported once each shipment has been shown to be safe. As expected, the repeated outbreaks and nationwide recalls of papaya from Mexico have affected the perception of the safety of the fruit coming from that country in the US market. This has created a challenge for producers from Mexico, while alternatively providing an opportunity for exporters from other countries who have not had issues with their product. This case study is of the approach that major producers of papaya in the Caribbean took to significantly minimize the risk to their customers of foodborne illnesses from their products while capitalizing on the high and increasing demand for their product.

Background information and description of the value chain

In the early 2000s, having effectively dealt with the advent of the Papaya ringspot virus (PRV) that wreaked havoc on papaya orchards worldwide in the late 1990s to mid-2000, with the help of a local private sector organization and the Ministry of Agriculture (Fig. 3.9),

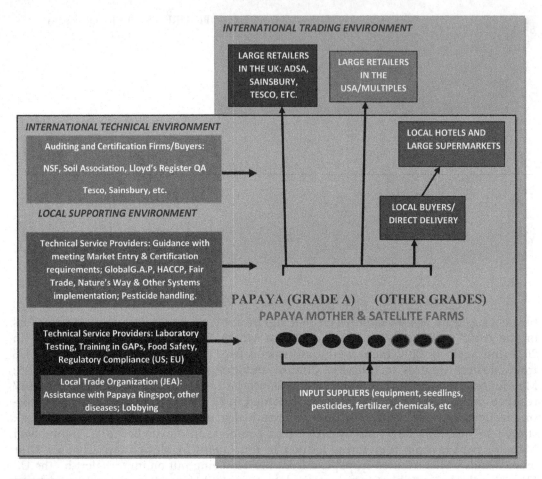

FIGURE 3.9   Summarized value chain for the production and export of high-quality solo variety papaya to the international market from a leading producer.

the Jamaican growers that survived began to expand production. One major producer, Advanced Farm Technologies Limited (AFTL) had production in both Mexico and Jamaica and exported its product to the US, Canada and the UK, the US and Canada being served in the 2000s largely from Mexico. The other major exporter of Solo papaya from the Caribbean is Valley Fruit Company Limited (VFCL), based in Jamaica, which exports mainly to the U.K. Both firms had begun the process to develop and implement a range of food safety and quality systems (FSQS) on their farms and in their pack houses in response to the demands of their buyers, mainly in the UK at the time, who were asking supplier firms to provide various kinds of certification that their market required. As a result of this, well in advance of what would become a global trend, the Jamaican firms began the development and implementation of HACCP-based food safety systems. At different times, they also implemented programmes to ensure compliance with other certification requirements such as GlobalG. A. P.,

Nature's Way (a Tesco[10] requirement) and Fair Trade, among others. Both firms also received assistance from the Jamaica Exporters Association (JEA) — managed technical support programmes, as well as other local support provided by the Ministry of Agriculture (Fig. 3.9). Both firms also retained the services of a regional technical service provider, Technological Solutions Limited (TSL)[11], at various times, to assist them with systems implementation for compliance with specific FSQS system requirements. AFTL, in particular, was first in developing and implementing a HACCP-based system in its pack house operations in the early-mid 2000s. The overall environment in which the value chains of which the firms were a part (and continue to be a part) is summarized in Fig. 3.9.

The value chain for papaya producers and exporters in the Caribbean country of Jamaica is based on the upstream planting, growing and delivery of high-quality Solo papaya by "mother" farms, supported, in some instances, by small "satellite" farms whose production augments the overall volumes sold locally and exported. All producers are supported at the primary production node of the value chain by input suppliers, the Jamaica Exporters Association (JEA), a local trade association that gave them a voice and assisted in mobilizing resources that successfully addressed Papaya ringspot virus — PRV (*Potyvirus*) in the late 1990s, and technical service providers (Fig. 3.9). In addition, the Ministry of Agriculture's (MOA) extension arm, RADA[12], and other areas of the Ministry helped with handling PRV and other disease-related challenges and generally provided technical and agronomic support. The farms explored the option of using genetically modified (GM) Solo papaya (Tennant et al., 1994) to address the PRV problem, but were dissuaded as their important UK buyers were adamant that no genetically modified organism (GMO)-based fruit would be allowed. Traditional approaches, including rogueing and careful field management were used instead. For a more detailed discussion of PRV and its control see Gonsalves (1998).

The trade association and the public sector technical support providers helped the farms successfully negotiate the PRV challenge in the late 1990s to early 2000s. With regards to compliance, other technical service providers assisted the Jamaican producers in achieving compliance with the regulatory requirements in both the US and EU. One such entity, TSL, provided consultancy, training and laboratory services support to both entities and worked with Advanced Farm Technologies Jamaica Limited (AFTJL), a subsidiary of AFTL, to implement its HACCP program for its packhouse (to which it is now certified), from 2002, and VFL to implement GlobalG.A.P. to which it is also now certified. The exporters also received technical support with compliance with other market-determined technical requirements, including traceability, this technical support being a key component of the local supporting environment of the sector (Fig. 3.9). Both AFTJL and VFL have achieved certification of their FSQ systems to a range of programmes that has allowed them to market their Grade A

[10] Tesco is one of the largest retailer multiples in the United Kingdom and has its own in-house certification systems, in addition to Global Food Safety Initiative-benchmarked systems such as GlobalG.A.P. and the British Retail Consortium's (BRC) Global Food Standard

[11] TSL is Dr. André Gordon's company provides consulting, training and auditing (against HACCP, GlobalG.A.P., Safe Quality Food (SQF) and other food safety audits) services and also operates an ISO 17025 accredited laboratory.

[12] RADA — Rural Agricultural Development Authority, Jamaica

papayas to the major multiples in the UK and, for AFTJL, those in the US also. Grade B and C papayas are sold to local hotels and supermarkets directly or through traders (Fig. 3.9). AFTJL owned or controlled all aspects of its specific value chain as also did VFL.

A pictorial overview of the process involved in the delivery of high-quality papaya to the international and domestic markets from the Jamaica Solo papaya value chain described in the value chain model shown in Fig. 3.9 is presented in Fig. 3.10. Papaya are

FIGURE 3.10 Stages of production for high-quality solo papaya to the international market. *Source: (E) Martha's Best.*

first planted in a carefully structured manner on lands about which the detailed agronomic history is known, as required by most Good Agricultural Practice (GAP) certification systems (as described in Chapter 4). The fruit is then grown under conditions in which inputs are recorded in detail, with application information and other relevant data also captured on the requisite hard copy or electronic forms. When fit, these are then carefully harvested by hand in an organized manner from pre-determined, labeled plots (sectors) by trained harvesters, taking care to protect the fruit from damage and keeping it from contacting the ground or any potential source of contamination. Pre-harvest intervals are strictly adhered to as this is required to ensure that fruit are free from residues of any pesticides (herbicides, fungicides or acaricides) that may have been used.

The fruit is then transported to the plant in field crates, the cleanliness of which is controlled by the pack house, while that of the vehicles is managed by the farm. On receipt at the pack house, the fruit is washed in chlorinated water (50–150 ppm hypochlorite), inspected, the stems removed and the stem area treated with an antifungal agent to prevent stem rot. The fruit is then dried and placed on a conveyor where they undergo grading to determine their disposition (Fig. 3.10). Grade A export fruit and others to be sold at retail are placed in lined carboard boxes to be sent for further inspection before final packaging. The cartons used are specially designed to cushion the fruit such that little or no damage occurs during handling and transportation to market. After visual inspection, the fruit are sleeved (if required) and the carton sealed and sent to storage. Throughout the entire process, records are kept and traceability of each batch of fruits back to the farm, the specific field from which it was harvested, and all of the inputs applied, is maintained. This is inclusive of the history of the husbandry of the trees and their fruit in the specific field or field segment (plot) in which it was grown.

For AFTJL, which retails in both the local and export markets under the brand Martha Best®, a detailed visual examination with a magnifying glass is done to ensure that the fruit is not bruised, shows no sign of damage which could lead subsequently to mold growth and is visually free from contamination. The fruit, once acceptable, is then sleeved, individually labeled with a branded sticker and repacked in the carton to be palletized and then taken to storage (Fig. 3.11). These final visual quality checks are an important, if simply aspect of quality assurance for the product. Along with the individual branding, this ensures that at the end of the value chain where the fruit is offered for retail sale, the best possible price is received and also that the consumer is able to associate their satisfaction with the fruit with a specific brand from a specific geographical location, Martha Brae in Jamaica where the fruit is from. This kind of branding also supports traceability and compliance for standards such as Fairtrade which include specific geographical considerations and seeks to provide a better return for farmers and workers at the beginning of the value chain who do not always get a fair return on their efforts (Fairtrade, 2019).

Food safety & quality considerations along the Jamaican solo papaya value chain

The supply of high quality, safe Solo papaya to the local hotels and major export markets in particular, required local producers to employ the full range of food safety and quality systems practices relevant to farming and pack house operations, in addition to a range of other requirements dictated by the certification requirements imposed by specific

FIGURE 3.11 Visual quality control checks on solo papaya prior to branding and packaging. *Source: (E) Martha's Best.*

markets. Among FSQS considerations (most of which will be discussed in greater detail in Chapter 4) were:

At the farm
- the quality of the water used for irrigation, washing farm utensil, sanitary purposes, mixing chemicals, etc.
- effective control of fungal, insect and other pests using appropriate, compliant pesticides
- the type, storage, mixing and use of chemicals and pesticides,
- personal hygiene of workers,
- adequacy of sanitary facility (including toilets and hand washing stations) on the farms
- pre- and post-harvest handling,
- product, process and systems knowledge of farm workers — training and capacity building requirements
- observation and documentation of the rules regarding manure and soil amendments
- observation and documentation of the required pre-harvest intervals for all crops
- post-harvest storage

As with other value chains, the management and workers on the farm had to be trained in a wide variety of areas including:

  i) the use, handling and storage of fertilizers, soil amendments, other nutrient and pesticides
  ii) proper field sanitation
  iii) contamination prevention during growing and harvesting
  iv) the use of appropriate harvesting containers (including keeping produce off the ground)
  v) post-harvest handling in the fields
  vi) pest identification and control

vii) Good Agricultural Practices (GAPs) and Good Food Handling Practices (GFPs)
viii) food safety (general), personal hygiene and food handling
  ix) traceability and recall;

At the pack house
- the quality of the water used for washing papayas, for mixing chemicals, restrooms and staff facilities, etc.
- management of the use of topical fungicide(s)
- the documentation, use and management of sanitation and other pack house chemicals
- attire, personal hygiene and GFHP practices of workers,
- adequacy of sanitary facility (including toilets and hand washing stations)
- handling of the fruit throughout the process,
- training and capacity building requirements for food safety and quality-critical functions
- suitability of the physical facilities, inclusive of food contact and non-food contact surfaces
- quality assurance-specific activities, including calibration, visual inspection, accuracy of labeling, application of labels, appropriate grading.
- effective pest control

At the pack house, the food safety team, management and pack house staff, as well as the persons working in storage, transportation, sales and marketing had to be trained in:

  i) food safety (general), personal hygiene and sanitation
 ii) traceability throughout the handling and distribution channel
iii) Good Food Handling Practices (GFHPs)
 iv) quality assurance
  v) Preventive Controls
 vi) pest control
vii) Food Defense, Food Fraud and Food Safety Culture
viii) Crisis management
    among others.

As far as possible, much of this training was done in-house, with organizations such as the Bureau of Standards Jamaica (BSJ), the Ministry of Health's Public Health Department, the Pesticide Control Authority (PCA) and, where required, TSL, providing training in specific areas of their expertise. These all were a key part of the supporting environment indicated in Fig. 3.9.

Other considerations The foodborne illness outbreak associated with papaya in the US in 2011 (Mba-Jonas et al., 2018) brought a new dynamic into the industry. It caused a renewed focus on food safety practices in the papaya industry and the need for exporters to demonstrate food safety along the entire value chain, from farm to delivery to retail. In the case of one major producer, Advanced Farm Technologies Limited (AFTL) that had production in both Mexico and Jamaica, the outbreak in 2011 led the company to make a decision that saw the curtailment of production in Mexico and the consolidation and expansion of the operations of in Jamaica. Advanced Farm Technologies Jamaica Limited (AFTJL) now had to supply both markets as well as Canada with all of their signature Strawberry variety of Solo papaya for which the company was known. Consequently, the current success of this value chain has been assisted by the increasing focus not just on the nutritional benefits, flavor, quality and wholesomeness of papayas, but critically as well, their safety. As such, this is an excellent example of a value chain, the success of which is closely intertwined with the food safety of the product on which it is based.

Case study 3: a fine flavored cocoa production and chocolate industry value chain

Introduction

Cocoa is one of the more important food crops grown in developing countries. Cote d'Ivoire, Ghana, Nigeria and Cameroon, in that order, are the major producers in Western Africa. In addition, Benin and Gabon with Indonesia and Malaysia in Asia, and several countries in the Americas including Brazil, Ecuador, Colombia, Trinidad and Tobago, Grenada and Jamaica are also involved in production and export of cocoa. Cote d'Ivoire, the world's largest producer produced an estimated 1.96 million tonnes of Africa's 3.50 million tonnes in 2018, with Brazil and Ecuador producing 204,000 and 285,000 tonnes, respectively, of the Americas 819,000 tonnes and Indonesia producing 240,000 tonnes of Asia and Oceania's 326,000 tonnes (International Cocoa Organization ICCO, 2019a). There are two basic types of cocoa, "fine" or "flavor" cocoa (also called fine flavored cocoa) and bulk or "ordinary" cocoa (International Cocoa Organization ICCO, 2019b). Most of the world supply of cocoa is bulk cocoa and is produced in Western Africa and Indonesia, with Africa producing 73%, the Americas 15%, and Asia and Oceania 12% of total global cocoa output (International Cocoa Organization ICCO, 2014). While the production of cocoa in the Caribbean region of the Americas is much less than elsewhere, this is where the majority of fine flavored cocoa, typically of the **Trinitaria** variety is grown, with Trinidad and Tobago, Grenada and Jamaica being major producers of this variety (International Cocoa Organization ICCO, 2019b). This cocoa is typically either a bright orange-yellow or purple-red in hue (Fig. 3.12) and produces cocoa when ground that has a sought-after red color and distinctive flavor, hence its name.

In many producer countries, cocoa is mainly processed only to the beans which are then exported, mainly to Europe and North America for conversion to chocolate and a range of other cocoa-based products (beverages, ice cream, sweets, etc.). This has meant that the food safety and quality systems (FSQS) that would be relevant to the production of cocoa would only have involved the aspects that dealt with primary production, i.e. those focused on GAP and good post-harvest handling and storage, including traceability. Over the last ten years, however, there has been a growing trend when the grade of cocoa grown is "fine-grade" for artisanal producers to take the cocoa beans all the way to fine flavored chocolate (Fig. 3.13). This has taken a cocoa/chocolate value chain that was transnational in scope (as is the case for the cashew nut value chain discussed later on) and localized it, bringing enhanced value to the producer countries. This has also introduced the requirement for these now localized (national) cocoa/chocolate value chains to implement within their operations the FSQS that would traditionally have been required of

FIGURE 3.12   Fine flavored cocoa on trees in the Americas. *Source: Gordon (2017).*

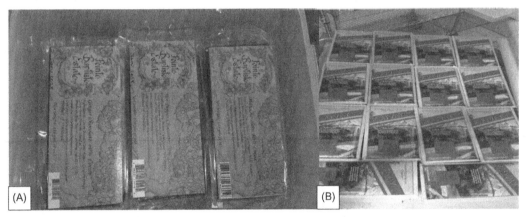

FIGURE 3.13   (A) Artesanal, hand-made fine flavored chocolate and (B) fine flavored Grenadian chocolate. *Source: (A) Pointe Baptiste Estate.*

manufacturers in the countries to which they exported their cocoa. Consequently, as cocoa producing countries have sought to move up the value chain, it has required a significant enhancement of their traditional practices to incorporate the required systems for the successful implementation and maintenance of compliant FSQS programmes. This case study outlines the FSQS considerations along cocoa/chocolate value chains based both on the traditional cocoa bean exporting-focussed production systems and the emerging cocoa-to-chocolate product production operations, particularly among fine flavored chocolate producing nations.

Background information and description of the value chain

The cocoa production/chocolate manufacturing process

Understanding the cocoa and chocolate industries, the food safety and quality systems considerations, how these impact production at different nodes in the value chain and how, ultimately, opportunities can be created for greater value addition along the chain requires a knowledge of the overall global market and cocoa/chocolate industry value chain. Traditional cocoa processing involves first harvesting the beans from the tree (Fig. 3.12), typically with a sharp cutting implement (e.g. a machete). The wet beans are then harvested from the pods by breaking these open to expose the pulp-covered beans which are removed and put to ferment, either in a pile covered with banana leaves, as is often done traditionally in some countries (International Cocoa Organization ICCO, 2012) or in a box (Fig. 3.14A) which is done traditionally as well, with natural microorganisms driving the fermentation process in both cases. The fermentation process which is critical to the development of the complex flavor and color of the cocoa (International Cocoa Organization ICCO, 2012) has been extensively studied, using the processes traditional to the Caribbean, West Africa and Indonesia and also under simulated conditions (Biehl et al., 1985; Ardhana and Fleet, 2003; Camu et al., 2007; 2008) Fermentation is first initiated by yeasts and fungi which are present on the pulp surrounding the beans. This starts the breakdown of the pulp, resulting in its dissolution and initiation of a cascade of interrelated microbial, enzymatic and chemical reactions that develop the delicate flavor and colors characteristic to cocoa of different geographical origins. A more detailed discussion of the microbial ecology of cocoa bean fermentation is presented by Thompson et al. (2013).

The next stage of the process involves drying which is traditionally done by sun-drying on barbeques or on wooden racks (Fig. 3.14), after which the cocoa beans are cleaned to remove extraneous material (International Cocoa Organization ICCO, 2013) and then roasted to develop the color and flavor of the beans (Fig. 3.15A). After roasting, a winnowing machine is used industrially and for cottage-industry scale manufacturing to remove the shell from the beans, leaving just the cocoa nibs. For artisanal producers, this is done by hand (Fig. 3.15B). The cocoa nibs are then treated with potassium carbonate, a process known as alkalization, to further develop the color and flavor of the cocoa and milled to produce cocoa liquor (Fig. 3.15C), which is a mixture of cocoa butter and suspended cocoa particles. Any mixing of liquor from different types of beans is then done to produce the

FIGURE 3.14   Fermentation chamber (A) and drying racks (B) used for cocoa. *Source: Gordon (2017).*

FIGURE 3.15 (A) roasting, (B) winnowing (done manually for artisanal processors) and (C) milling of cocoa. *Source: Gordon (2017)*

characteristics that are desired for the final product before pressing of the liquor to extract the cocoa butter and create a solid mass called a "cocoa presscake" is done. For fine flavor chocolate, often only one type of bean is used and so the mixing step is not required. The cocoa presscake is broken into smaller pieces and ground to produce cocoa powder, the composition of which is dependent on the mixture of beans used. The cocoa liquor remaining is used to make chocolate by mixing in additional cocoa butter, sugar, dairy products and other ingredients and then refining the chocolate mixture by passage through a series of rollers that improve the texture (Fig. 3.16). In artisanal manufacture, purpose built small-scale equipment is used for this purpose (Fig. 3.17).

The next stage involves the kneading and smoothing of the mixture to further develop the texture and enhance the flavor, a process known as "conching" (Fig. 3.18A). An alternative to conching is an emulsification process that achieves the same objective. The next

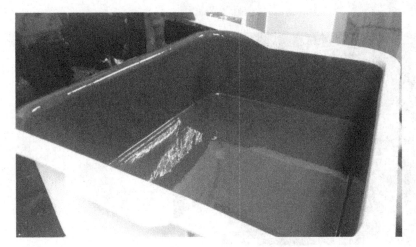

FIGURE 3.16 Cocoa liquor mixture after refining at a cottage-scale manufacturer. *Source: Gordon (2016).*

FIGURE 3.17 Refining of cocoa liquor mixture in artisanal manufacture of chocolate from fine flavored cocoa. *Source: Gordon (2017).*

FIGURE 3.18   (A) Conching and (B) Tempering using cottage-scale equipment at a cottage-scale manufacturer of chocolate. *Source: Gordon (2016).*

FIGURE 3.19   Handling and storage of molds and containers.

step is tempering of the chocolate (Fig. 3.18B), a cyclical process of heating/cooling/reheating to prevent discoloration and the development of fat blooms by inhibiting crystallization of cocoa butter fat (International Cocoa Organization ICCO, 2013). The final stages of processing involve pouring the chocolate mixture into molds that create the desired shapes (Fig. 3.19) or using the chocolate for enrobing[13] (e.g. when making milk chocolate containing nuts), cooling to harden the chocolate, cutting to shape and size (if required) and then packaging for distribution and retail (Fig. 3.13).

Description of the value chain

More than three quarters of global production of cocoa originates in West Africa, Cote d'Ivoire to be specific, but very little grinding (the initial step of further processing after winnowing) is done there. While the post-harvest processes of fermentation and drying are localized in the producer countries (upstream), grinding and further processing has traditionally been located downstream in "buyer" countries where the main consumer

[13] Enrobing is a process by which chocolate is used to completely cover other ingredients to create a chocolate exterior surrounding other material (e.g. caramel, nuts, etc.).

base of the value chain is located. This has traditionally been done by companies which dominate the retail market for cocoa products and chocolate, the latter being the single largest product in which cocoa is used. Companies such as Mars Wrigley Confectionary — US (US$18 billion), Ferrero Group — Luxembourg/Italy (US$12.39 billion), Mondelez International — US (US$11.79 billion), Meiji Co. Ltd. — Japan (US$9.62 billion), Hershey Co. — US (US$7.78 billion) and Nestlé — Switzerland (US$6.14 billion), the top six in global chocolate sales in 2018 (International Cocoa Organization ICCO, 2019a), command the lion share of global cocoa supply. These, and others are the firms who typically undertake the processing steps after drying as they are the entities at the end of the value chain where the interface is directly with the consumer. These firms are also the ones that would have the most stringent and developed FSQS designed to protect their investment and their brands and which any supplier of intermediate processed cocoa to them would have to meet in order to supply further down the value chain. It is in this context that producer firms in countries where the cocoa is grown have increased the percentage of the bean that is further processed to derive greater benefits from the industry value chain overall. As such, producers in countries such as Cote d'Ivoire and West Africa, as well as Indonesia which has been among the more dynamic countries in extending its industry into processing (International Cocoa Organization ICCO, 2014), typically focussed on doing more milling (grinding) in-country. In the Americas, the focus has been more on extending involvement to the consumer end of the value chain by producing cottage-scale or artisanal, high value fine flavored chocolate, requiring all of the stringent FSQS systems as for any other firm delivering chocolates to a consumer market.

Food safety and quality management along the cocoa industry value chain

The FSQS requirements for the production of cocoa revolve around GAP and traceability requirements covering the nut from tree to drying and then bulk packaging for shipment. The issue of traceability would therefore be the most difficult of the requirements to comply with, the cocoa being still in the shell when shipped, with the significant kill steps and potential for contamination occurring further down the chain. The robust nature of the unprocessed cocoa still in its shell and, consequently, the easy-to-comply-with basic requirements for handling and GAP as compared to a fruit like papaya or fresh vegetables make the delivery of dried bulk packaged cocoa beans for export shipment or domestic processing relatively uncomplicated. This accounts for the high compliance rates that obtain in the Cote d'Ivoire and also in the production of fine flavored cocoa in the Caribbean. The complexity of the value chain for cocoa products, including chocolate, on the other hand, is much greater and with this is a corresponding increase in the number and nature of the FSQS required. The FSQS requirements for cocoa powder, for example, would end after the grinding of the presscake to produce cocoa powder, which would then be packaged. This product would still additionally require that the facilities implement GMPs, traceability, good sanitation and hygiene, worker training to ensure competence, HACCP-based preventive controls programmes and all of the other food safety systems and protocols required for manufacturing, as discussed in detail in Gordon (2016). This includes ensuring proper sanitary design and also proper provisions for hand washing (Fig. 3.20), proper cleaning, handling and storage of equipment and utensils (Fig. 3.19), adequate pest control and training in appropriate quality control and assurance

FIGURE 3.20 Sanitary design of facilities and provision of hand washing stations.

procedures to ensure that the product meets the requirements of the markets targeted. Chocolates, including enrobed products, have much more involved requirements for FSQS, the complexity of the systems required, particularly on the quality assurance side, increasing with the number, nature and type of ingredients included in the product.

FSQS for chocolate products, therefore, cannot be generalized but have to be developed specifically to provide the appropriate quality assurance for meeting specifications and customer requirements. As well, they must include specific food safety systems to control the risks (including allergen cross contact) from all raw materials and inputs, in process materials (such as the allergen risk, where applicable, from rework), the risks from the process itself and risks from the environment, sanitary handling and sanitation and the human factor. Consequently, while FSQS systems for products derived from cocoa earlier up the value chain may be similar to those of other production systems, significant work will have to be invested in developing FSQS for the various types of chocolate products that a large plant may make, particularly the specific quality assurance requirements in addition to the food safety systems required for all compliant manufacturing facilities.

Case study 4: the Indo-Ivorian cashew nut value chain: influence of FSQS considerations

Introduction

The global snack food industry and foods based on the inclusion of nuts continue to show robust demand, despite the greater focus now being placed in global food safety systems on allergens, including those due to tree nuts. Snack foods and breakfast cereals are two of the main categories in which a product like cashew nuts are used, although cashews are already an important part of some cuisines, such as Chinese, Korean and Thai cuisines that have established a growing global footprint. The transition of these cuisines mainstream has also meant increased demand for ingredients such as cashew nuts which are an important component of some dishes. These trends have increased the demand for cashew nuts globally and have been among the drivers of the expansion of the transnational value chain in cashew nut. This case study examines the Indo-Ivorian global cashew nut value chain, demonstrating the linkages between small farmers in rural communities in developing countries, the major international importers/manufacturers of nut-based snack and other foods, and consumers in developed markets. It examines how food safety and quality considerations influence value creation along the nodes of the value chain both upstream and downstream.

Background information and description of the value chain

Cashew nuts are supplied globally mainly from the African and Asian continents. In 2017, Côte d'Ivoire was the world's largest producer of raw cashew nuts supplying 711,000 tonnes or 23% of the world's total production. Côte d'Ivoire was also the world's largest exporter, with the export value of cashews estimated at approximately US$800 million (World Bank, 2018). Cashew is therefore one of Côte d'Ivoire's more important non-traditional food crops and has gained greater importance with the increasing demand for differentiated cashew food products in the global market. Other significant cashew nut producers are Vietnam, India, Brazil, Benin, Guinea-Bissau, Tanzania, Mozambique and Indonesia.

There are approximately 450,000 farmers growing cashew in Cote d'Ivoire, mainly in the northern part of the country. Cashew is the most important source of rural cash income and is harvested during the dry weather from February through June, a period known as the "money harvesting season" for cashew farmers. Once the fruit is harvested, the nuts are separated from the fruits and sun-dried for 2−3 days before they can be processed. Cashew kernels are consumed in a variety of ways including as a snack, raw, salted, seasoned, flavored, as a major ingredient in sweets, chocolates and cookies. The nuts are also used in baking, cooking, as a topping and as a major ingredient in Asian cuisine. The demand from global manufacturers and consumers for differentiated and value-added cashew products has necessitated the implementation of strict food safety and quality systems (FSQS), GMPs and HACCP-based food safety programs along all nodes of the cashew nut value chain. In Côte d'Ivoire, cashews are grown and harvested by small farmers in accordance with the industry standards for compliance with food safety regulations, especially those set out by the Global Cashew Council of the International Nut and Dried Fruit Council (INC). This has helped to ensure that consistently high-quality cashew kernels are exported from Côte d'Ivoire, an achievement for the sector which has been recognized internationally (International Nut and Dried Fruit Council (INC), 2015).

Despite being the world's largest cashew nut producer, the Ivorian end of the global cashew nut value chain is poorly developed, with only 7% of global processing carried out in Côte d'Ivoire itself (World Bank, 2016). In fact, almost 99% of cashew nuts are exported unprocessed to India and Vietnam for further value-added activities. The World Bank has estimated that the raw cashew accounts for only 34% of final product value (World Bank, 2018). This is important to note, since significant value addition in the cashew nut value chain is realized through processing, with roasting, shelling and salting and packaging adding another 24% to the final product value (Tessmann, 2017). Existing installed cashew nut processing capacity in Côte d'Ivoire is estimated to be around 109,000 tonnes (less than 15% of annual production), with 50% of the processing carried out by three large industrial factories in the central Ivorian city of Bouaké. Côte d'Ivoire's low installed processing capacity limits the vast potential for value addition, job creation and additional earnings at this end of the chain. A World Bank review of the cashew industry has noted that processing has been constrained by access to finance, processing infrastructure and technology, and poor rural infrastructure. This study further indicated that if just 50% of the cashews harvested in Côte d'Ivoire's were processed locally, more than 90,000 new jobs and approximately US$100 million in additional income could be generated annually for the local economy (World Bank, 2016). The lack of investment in processing translates to limited value addition, low margins to the processors and even lower margins in Côte d'Ivoire for the cashew growers upstream in the global cashew value chain.

The Côte d'Ivoire's end of the cashew value chain involves several traders that link local farms to Indian processors in what has been described as the Indo-Ivorian value chain (Tessmann, 2017). The distribution channels for raw cashew nuts in Côte d'Ivoire are complex and involve numerous actors (Fig. 3.21), including buyers working for export companies, as well as independent buyers with their own collection and distribution channels (African Cashew Initiative ACI, 2010). The raw nuts are usually collected at the farm gate by local traders or middlemen acting on behalf of exporters requiring large volumes from several supply sources. Since cashew must be processed soon after harvesting in order to maintain its properties and quality, processors must be able to access and purchase sufficient cashew nut stock during the harvesting season to be able to process continually over ten months of the year. Although the middlemen play an important role as intermediaries between growers and buyers, their involvement also reduces the margins that accrue to the farmers.

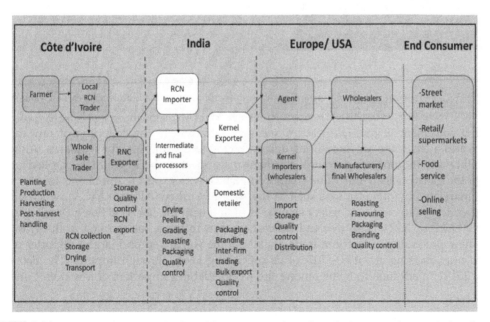

FIGURE 3.21 Main participants and value adding activities along the Indo-Ivorian cashew nut value chain.

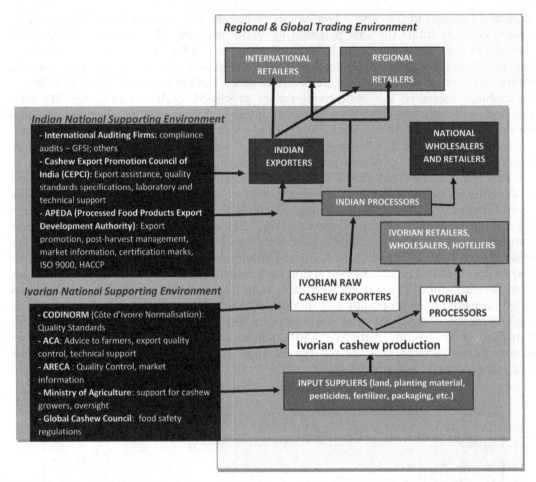

FIGURE 3.22 Transnational supporting and regulatory environment for the Indo-Ivorian cashew value chain.

The involvement of multiple traders and subcontractors in the collection of raw cashew nuts creates a challenge for monitoring and enforcing safety and quality standards. Although traceability specifications and control mechanisms appear to be less strictly applied at this node of the global cashew nut value chain, compliance with export requirements for raw cashews is mandated by organizations such as the African Cashew Alliance (ACA) which verifies adherence to quality and food safety standards in line with the US Food Safety Modernization Act (FSMA) (Fig. 3.22). High demand for cashew nuts in India has resulted in the growth of the cashew processing industry in India, making it the world's second largest importer of raw cashew nuts, meeting about 60–70% of its processing needs through imports (The Economic Times, 2019). In addition to being among the world's largest importers of raw nuts[14], India is

---

[14] India imports most of Côte d'Ivoire and Mozambique raw cashew nut production.

also among the world's largest producers of raw cashew nuts (45% of global production)[15] and the largest exporter of processed cashew nuts (World Bank, 2016).

The processing of cashew nuts in India and their subsequent export to the USA, UAE and European markets (mainly Germany and the Netherlands) are critical activities in which value is realized in the global cashew nut value chain that originates downstream in Côte d'Ivoire. India is highly dependent on imported raw cashew nuts for value addition, with approximately 38% of the raw kernels processed imported from Côte d'Ivoire. India's cashew processing capacity is approximately 2 million tonnes, with processing carried out by over 3500 small and medium sized processors with much inter-firm trading between the major processors that ship to the international markets under their own brands (The Economic Times, 2019). India's cashew industry is also largely export-oriented, providing an important source of income for India's small farmers and for the more than 500,000 persons, 95% of whom are women, employed in the processing factories (World Bank, 2018). A large part of the value addition, job creation and business opportunities in the cashew nut value chain is therefore realized in India, specifically through processing (roasting, salting), inter-firm trading, packaging, branding and bulk export to markets in the USA, the EU and Asia. The participants/stakeholders involved at different stages of this transnational cashew value chain are shown in Figs. 3.21 and 3.22.

Cashew nuts are made up of two main components, the fruit and the kernel, with the kernel accounting for more than 80% of the value of the nut. The processing of the cashew nuts into marketable kernels involves roasting or steam-boiling, shelling and drying of the nuts, then peeling and grading, followed by packing. This process takes about 5–7 days and involves strict quality controls under clean, hygienic conditions. The kernels are of different sizes and colors and can be broken into pieces of different sizes (Fig. 3.23). As such, there are up to 25 different grades of cashew nut handled by the cashew nuts processing industry, with the bigger, whole, white edible kernels attracting the highest prices and being preferred for export (Fig. 3.22). Appropriate packaging of the raw cashew nuts and optimal storage are critical for maintaining product quality by avoiding damage, contamination and deterioration.

The three main cashew products exported from India are raw cashew nuts, roasted cashew kernels (whole and broken) and cashew nutshell liquid. India's Cashew Export and Promotion Council (CEPCI) specifications have been adopted by the industry for grading cashews. The kernels are packed in flexible polymers or metallized vacuum packs for export to international markets. International market prices are mainly based on the kernel size and the percentage of broken kernels, as per ISO specifications (Tessmann, 2017). Whole kernels (Fig. 3.23) fetch higher market prices. It is at this node in the global cashew value chain, the packaging of the nuts for export, that much importance is placed on food safety and quality. Despite this, as is evident from the transnational complexity of the value chain presented here, that it takes considerable coordination, the application of regulatory controls, technical support and collaboration along this Indo-Ivorian value chain to bring cashews from primary production in Cote d'Ivoire to the international end-user markets (Fig. 3.22).

---

[15] India's major cashew nut producing states in India are Maharastra (32.3%), Andhra Pradesh (16.15%), Orissa (13.7%) and Karala (10.76%).

FIGURE 3.23 Whole cashew nut kernels. *Source: Photo taken by D. Gordon in 2019.*

Food safety and quality systems in the Indo-Ivorian cashew nut value chain

As indicated previously, export standards are mandated and enforced at the Ivorian end of the Indo-Ivorian cashew nut value chain by the ACA, with traceability being required of exporters by the importers in India. While this remains a challenge, efforts are being made to improve traceability back to at least collecting areas, as this would help to control the risk arising from any potential mis-handling or poor storage of nuts. This is essential for Indian importers and manufacturers as their customers, importers and manufacturers in the international market, require periodic audits of processing and storage facilities in India in order to ensure compliance with food safety requirements. These audits are managed by representatives of international roaster/processing companies and their importers. There are increasing stringent demands for traceability, safety, quality and third-party verification (GFSI-benchmarked FSQS certification) at this node of the cashew value chain (Fig. 3.22), with particular requirements for supply-chain visibility i.e. understanding and being able to follow the movement of nuts along the supply route, if necessary. Special concerns arise at this node regarding insect infestation and microbial contamination. This has led to demands for pasteurization in cashew processing to ensure protection against microbiological contamination (Tessmann, 2017).

Quality measurements are carried out in accordance with agreed procedures to assess the foreign matter content of the nuts, the moisture levels (to eliminate the risk of mold growth), the nut count per kilogram (a measure of the size of the kernels) and the kernel outturn ratio (the quantity of acceptable kernels in an 80-kilogram bag after shelling).The export of processed kernels to the EU requires compliance with increasingly strict EU food safety, quality and traceability standards which require documentation both upstream and downstream and cooperation between kernel exporters and their buyers to guarantee a management system for food safety and risk management. Likewise, under the US Food Safety and Modernization Act (FSMA), the FDA requires American kernel importers to verify the implementation of a food safety system in their supply base as per the Preventive Controls for Human Foods Rule (PCHFR) and the Foreign Supplier Verification Program (FSVP), both discussed in Chapter 1. This compliance is often verified by external auditing and certification schemes applied both upstream and downstream before and after shipment by international auditing and certification bodies (Fig. 3.22)

The cashew grown by Côte d'Ivoire's small farmers and the cashew kernels processed in India and exported to the USA and Europe are purchased directly from importers or from wholesalers, and further processed by a small number of leading manufacturers such as Planters (Fig. 3.24), United Biscuits (KP Nuts) and Percy Daltons. Most of the final processing is carried out by these companies that determine the terms of trade, export conditions and with their extensive marketing and distribution channels, influence price and capture most of the revenue. It has been suggested by Harilal et al. (2006) that a serious constraint to the cashew nut processing industry is that of meeting traceability requirements, especially with raw cashews sourced from so many different countries. With a few large international buyers prescribing increasingly stricter quality and traceability certification requirements, there is limited room for further value addition through production activities at the midstream node of the global value chain (in India, mainly). This is because these processors would then have to take the full risk of not only developing an effective reach into the consumer markets, but also managing a much more complicated forward and backwards traceability scenario in order to effect a quick market withdrawal

FIGURE 3.24 Branded cashew nuts on sale in a developing country supermarket. *Source: Kraft Foods, Inc.*

or recall should the need arise. The international trade in cashews, therefore, unlike other value chains discussed previously, presents an opportunity to explore future innovative solutions to ensuring fully compliant, sustainable FSQS in developing countrie, producing the primary materials and extending these directly to the consumer market. This is a requirement if producers at the upstream and mid-stream nodes of the value chain are to derive greater rewards from the value harvested at the consumer end.

References

African Cashew Initiative (ACI), 2010. Analysis of the cashew sector value chain in Côte d'Ivoire. African Cashew Initiative (ACi), Accessed on 2 June 2019 from http://www.africancashewinitiative.org/files/files/downloads/aci_cote_d_ivoire_gb_150.pdf.

Ardhana, M., Fleet, G., 2003. The microbial ecology of cocoa bean fermentations in Indonesia. International Journal of Food Microbiology 86 (1-2), 87–99.

Biehl, B., Brunner, E., Passern, D., Quesnel, V.C., Adomako, D., 1985. Acidification, proteolysis and flavour potential in fermenting cocoa beans. J. Sci. Food Agric. 36 (7), 583–598.

Camu, N., De Winter, T., Verbrugghe, K., Cleenwerck, I., Vandamme, P., Takrama, J.S., et al., 2007. Dynamics and biodiversity of populations of lactic acid bacteria and acetic acid bacteria involved in spontaneous heap fermentation of cocoa beans in Ghana. Appl. Environ. Microbiol. 73 (6), 1809–1824.

Camu, N., De Winter, T., Addo, S.K., Takrama, J.S., Bernaert, H., De Vuyst, L., 2008. Fermentation of cocoa beans: influence of microbial activities and polyphenol concentrations on the flavour of chocolate. J. Sci. Food Agric. 88 (13), 2288–2297.

Centres for Disease Control and Prevention (CDC), 2017. Multistate outbreak of Salmonella infections linked to imported Maradol papayas (Final update). Accessed on 1 May 2019 from https://www.cdc.gov/salmonella/kiambu-07-17/index.html.

Coltrain, D., Barton, D., Boland, M., 2000. Value Added: Opportunities and Strategies. Arthur Capper Cooperative Center, Department of Agricultural Economics, Kansas State University.

De Silva, D., 2011. Value Chain of Fish and Fishery Products: Origin, Functions and Application in Developed and Developing Country Markets. FAO Value Chain Project Reports. FAO, Rome, 63 pp. Also available at www.fao.org/valuechaininsmallscalefisheries/projectreports/en/.

Dyer, J.H., 2000. Collaborative Advantage: Winning Through Extended Enterprise Supplier Networks. Oxford University Press.

Ensign, P.C., 2001. Value chain analysis and competitive advantage: assessing strategic linkages and interrelationships. J. Gen. Manag. 27 (1), 18–42.

Fairtrade International. 2019. Fairtrade standard for small-scale producer organizatons, v2.2, 3 April 2019. Accessedon18 September 2019 from https://files.fairtrade.net/standards/SPO_EN.pdf

FAO, 2014. Developing sustainable food value chains: Guiding principles. Accessed at http://www.fao.org/sustainable-food-value-chains/library/details/en/c/265156/.

Gereffi, G., Kaplinsky, R., 2001. The value of value chains: spreading the gains from globalisation. IDS Bull. 32 (3).

Gereffi, G., Fernandez-Stark, K., 2011. Global Value Chain Analysis: A Primer. Center on Globalization, Governance and Competitiveness (CGGC). Duke University, Durham, North Carolina, USA.

GlobalEDGE, 2019. Jamaica trade statistics. Accessed on 28 May 2019 from https://globaledge.msu.edu/countries/jamaica/tradestats.

Gonsalves, D., 1998. Control of ringspot virus in papaya: a case study. Annual Review of Phytopathology 3 (6), 415–437.

Gordon, A. (Ed.), 2016. Food Safety and Quality Systems in Developing Countries: Volume II: Case Studies of Effective Implementation. Academic Press.

Gordon, A., 2018. Rural futures: Cultivating sustainable livelihoods in the Caribbean. Presented at "Sustainable Futures for the Caribbean. Critical interventions and the 2030 agenda", SALISES 19th Annual Conference, 25–27 April 2018. Holiday Inn Resort.

Gordon, D., 2019. The JBI/JSIF Water Harvesting and Greenhouse Custer Project. Working Document, Jamaica Bauxite Institute.

Harilal, K.N., Kanji, N., Jeyaranjan, J., Eapen, M., Swaminathan, P., 2006. Power in global value chains: implications for employment and livelihoods in the cashew nut industry in India. Published by the International Institute for Environment and Development (IIED). Printed by Oldacres Ltd, London, UK.

Hendrikse, G., Bijman, J., 2002. Ownership structure in agrifood chains: the marketing cooperative. Am. J. Agric. Econ. 84 (1), 104–119.

Humphrey, J., Memedovic, O., 2006. Global value chains in the agrifood sector (working paper). Available at: http://tinyurl.com/y8qh4obd.

International Cocoa Organization (ICCO), 2012. Harvesting and post-harvest processing. Accessed on 21 May 2019 from https://www.icco.org/about-cocoa/fine-or-flavour-cocoa.html.

International Cocoa Organization (ICCO), 2013. Processing Cocoa. Accessed from https://www.icco.org/about-cocoa/fine-or-flavour-cocoa.html.

International Cocoa Organization (ICCO), 2014. The Cocoa market situation. EC-4-2 Cocoa Market Situation, TheICCO Economics Committee Fourth Meeting, 16–18 September 2014. Wembley, London, United Kingdom.

International Cocoa Organization (ICCO), 2019a. ProductionQBCS XLV No. 1. February 2019.

International Cocoa Organization (ICCO), 2019b. Fine or flavour cocoa. Accessed on 21 May 2019 from https://www.icco.org/about-cocoa/fine-or-flavour-cocoa.html.

International Nut and Dried Fruit Council (INC), 2015. Cashew. Global Cashew Council. Accessed at: https://www.nutfruit.org/files/tech/FITXES-CASHEW-10-low.pdf.

Jamaica Bauxite Institute (JBI), 2016. The JBI Water Harvesting and Greenhouse Cluster Project. JBI, Hope Gardens, Kingston, Jamaica.

Jamaica Social Investment Fund (JSIF), 2016. 20th Annual Report 2015-2016, JSIF, Kingston, Jamaica.

Kampen, K., 2011. Financial Analysis of Three Value-added Dairy Enterprises in Vermont, Wisconsin, and New York (PhD thesis). California Polytechnic State University, San Luis Obispo, CA, USA.

Kaplinsky, R., 2000. Spreading the gains from globalisation: what can be learned from value chain analysis? J.Dev. Stud. 37 (2).

Kaplinsky, R., Morris, M., 2001. A handbook for value chain research, prepared for the International Development Research Centre (IDRC), p. 4–6 (emphasis added) (Accessed 22/09/17) Sustainability, UNEP and UN Global Compact.

Larsen, L., 2019. Mexican papayas still on import alert for Salmonella contamination. Food Poisoning Bull. Accessed on 30 April 2019 from https://foodpoisoningbulletin.com/2019/mexican-papayas-import-alert-salmonella/.

Mba-Jonas, A., Culpepper, W., Hill, T., Cantu, V., Loera, J., Borders, J., et al., 2018. A multistate outbreak of human *Salmonella agona* infections associated with consumption of fresh, whole papayas imported from Mexico—United States, 2011. Clin. Infect. Dis. 66 (11), 1756–1761.

Mentzer, John, et al., 2001. Defining supply chain management. Journal of Business Logistics 22 (2). Available from: https://docplayer.net/5820322-Journal-of-business-logistics-vol-22-no-2-2001-1.html.

Porter, M.E., 1985. Competitive Advantage: Creating and Sustaining Superior Performance. The Free Press, New York.

Reardon, T., Berdegué, J.A., 2002. The rapid rise of supermarkets in Latin America: challenges and opportunities for development. Dev. Policy Rev. 20 (4), 317–334.

SGS, 2014. Managing risk in global supply chain. Accessed on 12 April 2019 from https://www.mygfsi.com/news-resources/news/news-blog/1748-importance-of-managing-risk-in-food-supply-chains.html.

Sporleder, T., 2006. Strategic alliances and networks in supply chains: knowledge management, learning and performance measurement. Quantifying Agri-Food Supply Chain 15, 161–171.

Taylor, F.W., 1919. The Principles of Scientific Management. Harper & Brothers Publishers, The Plimpton Press, MA, USA.

Tennant, P.F., et al., 1994. Differential protection against papaya ringspot virus isolates in coat protein gene transgenic papaya and classically cross-protected papaya. Phytopathology 84, 1359–1366. Available from: https://www.apsnet.org/publications/phytopathology/backissues/Documents/1994Articles/Phyto84n11_1359.PDF.

Tessmann, J., 2017. Governance and upgrading in south–south value chains: evidence from the cashew industries in India and Ivory Coast. Accessed on 30 May 2019 from https://www.comcashew.org/imglib/downloads/TESSMANN-2017-Global_Networks.pdf.

The Economic Times, 2019. Accessed at https://economictimes.indiatimes.com/.

Thompson, S.S., Miller, K.B., Lopez, A.S., Camu, N., 2013. Cocoa and coffee. Food Microbiology. American Society of Microbiology, pp. 881–899.

University of Cambridge Institute for Manufacturing (IfM, 2019. Porter's Value Chain. Management and Technology Policy, University of Cambridge. Accessed on 30 May 2019 from https://www.ifm.eng.cam.ac.uk/research/dstools/value-chain-/.

US Food and Drug Administration (FDA), 2017. FDA investigates multiple Salmonella outbreak strains linked to papayas. November 4, 2017. Accessed on 25 April 2019 from https://www.fda.gov/food/outbreaks-food-borne-illness/fda-investigates-multiple-salmonella-outbreak-strains-linked-to-papayas.

World Bank, 2016. Cashew Value-Chain Competitiveness Project (P158810): Project Information Document/Integrated Safeguards Data Sheet (PID/ISDS). Accessed on 30 May 2019 from http://documents.worldbank.org/curated/en/128721473898123367/pdf/ITM00184-P158810-09-14-2016-1473898120922.pdf.

World Bank, 2018. How to Cash-in on Cashews? Supporting Côte d'Ivoire to Create More Jobs and Business Opportunities in Agribusiness with an IBRD Enclave Loan Using the Future Export Revenue of Cashew Nuts. The World Bank, Washington, DC, Accessed on 23 June 2019 from http://pubdocs.worldbank.org/en/347461532029238904/case-study-IBRD-enclave-Ivory-coast-2018.pdf.

Further reading

Cucagna, M.E., Goldsmith, P.D., 2018. Value adding in the agri-food value chain. Int. Food Agribus. Manag. Rev. 21 (3), 2018. Available from: https://doi.org/10.22434/IFAMR2017.0051.

Diamond, A., Tropp, D., Barham, J., Frain Muldoon, M., Kiraly, S., Cantrell, S., 2014. Food Value Chains: Creating Shared Value to Enhance Marketing Success. U.S. Dept. of Agriculture, Agricultural Marketing Service, May 2014.

International Cocoa Organization (ICCO), 2019c. The chocolate industry. Accessed on 21 May 2019 from https://www.icco.org/about-cocoa/fine-or-flavour-cocoa.html.

Kone, M., 2019. Analysis of the cashew sector value chain in Côte d'Ivoire. F1000Res. 3.

Krishnakumar, P.K., 2019. Tight global supplies hit cashew industry in India. Economic Mark Accessed at: 4.15 p.m. 17 June 2019 at https://economictimes.indiatimes.com/markets/commodities/news/tight-global-supplies-hit-cashew-industry-in-india/articleshow/58510169.cms?from = mdr.

Simatupang, T.M., Piboonrungroj, P., Williams, S.J., 2017. The emergence of value chain thinking. Int. J. Value Chain Manag. 8 (1), 40.

United Nations Conference on Trade and Development (UNCTAD), 2013. Global Value Chains and Development: Investment and Value-Added Trade in the Global Economy.

CHAPTER

4

Supplier quality assurance systems: important market considerations

André Gordon[1], Carolina Mueses[2], Helen Kennedy[3] and Rickey Ong-a-Kwie[4]

[1]Chairman & CEO, Technological Solutions Limited, Kingston, Jamaica, West Indies
[2]AgroBioTek Dominicana, Santo Domingo de Guzmán, Distrito Nacional, República Dominicana [3]Technical & Consulting Services, Southern Caribbean, Technological Solutions Limited, Port of Spain, Trinidad and Tobago [4]Principalis Consultants N.V., Paramaribo, Suriname

OUTLINE

Introduction

As the global food supply chain becomes more inter-connected and reliant on production in countries and locations far removed from the point of consumption, the role of global value chains, such as those discussed in the previous chapter, as well as the ability of the food system to assure safety, quality and satisfaction of customer and consumer demands become more important. This has led to the development and application of a range of *Supplier Quality Assurance* (SQA) systems which are based on producers conforming to agreed food safety and quality system (FSQS) requirements and delivery of products to

FIGURE 4.1   A cereal product from a developing country contract packed for Carrefour. *Source: André Gordon (2017).*

meet pre-determined, communicated specifications of the buyers. These SQA systems have become the norm in all areas of the food industry, from the production of raw agricultural items (fruits, vegetables, produce), meat, poultry and dairy products, to fabricated and manufactured foods. The increasing complexity of the markets, including the kinds of segmentation that are discussed in various parts of this volume, have resulted in a plethora of SQA programs and systems, several of which have become formal certification programs. Others became a pre-requisite to supply major buyers, including specific requirements for the retail trade for firms such as Carrefour (Fig. 4.1), Walmart, Tesco, ASDA, Safeway, Albert Heijn and others, as well as manufacturing groups (e.g. Nestlé, Unilever, Kraft-Heinz, InBev). Transnational food service groups have also developed their own requirements as they seek to mitigate the risk of foodborne illness, poor quality and other brand-damaging events that could occur without a system in place among suppliers to prevent this.

The result of this ongoing evolution in the area of food safety, quality systems and supply chain management to guarantee standardization across markets using a global supply chain and mitigate risks from sub-standard quality and unsafe food and food ingredients has been the universal application of SQA programs and attendant specifications. Whether the supplier is an agricultural (Figs. 4.2 and 4.3), processing (Fig. 4.4), food service (Fig. 4.5) or distribution (Fig. 4.6) operation, there are specific requirements that must be met to supply global food value chains. As discussed in Gordon (2015a), oftentimes buyers insist that their suppliers meet the requirements of the Consumer Goods Forum's (CGF's) Global Food Safety Initiative (GFSI)-benchmarked set of standards which now cover all of these areas, with some, such as Safe Quality Food (SQF) having benchmarked standards for all areas of production and distribution. Others, such as the British Retail Consortium (BRC) and Food Safety System Certification (FSSC) 22000 cover several areas of the food industry

FIGURE 4.2 Agricultural practices that support the implementation of good agricultural practices (GAPs). *Source: André Gordon (2015).*

FIGURE 4.3 Produce being prepared for export in two developing country pack houses. *Source: André Gordon (2014, 2015).*

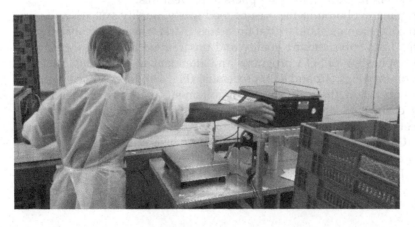

FIGURE 4.4 A food manufacturing operation in South America. *Source: André Gordon (2019).*

FIGURE 4.5 A food service operation in a developing country. *Source: André Gordon (2019).*

FIGURE 4.6 World-class distribution operations in two developing countries. *Source: André Gordon (2018, 2019).*

but may, at the moment, not have benchmarked standards for some areas (e.g. food service and agriculture for both BRC and FSSC 22000). In these instances, buyers may require other non-GFSI standards or have their own specific addenda (e.g. Costco and Restaurant Brands International (RBI) — Burger King and Popeyes). Other buyers, such as Tesco in the UK, may insist that suppliers also must specifically meet or be certified to their own or other buyer-approved standards such as Tesco's Nurture 10, or the Rainforest Alliance's Fairtrade standard. In addition, market trends and changing demographics are driving a rapid expansion in the demand for organically certified foods, as well as *kosher* and *halal* certification. All of these require that food industry participants, practitioners, industry support organizations and regulators in developing countries understand the changing landscape and the nature of the requirements to be successful in supplying developed country buyers, whether located domestically or in third country markets.

This chapter examines the general requirements of SQA programs, as well as those of other types of certification that developing country suppliers now have to deal with. It will look at the basic requirements for compliance with Good Agricultural Practice (GAP), Good Manufacturing Practice (GMP), Good Distribution Practice (GDP) and Good Food Handling Practice (GFHP) for foodservice operations. It will present selected examples of some of the additional requirements that selected buyers in food service, manufacturing and distribution have, and will also discuss the requirements for attaining organic, halal and kosher certification. From among the GFSI-benchmarked standards, it will briefly examine three standards as examples. These are GlobalG.A.P., a standard the covers primary production, FSSC 22000 which covers manufacturing and SQF which covers from primary production, through manufacturing and distribution and which has a (non-benchmarked) standard for food service as well. Overall, the chapter will provide a clear understanding of where the food industry is with regards to the market and certification requirements for growing business with third country buyers. It will also clarify the future direction of food safety and quality systems and certification in the development of the global food industry value chains.

The importance of specifications in food safety and quality in a supplier assurance program

Any program in which achieving specific target characteristics and values for specific food safety or quality parameters is required (e.g. microbial numbers or mass, respectively) will depend on the ability of suppliers to consistently meet these targets. As such, the role of specifications is critical when seeking to establish criteria with which all parties along a supply chain can agree, understand and by which they seek to manage products and commercial arrangements. The language outlining what is required to meet the buyers needs should be clear, with unabiguous descriptions of what is required being at the core of the description of characteristics or parameters (Fig. 4.7). Clear, unambiguous descriptions and/or specific measureable criteria allow for buyers and suppliers to speak the same language and understand specifically what is being offered and what are the expected compliance limits, compositional criteria and measuable parameters or values. They facilitate uniformity in measurement or assessment of products by buyer, seller, regulators or, in the case of disputes, an independent third party. They allow an input such as packaging to be sourced and delivered to meet very specific criteria, an example of which is shown in Fig. 4.7.

EXAMPLE 1

■ A clear, transparent 8 oz (250mL) glass bottle, with a height of 6" (15 cm) with a screw-on type cap with lug markings parallel to the base. The body of the bottle should have a diameter of 2 ½" (6.25cm), with a neck diameter of 1" (2.5cm). The thickness of glass should be a minimum of 4.5 mm. The bottles should be shipped packed in cartons, 24 per case with a total stack height of 10 cartons per pallet.

EXAMPLE 2

■ Clear glass bottle; transparent
■ Screw on cap with lug markings parallel to the base
■ Capacity - 8 oz (250 mL)
■ Height – 6" (15 cm)
■ Diameter of Body – 2 ½" (cm)
■ Diameter of Neck – 1" (2.5cm)
■ Thickness of glass – 4.5 mm (min)
■ Packed in cartons, 24 per case
■ Stack height – 10 cartons

FIGURE 4.7 Two examples of summary specifications for a glass bottle.

Note that the format of the specifications may vary, depending on seller/buyer preference, whether any particular format is standard in an industry subsector or in keeping with a predetermined documentation style required by the buyer or which is typical in the industy. Specifications may be written in a paragraph which communicates the details of what is required, as shown in Example 1 (Fig. 4.7) or they may be arranged in a bullet point summary format (Example 2) or any other convenient format that unequivocally communicates requirements to all parties involved. Whatever the format used, the purpose is to communicate what is required in an easy to understand manner.

Alternatively, in developing countries where literacy, a language barrier or an inherent difficulty in easily describing the characteristics required of a product or service exists, pictorial reperesentation of the characteristics being sought can be useful. Examples of this are often seen where the trade is in produce, fruits and vegetables and there are difficulties describing what is required in a clear and unambiguous manner. In these cases, it is often helpful to use visual representations of what is required, an example of which is the pictorial descriptions used in areas of the spice trade in some developing countries (Fig. 4.8). These visual guidelines allow for uniformity in the sourcing, preparation, handling and purveying of traditional crops for which readily documented, standard format specifications or criteria may not be available.

Whatever the mechanism used to communicate the specifications being targeted, it is important that the buyer (the importing firm for transnational supply/value chains) and the supplier clearly understand what is being sought and therefore what is meant by each term if written specifications are used. This should be confirmed by the buyer in writing, a process which could also include a range of tests and analyzes done on the initial and subsequent samples of product supplied, with communication of the findings back to the supplier for their action. This is critical if supplier/buyer relationships are to operate seamlessly and remove the element of uncertainty in commercial transactions and in the

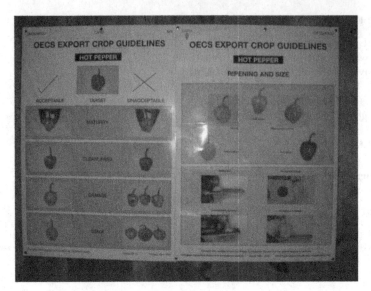

FIGURE 4.8 Example of pictorial specifications: peppers in the OECS region. *Source: A. Gordon (2014).*

determination of the levels of compliance with the agreed criteria on the basis of which the transaction is being done. This applies equally to all types of food items.

Combating food fraud: technical approaches to brand protection

Food fraud has been defined as *"the act of purposely altering, misrepresenting, mislabelling, substituting or tampering with any food product at any point along the farm—to—table food supply—chain"* (Smith, 2016). It can impact the producer if they are supplied with ingredients that have been misrepresented, enhanced using unapproved additives (e.g. colorants) or which have been substituted or had components substituted without their knowledge, thereby causing them to unknowingly also perpetrate a fraud on the consumer. In other cases, the producer or distributor may be involved in fraud if they misrepresent the country of origin, misbrand the product or infringe the intellectual property rights of another entity. Whatever the reason, the person who suffers financial loss or deception may, in the worse-case scenario, also suffer intended or unintended public health consequences (Smith, 2016; Spink and Moyer, 2011).

Any discussion of food fraud should include the motivation for the fraud, including economically motivated fraud or fraud intended to inflict harm on the consumer or the brand. Other considerations include criminal liability for the fraud, the ethical concerns and intended or un-intended private or public health consequences (Smith, 2016; Spink and Moyer, 2011). An area receiving increasing attention today as one of the requirements for any Global Food Safety Initiative (GFSI) benchmarked certification programs, prevention of food fraud is now a major focus of any robust food safety and quality system program. While it can occur with specific inputs, including raw materials and packaging, it also occurs with the finished product being marketed to consumers and may, at this stage, also include the food's packaging. If the fraud delivers a product that is of noticeably inferior quality, or causes injury or illness to the consumer, the manufacturer or brand owner's brand equity can be damaged (Spink et al., 2016). Alternatively, the fraud may misrepresent the nature of the product and suggest that it is organic, halal, kosher or made in a plant that is certified to a GFSI-benchmarked program (all discussed later on in this chapter) when it is not.

Another type of food fraud is the counterfeiting of food products which has become a billion-dollar business and is a major concern in the food industry. It is typically targeted at major brands or products which, by virtue of the properties associated with their name, brand or origin, command a premium price in the marketplace or enjoy significant commercial success. Counterfeiting is a type of food fraud that involves *intellectual property rights infringement* and is very hard to prevent (Spink, 2019) and often even harder to prove. Obtaining proof requires careful planning and application of various aspects of food science as well as a knowledge and understanding of the relevant regulations in the jurisdiction in which the fraud is being perpetrated. This has been the challenge that has faced several firms from developing countries who find counterfeit products being passed off as theirs in their major markets. These firms are often only made aware of the problem by customer complaints in which the consumer complains about changes in taste, color or other characteristics of the product. The diligence of the firm and their use of a meticulous approach to dealing with the challenge to their brand and intellectual property can result in a successful prosecution of the perpetrators of the fraud and force them to terminate their fraudulent activities. Counterfeiting can, however, be difficult even for firms from developed countries to deal with and so firms from developing countries with increasingly successful brands and their import partners should be on the lookout for this.

Basic requirements for agricultural production systems

In a food value chain, an important link is undoubtedly agricultural production. Without proper control of the activities at this point in a value chain, achieving a finished product that complies with food safety and quality requirements will be difficult. What every primary food producer wants is for all of their products to be sold at the best possible price. Oftentimes, due to a limited local market, to accomplish this will require the firm to sell products to the local as well as the international market. A number of conditions must therefore be met to guarantee the safety of the products for both the local and international market.

Among these are effective management of the value chain, which requires having a functioning food safety system implemented in the primary producers' operations. By having a food safety system implemented (certified or not), an agricultural production company is able to demonstrate that all processes within the company are controlled. It is important that all of these facets of agricultural production that make the difference between success and failure for the farmer are known and properly delineated beforehand. A simplified flow chart of the production process on a farm for fruit and vegetable products is shown in Fig. 4.9. The sections that follow discuss the most important considerations for agricultural operations.

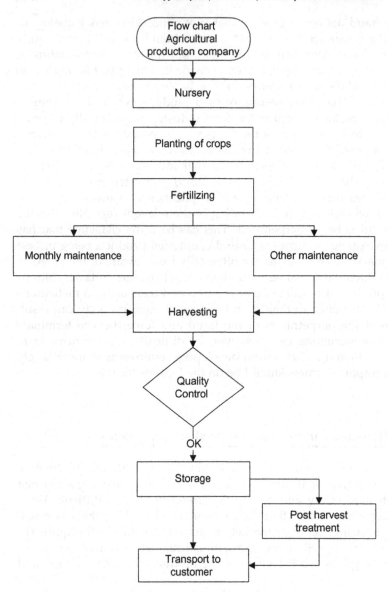

FIGURE 4.9 Simplified flowchart of agricultural production. *Source: Rickey Ong A Kwie (2019).*

Selection of the location of the farm

The history of the area and the immediate surroundings of the farm must be known and documented. The main reason for this is that if there have been activities in the area in the past that may have caused long-term pollution of the soil by heavy metals, chemicals, pesticides, etc. (e.g. by industrial activities such as gold or bauxite mining, sugar cane or other monocrop production), this can have significant consequences that have to be taken into

FIGURE 4.10 Area with soil compaction. *Source: Rickey Ong A Kwie (2019).*

consideration. A possible consequence of past history of the farming location is that the yield of the crops is low due to the pollution of the soil, a fact which is also likely to lead to contamination of the food grown by these pollutants. This may occur because there is a chance that the pollutants, often chemical in nature, can be absorbed by the plant and thus be introduced into the human food chain. Because of this, soil analysis prior to start-up is very important. The results of the analyzes will help to advise on any possible pollution concerns; it will also advise the producer which crop the soil is most suitable for growing and which (micro) nutrients will possibly need to be added to have optimal growth.

Another reason for assessing the soil prior to commencement of a farming operation is to assess the extent of compaction of the soil. Soil compaction (Fig. 4.10) and erosion should be avoided as much as possible because compacted soil has a reduced rate of both water infiltration and drainage and slower exchange of gases, causing an increase in the likelihood of aeration-related problems. Finally, compacted soil also requires roots to exert greater force to penetrate the compacted layer. All of the above-mentioned issues may result in slower growth (deJong–Hughes et al., 2018) as shown in Fig. 4.10 where soil compaction has resulted in inferior growth rates. This often leads to excessive use of fertilizers and plant growth stimulators to counteract the slow growth rates, options that are not in keeping with Good Agricultural Practice (GAP).

Water quality and water management

It is difficult to achieve good quality, safe agricultural crop production and with good yields without a proper water supply and adequate drainage. During the growth phase of production on the farm, water used may come from sources like the national domestic water supply system, rivers, creeks or from rainwater harvesting reservoirs or tanks (see Chapter 3) in countries where adequacy of water for irrigation is sub-optimal. Whatever the origin, no water should be used that comes from sources known to be polluted by drainage systems, spill water systems or run-off from other farms or operations, for example, that flow into these water sources. It is therefore critical that the chemical and microbiological quality of the water being used is known and monitored on a pre-determined basis. This is also a basic requirement of most SQA systems for primary (agricultural) production.

The water used in an agricultural production operation includes:

- Agricultural Water (for irrigation/fertigation and field preparation)
- Cooling Water (used to remove field heat from some crops)
- Wash Water (used in packing/processing to clean harvested produce)
- Processing Water (used in the pack house/packing operation, generally)
- Water used for sanitary purposes (for hand washing on the farm, in the field or pack house)
- Water used for preparation of antimicrobial chemicals/pesticides/fertilizers

Water used for irrigation or fertigation must be free from harmful contaminants and adequate measures must be taken to prevent water flowing into the fields from undesirable sources or being used for irrigation after being contaminated by sources such as landfill areas, industrial sites, dump sites and hospitals. It is important, therefore, to have an understanding of the microbial and chemical hazards that may be associated with water for various uses in farming operations and to have a record of the nature of the water being used.

While there are different requirements for the water used at different phases of the operations in which water is used, the water used for sanitary purposes and post-harvest activities (including cooling and wash water) should be potable water (of drinking water quality). This is critical because in several of the instances in which foodborne illness outbreaks have been associated with agricultural production such as produce, fruits and vegetables, breakdown in sanitary practices or the use of contaminated, non-potable water has been the cause. Where human enteric pathogens such as *Hepatitis A*, *Salmonella* and pathogenic *E. coli* have been implicated, poor personal hygiene and sanitation exacerbated by the lack of availability of potable water, effective cleansers (soap/detergent) and sanitizers have often been the cause.

Planting material

The history, immediate source, origin, quality, freedom from disease and genetic composition of planting material is often required information for a good primary production operation. This information not only provides the owners with valuable insights as to overall productivity of their farm, it helps with selection of seeds, seedlings, cuttings and other planting material for optimal returns and, importantly, is also a requirement of GAP and several SQA systems. Other requirements related to planting material are important for fruits, vegetables, produce and other materials of plant origin (e.g. spices, etc.) being exported to or sold in the EU, Japan, Australia/New Zealand and selected other third country markets. An example of this is the non-genetically modified organism (GMO) status of the product, which is seen as critical for success in these markets. As such, the value chain delivering the product to market has to be able to clearly demonstrate by effective *traceability* back to the origin of the planting material that the food being offered for sale is non-GMO.

Among other reasons for control of the origin of planting material include the risk of introduction of diseases and pests on plant material originating from infected countries, areas or nurseries and the use of inferior planting material from plant propagation facilities (nurseries) with less than optimal practices. This often makes the difference between a successful and a less successful operation, in addition to presenting a potential risk to both the safety of the product and its pest and disease status depending on the growing areas. A fruit that is very highly prized and in high demand in North America such as *guanábana* (soursop), also known as *graviola* in parts of the Americas (Fig. 4.11), for example, has had restrictions put on its import into the US because of pests associated with the fruit originating from some countries. This makes the origin of the fruit and source of the original planting material important to the commercial success of an export venture for the fruit. Other advantages of using carefully selected, suitable plant material is reduced use of fertilizers and pesticides (particularly in the early stages of plant propagation), improved plant growth, better product yields and also better product quality.

Other important aspects of primary production operations, including practices and requirements for compliant operations on the farm, during harvesting, transportation and in the packhouse are summarized under Good Agricultural Practice (GAP) discussed further, below.

FIGURE 4.11 Guanábana (Soursop – *Annona muricata*) in a pack house in Latin America. *Source: André Gordon (2019).*

Good agricultural practices (GAPs)

GAPs are defined by the Food and Agricultural Organization (2003) as "a collection of principles to apply for on-farm production and post-production processes, resulting in safe and healthy food and non-food agricultural products, while taking into account economic, social and environmental sustainability." GAPs provide guidance for employing best management practices to help reduce the risks of microbial contamination of fruits and vegetables. In order to minimize microbial food safety hazards, the growers, packers and shippers of fresh produce must use good agricultural practices in the areas over which they have control.

Factors to be controlled

For Good Agricultural Practice, the following must be controlled in addition to the site selection and history, water management and planting material already discussed:

Worker hygiene and health. During the production, harvesting, sorting, packing and transportation of fresh produce, worker hygiene and sanitation practices play a critical role in minimizing the potential for microbial contamination. All workers should be trained in personal health and hygiene practices for minimizing microbial contamination. Written instructions on personal hygiene should be provided and prominently displayed.

Sanitary facilities. Sanitary, well-ventilated and secure toilets and hand washing facilities must be available to all workers. These must be maintained in a hygienic condition. Importantly, also sewage must be disposed of in a manner that precludes contamination to produce.

Water quality. As previously indicated, water is critical at all stages in the production and handling of food. The source and quality of water that comes into contact with fresh produce (whether in production or harvesting or processing) determines the potential for contamination.

The use and handling of manure. Human and animal feces and composted manure are major sources of microbial contamination. All workers who are required to use manure should be trained in the use and handling of the manure. The use of manure and other soil amendments, additives and chemicals should be closely managed to minimize contamination of fresh produce.

The use and control of fertilizers. All farming operations except for organic farming (discussed later in this chapter) use fertilizers and other plant growth stimulators to increase the rate of growth, foliage and productivity of crops being grown. Overuse of fertilizers can result in the soil or crops being grown becoming contaminated with chemical residues (e.g. nitrates or phosphates) which can be harmful in too high concentrations. Similarly, misuse of fertilizers (poor or wrong application) can see foliage or crops directly being contaminated with fertilizer residues at harvest, resulting in a risk to consumers from residual fertilizer on these fruits, vegetables or produce.

Use and control of pesticides. Pesticides and other chemical are required to be used in accordance with their usage instructions. Where pesticides are used, the appropriate *pre-harvest interval* (PHI)[1] is expected to be scrupulously observed to eliminate the risk of residual pesticides being on the produce and creating a food safety risk for consumers.

Field Sanitation. General harvest conditions should be sanitary. Produce should not be placed directly on the soil, be contacted by a potentially unclean surface (Fig. 4.12A and B) or put on the floor of the transportation or of the handling, packing or storage areas (Fig. 4.12C).

Equipment, containers and materials. The general requirement is for ensuring that equipment and containers used on the farm and in the packhouse are checked for soundness, food safety status (i.e. they do not contain toxic components or residues) and fitness for use. As such, containers such as field crates are required to be stackable (Fig. 4.12D) in a manner such that produce can be handled without being contaminated by unclean surfaces (e.g. the bottom of a field crate), cleaned before use and cleaned and sanitized periodically, as well as being repaired or discarded, as required.

Chemicals. Among the chemicals used across the operations that need to be controlled under the chemical management protocol are the agro-chemicals i.e. those used for cleaning, maintenance chemicals, pesticides (herbicides, acaricides, fungicides, pre- and post-emergent herbicides, whether contact or systemic), fertilizers and sanitizers (used in the pack house operations). Special areas should be constructed and properly outfitted for the storage of agro-chemicals. Non-agrochemicals such as grease, fuels and oils that are also required for other purposes on the farm must also be controlled. Workers should be trained in agrochemical/fertilizer management and use protective gear in their application. Equipment for applying chemicals must be maintained in good working condition and checked on a scheduled basis by technically competent persons. For chemicals, consideration should also be given to receiving controls, including the location and timing of receivals, the location and suitability of the area for immediate storage and whether the chemical is food-grade (approved for use with food) or not. Storage standards for chemicals should include ensuring appropriate segregation of food-grade and non-food grade chemicals, adequate ventilation, labeling of all individual units, including dispensing containers, and the segregation, sanitation and labeling of field containers used for different kinds of chemicals.

Harvesting. This should be done in an organized, planned manner, by pre-designated lots or fields to facilitate ease of traceability and management. Harvesting should be done in a sanitary manner with produce being placed directly in sanitary containers such as field crates (Fig. 4.12), where possible, and kept from direct contact with soil and the ground.

Post-harvest handling. Handling of produce when harvested is critical as any contamination at this stage may be likely to end up on the product delivered to the consumer, depending on what is involved in the post-harvest handling and treatment of the product on the farm, during transportation to, and storage in, the packhouse and

---

[1] The pre-harvest interval is the time between application of the pesticide and harvest of the crop that is required to ensure the complete removal of the pesticide from the plant prior any likelihood of it being consumed.

FIGURE 4.12 Collection and handling of produce: (A & B) bottom of field crates in contact with product; (C & D) nesting of clean crates to prevent direct contamination of product. *Source: André Gordon (2007, 2019).*

subsequent handling. It is important that all harvested products are handled in a sanitary manner by staff who are trained in proper sanitary handling of produce.
Storage and transportation. Produce must be stored and transported separately from other items that are potential sources of chemical, biological or physical contamination. A general consideration is that vehicles for transporting food products must be kept clean, in good condition, free from chemical contamination and free from pests. Produce should be held in secure, purpose built vehicular compartments, including refrigeration units, where required, and kept under appropriate environmental conditions. Overloading of vehicles should be avoided.
Sanitary handling in the packing facility (packhouse). Food products should be handled, packed and stored under appropriate and hygienic conditions. This requires the development and implementation of a validated, effective sanitation program for the packing facility. Good handling practices, as documented in standard operating procedures (SOPs) and an appropriate pest control program as developed by a competent pest management professional (PMP) are critical for preventing cross contamination of fruits, vegetables and produce being handled from packing

FIGURE 4.13 Variation in practices: poor (A) and good (B) sanitary and food handling practice in a packhouse. *Source: André Gordon (2010).*

equipment, pests and other sources of potential contamination. Domestic animals are strictly prohibited in the pack house and environs of the facility.

In the packhouse, sanitary handling is critical to ensure that the good practices which resulted in good quality, safe and sanitary fruits, vegetables and produce being delivered to the packing environment are not undone by poor sanitary practice (Fig. 4.13A) but is enhanced by good practice (Fig. 4.13B). In order to ensure this, training of the packhouse staff in Good Food Handling Practices and Sanitation and Hygiene is recommended. This should be supported by appropriate signage and documentation of procedures (SOPs) in a packhouse operations manual or similar document. These procedures can then be encapsulated and issued to each work area as targeted work instructions.

Traceability. Accurate records regarding harvest, storage and processing of produce must be maintained. Detailed records should be kept of the areas from which produce was harvested, each of which should have unique identification by name, number or code. Packed produce should be clearly labeled or marked and identified to enable traceability to the source farm and field/plot/area on the farm on which they were grown. A critical aspect of traceability for safety is routinely adhering to, and recording, pre-harvest intervals for the application of each kind of pesticides for all crops. This traceability should extend from the farm through handling and packing operations at the packhouse and to final storage prior to distribution (Fig. 4.14), in which storage best practices should be followed, including storage of produce at the appropriate (recommended) temperatures.

Summary of important control points & compliance criteria for GAP

In summary, Good Agricultural Practices (GAPs) require that farming operations follow basic approaches in order to comply with the supplier quality assurance requirements of quick serve restaurants (QSRs), hotels (local or transnational chains), manufacturing

FIGURE 4.14 Best practices: appropriate storage and labeling of product in facilities for traceability, inspection and cleaning. *Source: André Gordon (2010).*

operations for which produce are an input and supermarkets and multiples (domestic or transnational, respectively). These are also the same approaches that result in general good primary production management and deliver the kinds of returns that make the farm sustainably viable as a key part of a value chain. The main control points for such operations which have been discussed and which are also important for operational efficiency and viability are listed below:

- Site history and site management
- Soil and substrate management
- Variety and rootstocks
- Fertilizer use
- Irrigation/fertigation
- Crop protection
- Harvesting
- Produce handling
- Waste and pollution management, recycling and re-use
- Record-keeping and internal self-inspection
- Traceability
- Worker health, safety and welfare
- Environmental issues
- Complaints documentation and management

Once an operation has at least basic programs in place for each of these, they should comply with most of the requirements of agricultural supplier quality assurance programs.

Basic requirements for warehousing & distribution operations

In most cases, the food production process includes transport and distribution operations. While there may be some warehousing and distribution companies in developing countries that believe that once the product is in the warehouse, nothing further is required to safeguard food safety and quality, the warehousing and distribution operations are fundamental steps in ensuring that the product reaches the final consumer as good quality, safe product.

Control of reception and handling in warehouses

Once ingredients have been purchased, adequate procedures must be in place for their reception, storage and management during warehousing. It is advisable to have specifications for all raw materials so as to have a point of comparison to determine acceptability when receiving (see Fig. 4.7). Many distribution center (DC) operators do not understand the need for specifications. One solution has been to develop a specifications document that visually indicates what is acceptable to be received or what is to be discarded to help to guide persons doing the receival. It is necessary to have records that verify that products received complied with specifications at the point of reception. The records provide documentation of the condition in which the raw materials and products are received as well as relevant information such as lot numbers, "Best Before" or "Use By" dates and the temperatures for frozen and refrigerated products.

Handling and storage in the warehouse are important to keep items received in good condition compliant. Good personal hygiene and Good Food Handling Practices (GFHP) are required to support a Good Distribution Practices (GDP) program. This involves ensuring that adequate GDP reminders are posted where staff can see them, such as those reminding staff to wash their hands (Fig. 4.15) or those reminding staff not to eat, drink or smoke in the warehouse area (Fig. 4.16) in the appropriate languages.

Another important aspect of GDPs is ensuring that items are stored in a manner consistent with good practice, that is off the ground on pallets or racks and away from the walls to allow for inspection. Any pallets being used must be in good condition. Unlike those shown in Fig. 4.17, pallets must not present a risk of contamination of the items being stored on them by wood slivers. The warehouse must also be clean and organized. Cleaning and housekeeping are important in warehouses but is one of the areas that is most often neglected, especially deep cleaning. In general, it is found that while corridors are swept and kept clean, more detailed inspections below the shelves and

FIGURE 4.15 Best practices: well-placed signage in appropriate language for hand washing. *Source: A. Gordon (2019).*

FIGURE 4.16 Best practices: signage providing guidance with appropriate practices. *Source: A. Gordon (2019).*

FIGURE 4.17 Examples of poor quality pallets. *Source: A. Gordon (2019).*

FIGURE 4.18 Unlabeled items stored in filed crates in a warehouse (left); vegetables in field crates (right). *Source: A. Gordon (2019).*

pallets reveal accumulated dust and dirt and areas that are not routinely cleaned. The cleaning of warehouses and all storage areas should be included in the Master Sanitation Plan for the facility.

All ingredients or products should have labels indicating the received date and "Use By" or "Best Before" dates. This facilitates operations in a manner that ensures that the first product that enters is the first to be dispatched, that is first-in-first-out (FIFO) rotation of inventory. Alternatively, first-expired-first-out (FEFO) can be used to manage inventory particularly where short shelf life products are being handled. Among the main problems found in DCs is that the staff often fail to label the products if they feel that the products will only be in the warehouse for a very short time. While this may be true for some

FIGURE 4.19   Properly labeled and coded items stored in a cold room in a warehouse. *Source: A. Gordon (2019).*

products, whether the shelf life is short or very long, not labeling products (Fig. 4.18) often leads to costly errors and prevents proper inventory management and traceability in case of a recall. A better example of acceptable labeling for stored product is Fig. 4.19.

Many distribution centers and storage facilities include areas for the storage of refrigerated and frozen items. Good temperature control is required for these areas and this requires ongoing monitoring of temperatures and effective corrective actions when temperatures approach set limits or move outside of the desired ranges. The organization of the cold rooms is important in facilitating efficient operations and uniform temperature distribution. Ideally, where products of different kinds are being stored, there should be separate rooms or temperature zones for each category of product. For example, for fruits and vegetables, dairy products, meats, poultry, fish and seafood, prepared foods (and leftovers/rework in the case of food service and manufacturing operations' storage areas), separate storage or temperature zones are desirable as not all fruits and vegetables nor prepared foods have the same storage temperature requirements. Established maximum storage times for all items is also desirable. As for items in a dry storage distribution centre (DC), all products in frozen/chilled storage should be properly labeled (Fig. 4.19), including prepared products, carry-over product (food service and manufacturing) and fruits and vegetables. This then allows proper stock rotation as mentioned above and facilitates ease of traceability during loading, whether done manually or using electronic means, such as bar coding (Fig. 4.20).

Loading

An important aspect of the warehousing and distribution operations is the process of loading the product onto vehicles destined for distribution. This first requires that a designated team member in the distribution operation verifies that the vehicle(s) in which the food is to be transported are in the appropriate condition. The inspection process for

FIGURE 4.20 Best practices: proper handling of product, inclusive of traceability information and bar coding for data capture during dispatch. *Courtesy of Sranan Fowru, South America. Source: André Gordon (2019).*

delivery vehicles is oftentimes deficient because the responsible warehouse staff do not carry out a detailed inspection to verify the effectiveness of cleaning, absence of pests, conditions of the vehicle and any other consideration that can affect the safety of the product prior to dispatch. At times, the records are completed mechanically without real or meticulous inspection, resulting in product being distributed being put at risk. Good DC operations therefore need to ensure a focus on this area as the final step before loading and dispatch of the delivery container or vehicle.

In some instances, an improper practice observed for warehouses handling frozen or chilled products is the loading of products that require refrigeration and or freezing into vehicles that have not reached the optimal temperatures required to maintain the cold chain. The puts the product at risk. Once the pre-determined pre-cooling temperature is reached, best practice is loading through a temperature-controlled loading bay (Fig. 4.21) or by staging product for less than five (5) minutes prior to loading into a vehicle that has been pre-cooled. Other improper practices include the placement of the product directly on the floor of the vehicle for space optimization and dispatch of vehicles without the required safety seals or locks to ensure that they are secured prior to leaving the facility. All materials should be kept off the floor or stored on pallets, racks, shelving or other appropriate storage systems that allow for ease of inspection, cleaning and that preclude contamination of the products being transported.

Transportation

Loading problems that occur extend to the transportation of products: lack of records, inadequate vehicles, temperature failures and failure to inspect the condition of the

FIGURE 4.21   A vehicle preparing for loading: backing up into a fully enclosed loading bay. *Courtesy of Sranan Fowru, Suriname, South America. Source: A. Gordon (2019).*

vehicles result in deficient product transportation which can trigger safety and quality concerns. Often, the personnel who carry out the deliveries are not adequately trained, and this compromises the condition of the products being delivered. Because inadequately trained personnel create problems for warehousing and distribution businesses, proper training for staff involved in receiving, storage, picking and staging, loading, transporting and delivery is critical to assure a high-quality service that protects the safety and quality of the food being handled.

Basic requirements for manufacturing

Many manufacturing companies in developing countries have the desire to achieve industry best practices in order to facilitate growth within their own domestic marketplace and also in the global market through exports. This will require adoption of the best practices discussed in this section, as well as ensuring that they avoid many of the pitfall and unacceptable practices that can affect their customers and potential customers through the quality and safety of their products. Manufacturers that seek to employ Good Manufacturing Practice (GMP) should start by ensuring that their layout and design not only is compliant with the regulatory requirements in their target markets, industry best practices and GFSI facilities standards, but also are able to deliver efficient and competitive production of the items being manufactured, using appropriate equipment (Fig. 4.22).

GMPs employed by manufacturers must include proper solid waste management (Fig. 4.23), ensuring an adequate, properly treated and properly protected water supply and effective pest control (Fig. 4.24) and excellent personal hygiene and handwashing practices (Fig. 4.25). They should also ensure that staff are attired in a

FIGURE 4.22 Best practices: use of stainless-steel equipment and food contact surfaces in acidified food/ sauce (A) beverage (B & C), fruit & vegetables (D) and poultry manufacturing (E & F) in developing countries. *Source: André Gordon (2012, 2017, 2019).*

manner appropriate to the products being handled so as to protect the food from contamination (Fig. 4.26).

The quality and safety of the products being manufactured can only be guaranteed if proper process controls have been implemented during manufacturing. The controls in the manufacturing process will be determined based on the biological, physical or chemical hazards that have been identified in the risk analysis of the process. While the majority of process controls are usually associated with the biological hazards, depending on the nature of the product being manufactured, control over physical hazards (e.g. metal — Fig. 4.27) and chemical hazards (e.g. for certain seafoods) may also be required. Process controls, along with other preventive controls discussed previously in Chapter 1, constitute a fundamental part of a Food Safety Plan. They must be implemented based on science and experience and there are established parameters and limits that must be controlled to ensure safety of the final product. In order to achieve this, manufacturers

FIGURE 4.23   Best practice: enclosed garbage bins & dumpster for factory solid waste. *Source: André Gordon (2019).*

FIGURE 4.24   Best practice: (A) secured (arrow) auxiliary potable water supply & (B) pest control bait station (arrows) protecting the production facility. *Source: André Gordon (2019).*

have to overcome one of the fundamental challenges they encounter having grown their business, which is that many started as small companies which at start-up were not thinking about food safety and so did not design their operation in a manner to make it compliant. Consequently, implementing GMPs and compliance with food safety and quality system requirements when they need to export, or access more sophisticated markets often requires a significant transformation of their operations. In this section, the requirements for implementation of best practices is put in the context of manufacturers in developing countries seeking to upgrade their operations. Below are some practical examples of challenges that manufacturers face in complying with supplier quality assurance (SQA) and other requirements as they grow their businesses and implement systems to improve the compliance and competitiveness.

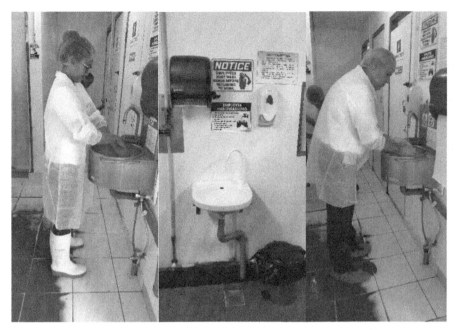

FIGURE 4.25 Best practices: appropriately located, adequately provisioned hand wash stations facilitating good handwashing practices. *Courtesy of Sranan Fowru, South America. Source: André Gordon (2019).*

FIGURE 4.26 Best practices: appropriate attire required for staff handling products in a plant. *Courtesy of Sranan Fowru, South America. Source: André Gordon (2019).*

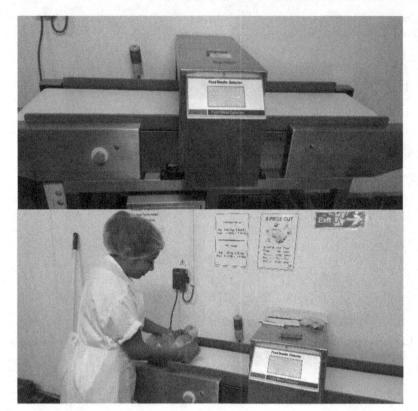

FIGURE 4.27 Best practices: metal detection used to manage metal contamination risk. *Source: André Gordon (2019).*

Let's take the example of a small bakery. The owner is interested in expanding the business and decides to offer their products to a chain of international restaurants. Upon receiving the customer's supplier assurance requirements, the owner is now faced with implementing a series of manufacturing controls that are quite costly. For example, the client requires a metal detector to be used to ensure the control of any potential hazard due to metal contamination. The owner of the bakery has probably never had a complaint about the presence of metal in the bread he makes, and when he carried out a hazard analysis, he never considered metal fragments as a risk. However, the chain of fast food restaurants has the metal detector as a mandatory requirement in its food safety guidelines. The small baker is then obliged to establish a manufacturing control that had not been originally contemplated and which has a cost that exceeds the current ability of his small business to finance it. Not installing the metal detector, however, limits the possibility of growth. Incorporating a metal detector in the operations means not just investing in the equipment but also redesigning the layout of the plant that has a completely manual process that does not include a conveyor for taking packaged bread through the metal detector. Consequently, a more pragmatic solution, such as using a stand-alone metal detector which operates off-line and through which each package can be manually put (Fig. 4.27) may be employed as a solution. This additional step may create a bottleneck in the

FIGURE 4.28 Best practices: posted whistle blower policy, appropriately located rest rooms and labeled pest control device locations. *Source: André Gordon (2019).*

process, but this would have to be accepted until a more automated solution that the bakery could afford can be implemented, if there is no other option.

Continuing with the example of the bakery, the bakery is also being required by this large new customer to implement new programs, improve their GMPs and also to bring their pest control program into compliance with their basic requirements. This means implementing some of the systems and practices shown in Figs. 4.22—4.25, inclusive of proper labeling and management of pest control fixtures and ensuring that its staff rest room facilities are fully compliant (e.g. Fig. 4.28). Further, when carrying out the risk analysis, the baker realizes that he handles several allergens in the process which he never considered before and for which he must now implement an *allergen control program*. He faces the disadvantage of not having a warehouse appropriately designed to separate the ingredients, or not having the logistics implemented to handle the allergens in the weighing area in a manner to prevent allergen cross contact. For the bakery, this may also present a challenge as the staff do not have sufficient knowledge regarding the seriousness of allergens and because of how the business developed, the company does not have an effective and proactive food safety culture. This bakery will therefore have to re-organize its production and process flows, re-train staff in food safety, particularly as regards the handling and storage of allergens to prevent allergen cross-contact and implement effective allergen preventive controls, supported by the appropriate records and documentation. They will also have to implement systems, such as implementing a whistle-blower policy (Fig. 4.28), that will help to build and reinforce the positive food safety culture.

Another example is the production of milk in a small dairy. A small cattle producer decided to add value to his product and acquired a pasteurizer that came with a set of established process parameters required to apply a compliant thermal process to the milk. This would also allow the processor to meet the customer requirements for implementing a HACCP-based Food Safety and Quality Management System. When he starts to implement the HACCP plan for his process, he realizes that the thermal processing step has become a fundamentally important part of the process because it is a Critical Control Point (CCP). A process that, in his opinion, had been working perfectly now must be controlled more strictly. He must define limits, corrective actions in case of deviations, validate if the process is effective and efficient, and probably at the end, he may have to invest in a new pasteurizer to guarantee that the milk is safe. In many cases he faces the reality that the process cannot guarantee consistency in the parameter he has established, because the equipment is old, inefficient or it does not have the capacity to consistently meet the required temperature for the time required, thereby guaranteeing safety.

The final example is that of a vegetable processing facility. There are many process controls that must be implemented in the process of producing cut and packaged vegetables, particularly after the foodborne illness outbreaks that have happened over the last several years (see Chapters 3 and 5 and Gordon, 2015). Buyers and processors of vegetables therefore have to exercise much greater control over their production processes and have to incorporate specifications for produce being sourced from suppliers. This will require, in most cases, training and re-orientation of suppliers who must now implement GAP in their operations. As with the bakery, many vegetable producers have largely or completely manual processes. These must now incorporate a step for effective cleaning and sanitizing of incoming vegetables, if this was not being done previously. This may be as simple as a manual dip or wash step or may incorporate an automated or semi-automated washing line and manual, semi-automated or fully automated packaging operation. Further, one of the final steps prior to warehousing may include the use of a metal detector. While none of these steps are themselves very complicated, the need for staff to monitor and control incoming materials, the use of a sanitizing step and/or metal detection as CCPs, as well as to adjust to best practice handling and data recording for these steps can turn the operation into a very involved process for a small producer not used to this level of controls. Invariably, this will require a retraining of staff and building of a food safety culture to assure the consistent supply of safe, quality foods to the customer and final consumer.

This section has sought to demonstrate the application of industry best practices and GMPs in the context of small manufacturers in a developing country setting, whether large of small, supplying locally to transnationals or exporting. These are now the norm, and manufacturers need to ensure that they comply with these requirements to participate in the global food industry.

Basic requirements of a system for food service

Developing a food safety and quality system (FSQS) in a Food Service operation is often more complex than it may appear initially. This is because a food service kitchen can handle many different products, including meat, poultry, seafood, fruits, vegetables, pastry and a variety of cooked and uncooked foods, presenting several opportunities for cross contamination and allergen cross contact, making the operation very complex.

The basic programs that should be developed and implemented in a food service operation are:

- Control of suppliers (Supply Chain Management)
- Control of reception, handling and storage of inputs

- Sanitation program
- Pest prevention program
- Good Food (Preparation and) Handling Practices (GFHPs)
- Prevention of cross contamination and allergen cross contact
- Good cold chain management practices (freezing, thawing and refrigeration)
- Maintenance of food safety during holding of ingredients and prepared foods
- Serving, delivery, dispatch and/or distribution

Control of suppliers & supply chain

Among the most important sources of risks in a food service operation are the raw material and ingredients use in the preparation of meals. Whether it is a large or small operation, many food service businesses do not have adequate controls over the inputs they use. Unlike businesses in more developed markets, inputs for food service businesses in developing countries may be sourced from multiple suppliers as a single provider is often unable to consistently supply all of the businesses' needs on a weekly basis, or as needed. As such, sourcing is done from distributors supplying imported or domestically produced items, wholesalers, supermarkets, markets, or directly from the producers themselves, including farmers. The supplies include a range of meats, vegetables and fruits, bakery ingredients, inputs or supplies, as well as packaging materials and preparation aids. They also include chemicals and cleaning agents and, depending on the nature of the operation (such as an airline caterer/flight kitchen (Fig. 4.29) or catering operation supplying multiple locations with meals), may also include the receipt, handling and cleaning and sanitation of trays, trolleys, lexans and holding/delivery equipment returned with the delivery vehicles. It is not unusual, therefore for the supplies to be received daily, sometimes from multiple suppliers and the inspection procedures at receipt are therefore critical to ensure that only safe, wholesome product of good quality are accepted.

Where there are no documented, effective receiving procedures or where the staff responsible for reception do not have the proper training, the process can become a major source of potential problems, particularly where physical, chemical and biological hazards are not effectively controlled. A combination of approaches is best applied in ensuring that what is received not only meets the requirements of the firm, but also complies with specifications and other requirements. These onsite receival checks will have to be augmented by a variety of other approaches to ensuring that the wide range of inputs that are combined to make the final product delivered to customers or consumers are not only of high quality, but also safe. These are similar to the approaches used to control reception of product in a distribution operation, discussed earlier, and are discussed in more detail in a section on the management of suppliers, which follows below.

Storage of raw materials, ingredients & inputs

Storage includes in the fresh state (as in the receipt and storage of produce, fruits and vegetables properly packed and protected in field crates – Fig. 4.18), as well as refrigerated and frozen states, all of which have particular requirements to optimize food safety

FIGURE 4.29 Prepared meals in a food service operation and airline catering establishment. *Source: A. Gordon (2019).*

and quality. It is necessary to control the temperature at which different meal ingredients inputs are stored and to take corrective action when they move out of the pre-set limits. In this regard, the management of the cold rooms (freezers and chillers/coolers) is important, inclusive of their organization, airflow, monitoring of temperatures, ensuring the accuracy and effective functioning of the temperature monitoring devices and installation of high (for safety) and/or low (for quality) temperature alarms. Ideally, different products should be separated in cold storage areas, with different rooms for fruits and vegetables, meats, poultry, fish and seafood, prepared foods and leftovers. While this is often a challenge or, in some cases impractical because of the space available, at the very least, different kinds of products with different temperature requirements (e.g. fresh vegetables and seafood) should be separated and uncooked products should not be stored with prepared meals. All products in cold storage units should be properly labeled and identified, and this is especially critical for prepared foods and leftovers, both of which have even more sensitive shelf life considerations than other products. The "Use by" storage times for all items should be established and each item in storage should be labeled with this use by or "Discard" date. While not always as critical, it is similarly a best practice that items in ambient temperature (dry) storage should also be labeled with unique identifiers that facilitate traceability and identify the date by which they should be used or disposed of (if relevant).

Sanitation programs

In a food service business, cleaning is fundamental for preventing cross contamination due to biological hazards. In general, cleaning is carried out in every kitchen after operations, but in many cases the procedures are not clear and are not properly documented. As in a food plant, there should be a cleaning and sanitation master plan and Standard Sanitation Operating Procedure (SSOP). The cleaning and sanitation regimes for incoming fruits and vegetables, produce and meats (where required) should be documented, based on industry best practices and be shown to be effective for the specific input being cleaned. Effectively cleaning soursop, also known as guanábana (*Annona muricata* – Fig. 4.11) to be used for making juice, for example, will be different from cleaning lettuce or tomatoes coming from a distributor. The personnel must be trained, and records must be kept. The kitchen should have control of the chemicals that are used and their method of preparation, as well as where they are stored after use. It is important to remember that the food will contain some microorganisms which can contaminate food contact surfaces, which can then become sources of cross contamination. As such, an effective cleaning and sanitation program, including deep cleaning, must be developed specific to each food service operation, validated as effective and systematically implemented. The sanitation program must also include an evaluation of the effectiveness of cleaning and sanitation.

Food preparation and handling practices for prevention of contamination

In the process of food preparation and handling, some practices like limiting the time that potentially dangerous foods are kept in the (temperature) danger zone, in addition to many other practices that must be carried out by employees, should be followed to prevent foodborne diseases. These include:

- Washing hands before preparing food;
- Using clean, sanitized equipment and utensils during food preparation;
- Separating raw foods from ready-to-eat foods;
- Using dedicated tables, knives and utensils, color coded (Fig. 4.30) if possible, to prevent cross contamination and allergen cross contact;
- Maintenance of temperatures for cold foods and hot foods (see below and endnote)[2];
- Discarding potentially hazardous foods after 4 hours post-preparation if not yet served;
- Use of gloves for handling ready-to-eat (RTE) food.

The use of only potable water that is adequately protected from contamination (intentionally or otherwise) is of paramount importance in ensuring that RTE foods and food being prepared for service and being generally handled in a food service setting is not exposed to potentially harmful organisms. As such, ensuring that an adequate supply of properly treated water with a minimum residual chlorine concentration of 0.2–0.5 ppm is

---

[2] It is typically recommended that chilled prepared meals or ingredients in a cooler be stored at a maximum of 4.4 °C (40 °F), while hot foods be kept at 60 °C (140 °F) or above.

FIGURE 4.30   Use of colored containers and utensils to differentiate products and stages of preparation in a food service operation. *Source: André Gordon (2019).*

available and that it is effectively secured (e.g. Fig. 4.24) is critical to overall assurance of the safety of the food being prepared and served.

Food must be handled at appropriate internal temperatures during the processes of cooking, holding, cooling and storage prior to plating or serving. A principal practice that should therefore be followed in food service is ensuring effective temperature control of ingredients and foods being prepared and handled. This is achieved by monitoring temperatures by taking at least 2 internal temperature measurements for the food at several stages of preparation and service, by inserting a sanitized thermometer into the thickest part of the product being served (e.g. meat). Records of these temperatures should be kept. Unfortunately, this is one of the practices that food service kitchens often fail to adhere to, requiring greater vigilance on the part of the management of these operations to ensure compliance. Foods known as "Potentially Hazardous Foods" need extra care as they can cause foodborne illnesses if not handled properly. Some of the foods considered potentially dangerous are meat, chicken, milk, egg, mashed and baked potatoes, fish, stuffed pies, soups, meat sauces, salads, prepared foods containing beans, rice and cooked vegetables, among others (see Chapter 5 for a full discussion on this).

Physical contamination of food with extraneous matter is the presence of any particle or object in the food that should not be there. Most of the time, physical contamination does not represent a food safety risk unless the physical contaminant exceeds 7 mm. However, the presence of a contaminants suggests a breakdown in the inspection or handling of the food and° its ingredients. Physical contamination is unpleasant and causes the customer to lose confidence in the food service outlet or food service provider. It is important therefore to identify the sources of physical contamination in order to establish preventive measures. Some possible sources of physical contamination are listed below:

These sources of physical contamination may include:

- Objects in the food from deficiencies in the hygienic practices of the employees, such as pencils and pen covers, false nails, hairs or other objects;

- Parts of tools or utensils used in the preparation of food;
- Pest that may be in food or in the environment such as flies, cockroaches, or larvae of insects that lodge in vegetables;
- Objects from the packaging of the products such as cartons, aluminum foil, staples, clips;
- Objects resulting from cleaning such as slivers of metal or glass from lamps in food preparation areas or other glass objects;
- Bones or splinters in filleted or meat saw sliced meats or fish.

Strict practices should be followed to prevent physical contamination of foods by these objects.

Another common task in a kitchen service is the process of defrosting and refrigerating food. Safe procedures for thawing meat, poultry and fish must be established. The most practical procedure for defrosting meat, poultry and fish in industrial kitchens is thawing under cold water, but there are several other approved procedures that are valid, including microwaving (if allowed), and thawing in the refrigerator/cooler.

Pest prevention programs

Pests need food, shelter and water and these three elements are found in a food service operation. Therefore, it is essential that a pest prevention program is implemented for each food service entity and is specific to the particular products being handled, pest burden and range of pests in the specific location and the layout and configuration of the operation, among other considerations. This program must consider controls to avoid entry of pests through doors, windows, other openings and through ceilings. Mechanisms must be established to ensure that measures for monitoring pest activity (both externally and within the interior) are effective. The most practical solution is to have a Pest Control Operator (PCO), that is trained and competent in developing, implementing and managing an effective pest control program. The PCO must perform the application of pest management solutions that comply with the requirements of local laws, which may vary from one jurisdiction to another, but in general, require permits from certain agencies, as well as evidence that the PCO is certified to perform the tasks. This is important for insurance against liability, should issues arise.

A pest prevention program in a food service establishment also requires the use of devices for prevention and monitoring, such as traps and related devices for rodents. In some locations and at particular times, the problem of excessive numbers of flies is exacerbated, requiring additional measures to be implemented to control them in a way that does not cause contamination of the product. This may require the deployment of additional devices and will certainly indicate the need for much tighter controls in the disposal of waste and waste management. A map of the location of all devices throughout the operation should be available. It is important that the pest prevention program is carried out correctly and steps should be taken to prevent pest control from becoming a potential source of chemical contamination during the process of application of the chemicals. Experience indicates that some food service operations in developing countries do not think that they should hire a person specifically for pest control but prefer to implement and manage the program themselves, although they often do not have the appropriate experience in the use of the correct chemicals. The inappropriate application of pest control chemicals can become a source of contamination and cross contamination.

Where there is any indication that rats have entered the facility and affected ingredients or prepared product, these should be discarded, and the access points should be clearly identified and sealed to prevent future incursions. It is important that inspections are carried out and that there is no evidence of excreta of rats in products, boxes, sacks, containers, of live pests inside the facilities, or decaying pests inside traps. While there may be the temptation in some developing countries to use cats to prevent the entry of mice, in food service operations no type of animal is allowed inside the facility, including cats or dogs. Where there is any indication of pest activity, the pest control supervisor and pest controller should be immediately informed, and the affected product discarded.

Pesticides must not be stored in preparation facilities and pest service providers must ensure that all pesticides used are food grade. For this, the PCO must supply a copy of the technical sheets and Material Safety Data Sheet (MSDS) of all the products used in providing the service. The MSDS are to be accessible in the event of an emergency with an employee.

Maintenance of safe temperatures for prepared food[3]

Once the food is prepared and ready for serving, it is imperative that it remains at safe temperatures and is properly handled. This is achieved through the use of warmers if it is to be served within a short period or, as is often the case, refrigeration if the food is to be held for several hours prior to being served. The hot-holding procedure must be carried out properly, ensuring that the holding equipment (heaters or warmers) are properly checked and at the holding temperatures before placing the food to be kept warm in or on them. The food must then be kept hot at or above 57 °C (135 °F) or above. Food temperature control should be maintained. In the case of cold foods, ice baths or commercial chillers are used, and these should be kept at a maximum of 5 °C (41 °F) or below.

The experience of the authors has shown that temperature abuse of hot- and cold-holding of foods (particularly hot holding) are among the most common areas of failure or noncompliance with food safety standards in a food service setting. Some problems include malfunctioning heaters/hot holding cabinets or coolers, resulting in hot foods not being kept hot enough and cold foods not kept at less than 41 °F. Food temperature control is among the areas most susceptible to failure in the food service industry.

Dispatch and distribution

Following on the safeguards associated with the preparation of food, it is essential to maintain these controls during transport. If the food is going to be served far from the area where it was prepared, it is essential to use containers that ensure the maintenance of the

---

[3] The temperatures mentioned in this section considered *critical temperatures (limits)* while the ones mentioned in footnote 2 above would typically be *operating temperatures (limits)*.

required temperature. These containers must also be sealable to protect the food and some businesses go as far as to seal each container with a tamper evident seal to provide additional assurance to the consumer that the product was protected from intentional or accidental contamination. When re-usable, containers should be impervious (i.e. not porous), be easily cleaned and should be included in the cleaning program. Vehicles used for transportation must also be properly cleaned and sanitized. If they are delivering frozen, chilled or otherwise temperature controlled meals, then the vehicles should be equipped with automatic temperature control devices, as well as calibrated temperature indicating device (s), the accuracy of which should be double checked at least twice each route day by the driver and/or his assistance. The temperature of all items on delivery should also be noted, as well as the time and condition of the product at the point of delivery.

While the implementation of the practices discussed in this section may present some challenges for foodservice operations in developing countries, these represent best practices and implementing them should be the aim of the operators of the business. Issues of food safety culture, the tendency not to want to document, or not to want to routinely measure critical limits (e.g. temperatures) throughout the workday are among the challenges that may exist. Other challenges include the unavailability of suitably knowledgeable and qualified trainers in food service food safety to assist with changing the culture and nature of the operations through training. Nevertheless, in multiple instances across many countries, the authors have seen where focussed, management-led programs designed to empower the staff while educating them on best-practice approaches and why these should be adopted have been successful in transforming operations.

Management of suppliers to the food manufacturing & food service industries

Introduction

One of the key elements that must be considered in order to ensure the quality and safety of food produced by a processor, manufacturer or food service operation is the formal approval of the suppliers of the ingredients, processing aids and packaging; both food contact and non-food contact. Suppliers may be primary producers, processors/manufacturers or distributors of food ingredients, processing aids and packaging. Once they supply items that are either integral to the food or come into contact with the food or food contact surfaces, they should be subjected to the scrutiny of the supplier approval process. Customers (processor, manufacturer or food service operation) must establish (develop and implement) a well-defined process not only for the approval of the suppliers, but also for the periodic monitoring of their performance.

Evaluation of suppliers

Before a supplier is approved to provide ingredients, packaging (both food contact and non-food contact) and processing aids to a processor, manufacturer or food service operation, it is necessary for an evaluation to be conducted by the customer to ensure that the supplier it capable of supplying safe food of appropriate quality. A questionnaire may be developed by the customer, which can be submitted to the supplier in order to obtain information on the company, the inputs which it proposes to supply and the company's Food Safety & Quality System (FSQS).

It is suggested that information on the following aspects of the supplier's operation be requested:

Company Information

    (i) Supplier Name
   (ii) Supplier Address
  (iii) Supplier Telephone/Fax Numbers
  (iv) Name, Position and 24 h Contact Information of Key Personnel (Internally and Externally)
    (v) Brief Company Overview

Food Safety & Quality System (FSQS)

    (i) Food Safety & Quality Management Organisational Structure
   (ii) Food Safety & Quality Management System Certification
  (iii) Food Safety & Quality Management System Standard

Elements of Food Safety & Quality System

      (i) Personal Hygiene
     (ii) Sanitation
    (iii) Supplier Management
    (iv) Foreign Matter Control
     (v) Chemical Control
    (vi) Allergen Management
   (vii) Traceability & Recall
  (viii) Receipt of Incoming Raw Materials
    (ix) Transportation & Storage
     (x) Building, Equipment & Premises Upkeep
    (xi) Water, Steam and Air Quality
   (xii) Preventive Maintenance
  (xiii) Human Resource Competencies & Training
   (xiv) Record Keeping
   (xv) Customer Complaints
  (xvi) HACCP
 (xvii) Food Fraud
(xviii) Food Defense

These are among the major aspects of FSQS programs required by most buyers. Another increasingly important requirement is for suppliers to have a specific program in place for *Food Safety Culture*. Existing suppliers may not have ever been subjected to such formal evaluation. It is recommended that they also be asked to complete the questionnaire when the program is implemented.

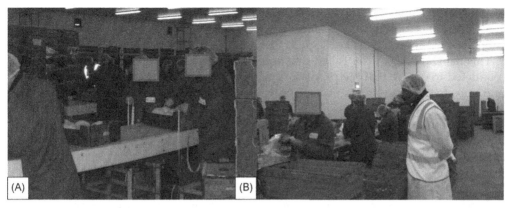

FIGURE 4.31   (A) Operations in a pack house (B) third party auditing of operations by visual observation of practices. *Source: A. Gordon (2014).*

An audit of the supplier's operation (Fig. 4.31) may also be conducted by the customer (second party) or an individual or firm (third party) contracted to provide the service on behalf of the customer. The audit should be based on a recognized food safety standard. The minimum requirements which should be met, in order for a supplier to be approved, should be determined by the customer and communicated to the supplier.

Following the receipt of the completed questionnaire and the audit, a review should be conducted by an internal team, to determine whether the prospective supplier should be approved. The supplier should be advised whether they have been approved, the specific items which they have been approved to supply and corrective actions which they will be required to implement, if any.

Documentation

The following documentation are amongst those often required to be maintained for approved suppliers.

1. Initial Supplier Evaluation Questionnaires
   This document provides information on the prospective suppliers such as the ingredient, processing aid or packaging material which it proposed to supply to the customer and details of its Food Safety and Quality Management System.
2. Annual Supplier Evaluation Questionnaires
   Over time, elements of the information may change. For example, the supplier may become certified to a GFSI standard or it may change its process for a specific item which may result in the elimination of a previously identified hazard or the introduction of one that was not considered in the Ingredient Hazard Analysis. These changes should be tracked and recorded.
3. Approved Supplier List
   i. Supplier Name
   ii. Supplier Address

iii. Supplier Telephone/Fax Numbers
iv. Name & Position of Key Contacts
v. Approved Ingredients/Packaging/Processing Aids

This list should be a controlled document and periodic assessments should be conducted to ensure that the processor/manufacturer or the food service operation continues to only use inputs from approved suppliers. This can take the form a formal audit in which a combination of high and low risk inputs are checked to confirm that they were purchased from the approved suppliers/distributors. The list should be updated when changes occur e.g. new suppliers are introduced, existing suppliers are either delisted or contact information or items which they are approved to supply are modified.

4. Product/Packaging Specifications

Purchases of ingredients, processing aids and packaging must always be based on established specifications. Relevant physical, chemical and microbiological parameters should always be included on the specifications. Where applicable, information on allergens should also be included. The current versions of these controlled documents must always be maintained on file.

5. Food Safety Certifications

Current versions of Food Safety and Quality Certifications for processors/manufacturers and distributors of ingredients, processing aids and packaging material should be maintained on file.

Monitoring of supplier's performance

Very clear criteria for the monitoring of suppliers' performance must be established. These should include product quality and food safety requirements for the ingredients, packaging and processing aids. It is suggested that the supplier's ability to conform to the following criteria should be considered in an annual evaluation of the supplier's performance:

a. On time delivery

The supplier should be able to deliver the items when the customer requires them. Late deliveries of ingredients, processing aids and packaging material could equate to lost sales.

b. Correct items delivered/no unauthorized substitution

The items which are delivered should be in accordance with the customer's request. If the customer requires cinnamon powder, then cinnamon sticks should not be delivered unless the customer approved the substitution.

c. Delivery of correct quantity of items

In order for the processor/manufacturer or food service operator to fulfill their customers' requirements, the quantity of items which were ordered should be delivered. Any variation of the quantity of items should only be made if the necessary approval is obtained from the customer by the supplier. The delivery of a smaller quantity of ingredients, processing aids or packaging material could negatively impact the customer's ability to meet their production targets while the delivery of a larger quantity could result in significant inventory challenges.

d. Condition of transport vehicles

Vehicles used to transport food ingredients, processing aids and packaging material should be designed and maintained for that purpose. They should always be clean, maintained in good repair, enclosed and locked/sealed. There should be no broken glass or other physical hazards and no pests or evidence of pests.

e. Condition of items on receipt

It is very important that the condition of ingredients, processing aids and packaging comply with the established specifications when they are received by the customer. Not only should an inspection be done to determine the condition of inputs on receipt, but the information should be documented and records should be maintained. The receipt of item which should be maintained in a frozen condition at a chill temperature, presents several food safety challenges. Similarly crushed, torn, punctured, wet/stained, leaking and open packages could result in levels of hazards in the inputs that could impact human health negatively.

f. Conformance of items to established quality and food safety specifications

On receipt, the ingredients, processing aids and packing material must conform to the food safety and quality specifications which would have been communicated to the supplier. Deviations, particularly those that are associated with food safety parameters and cannot be controlled by the facility's processes, can negatively impact the health of consumers.

g. Submission of required documentation e.g. COAs, Letters of Guarantee, etc.

It is necessary for deliveries of the items to be accompanied by the agreed documentation as in many operations, approval or rejection of the shipment is based on the submitted documentation, as extensive laboratory analysis is not conducted in house.

h. Conformance of cost of items to agreement

The agreed cost of the items ordered should not be changed without a discussion with the customer, whether it is a decrease or an increase. A decrease in price of an ingredient could result from improvements in the manufacturing facility which have positively impacted efficiency. It could also mean that the ingredient was sourced by a distributor from a primary producer, processor or manufacturer that employs questionable food safety and quality practices in order to maintain low cost of production.

i. Performance on Food Safety Audit i.e. achievement of pass mark

In addition to the submission of the supplier questionnaires on an annual basis, suppliers should be audited, at least annually. Audits may be conducted by a reputable auditing firm (third party) or by the customer (second party). The pass mark should be clearly defined to facilitate the supplier's maintenance of its Food Safety & Quality System (FSQS).

j. Responsiveness to customer complaints

Notwithstanding the implementation of a robust FSQS, unrecognized deviations may occur which can result in customer complaints. It is expected that the supplier would treat with the complaint expeditiously and would implement appropriate corrective action to prevent its reoccurrence.

GFSI-benchmarked certification programs

Background

The Consumer Goods Forum (CGF) initiated the process in 2000 which has led to the creation of the Global Food Safety Initiative (GFSI) set of benchmarked standards. This was, among other reasons, specifically to address the issue of a multiplicity and ever-increasing number of SQA programs with which suppliers had to contend. The GFSI was also intended to bring coherence and standardization through equivalence to different SQA standards, all of which were seeking to do the same thing: assure the safety of the food being supplied ultimately to the consumer. In lieu of brand-owned or administered SQA programs, therefore, companies, other businesses, consumer groupings and organizations globally have a set of benchmarked standards from which to choose to assure that their requirements are met by existing and potential suppliers. 'In this context, the GFSI is playing a critical role in improving food product quality and, more importantly, safety throughout the global food supply system. While an exhaustive treatment of all of the GFSI standards is beyond the scope of this chapter, the information presented here seeks to give a summary of three of the benchmarked GFSI standards. These cover primary production (GLOBALG.A.P.), manufacturing (Food Safety Systems Certification 22000 − FSSC 22000) and the complete supply chain from primary through to distribution (Safe Quality Food - SQF).

FSSC 22000 certification

FSSC 22000:2018 (Version 5) was published in May 2019. Owned by the Foundation for Food Safety Certification (FFSC), it is based on ISO 22000:2018 and is structured as follows:

Annex 1 | CB Certificate Scope statements
Annex 2 | CB Audit report template (FSSC 22000)
Annex 3 | CB Audit report template (FSSC 22000-Quality)
Annex 4 | CB Certificate templates
Annex 5 | AB Accreditation certificate scope
Annex 6 | TO Course specifications
Annex 7 | TO Training certificate templates

This standard specifies the requirements for a food safety management system (FSMS) and is applicable to all organizations in the food chain from manufacturing to distribution, regardless of size and complexity. In order to comply with the standard, organizations need to demonstrate their ability to consistently control food safety hazards in order to ensure that food is safe at the time of human consumption. The means of meeting any requirements of this International Standard can be accomplished through the use of internal and/or external resources.

The standard consists of the following:

- Annex A
- Cross references between ISO 22000:2005 & ISO 9000:2000
- Annex B
- Cross references between HACCP & ISO 22000:2005
- Annex C – Codex References

Providing examples of control measures, including prerequisite programs & guidance for their selection & use

This standard specifies requirements to enable an organization:

a. to plan, implement, operate, maintain and update a food safety management system aimed at providing products that, according to their intended use, are safe for the consumer

b. to demonstrate compliance with applicable statutory and regulatory food safety requirements

c. to evaluate and assess customer requirements and demonstrate conformity with those mutually agreed customer requirements that relate to food safety, in order to enhance customer satisfaction

d. to effectively communicate food safety issues to their suppliers, customers and relevant interested parties in the food chain

e. to ensure that the organization conforms to its stated food safety policy

f. to demonstrate such conformity to relevant interested parties

g. to seek certification or registration of its food safety management system by an external organization, or make a self-assessment or self-declaration of conformity to this International Standard

Organization

The standard is organized as follows:

Section 1: *Scope*
Section 2: *Normative References*
Section 3: *Terms and definitions*
Section 4: *Food Safety Management System*
Section 5: *Management Responsibility*
Section 6: *Resource Management*
Section 7: *Planning & Realization of Safe Products*
Section 8: *Validation, Verification & Improvement of the Food Safety Management System*

FSSC 22000, Version 5, in addition to being equivalent to other GFSI benchmarked standard, provides firms in the food manufacturing and distribution business an organized,

ISO 9000-based, ISO 22000 platform for the implementation and delivery of food safety and quality throughout their operations.

Safe quality food certification

Safe Quality Food (SQF) certification covers the food industry from the farm to the retail stores. Run by the SQF Institute (SQFI) for the Certified Programme Owner (CPO) the Food Manufacturing Institute (FMI), SQF is recognized by the GFSI, brand owners, retailers and food service providers worldwide. SQFI indicates that the programs are designed to provide a rigorous and credible food safety and quality system that assures compliance with industry, customer and regulatory food safety and quality systems requirements for all sectors of the food supply chain. The SQF scopes of certification to their fully GFSI-benchmarked standard (Version 8) includes primary production, manufacturing and storage and distribution. Their non-GFSI benchmarked programs include *Food Safety Fundamentals* and *SQFI Food Service Program*, done in conjunction with the United States-based National Restaurants Association (NRA). Summarized description of the SQF programs are presented below.

SQF food safety codes

The SQF Food Safety codes available for fully GFSI-benchmarked certification are equivalent to those offered by other GFSI programs and are structured in a manner to meet the core requirements for *food safety management, management responsibility, complaint and crisis management, regulatory compliance, supplier approval, traceability and recall management, allergen management, food fraud, food safety culture, food defense*, physical facility and prerequisite program requirements, among others. These are detailed under modules 2 and 11, respectively. Details can be accessed at https://www.sqfi.com/what-is-the-sqf-program/which-program-is-right-for-me/.
The specific areas under each code are as follows:

SQF Food Safety Code for Primary Production, Version 8.1[4]
Scopes Covered

- **AI** Farming of Animals
- **BI** Farming of Plants

SQF Food Safety Code for Manufacturing, Version 8.1
Scopes Covered

- **C** Animal Conversion
- **D** Pre-Processing Handling of Plant Products
- **EI** Processing of Animal Perishable Products
- **EII** Processing of Plant Perishable Products
- **EIII** Processing of Animal and Plant Perishable Products (Mixed Products)

---

[4] SQF Version 9.0 is currently being prepared as an update to Version 8.1

- **EIV** Processing of Ambient Stable Products
- **F** Production of Feed (SCOPE EXTENSION)
- **L** Production of (Bio) Chemicals
- **M** Production of Food Packaging

SQF Food Safety Code for Storage and Distribution, Version 8.1
Scope Covered in Application

- J Provision of storage and distribution services

In addition to its fully GFSI-benchmarked programs, SQF certification for the food industry value chain is structured to allow firms to access certification at various levels, depending on their current circumstances, needs and capabilities. SQF allows producers and retailers to access certification at a level that is not equivalent to, nor therefore benchmarked against GFSI. This certification is, however, more than sufficient to meet the requirements of most buyers SQA programs and allow firms to begin the journey towards full GFSI-benchmarked SQF certification (to SQF Food Safety Code for Manufacturing, Version 8.1, for example). This initial level of systems implementation and certification is called SQF Fundamentals and is available at the "Basic" and "Intermediate" levels.

As indicated above, SQF also facilitates certification of primary production operations such as the one shown in Fig. 4.32 at the Fundamentals and fully benchmarked levels, depending on the readiness of the operation and its current market determined needs. In this regard, along with Canada GAP, Asia GAP and Primus, SQF Code for Primary Production facilitates certification of farming operations in a manner equivalent to GLOBALG.A.P., discussed below.

FIGURE 4.32 Protected agriculture operations – certifiable under a GFSI good agricultural practice certfication program such as SQF or GLOBAL.G.A.P. *Source: Dr. Dianne Gordon (2018).*

GLOBALG.A.P. certification

Background

Having looked at Good Agricultural Practice (GAP) earlier in this chapter, this section examines what is specifically required for GLOBALG.A.P. certification. GLOBALG.A.P. started as EUREPGAP in 1997 before being renamed in 2007 after 10 years in existence and significant expansion in its reach globally. It is run by the Euro-Retailer Produce Working Group, an organization of growers and buyers with the focussed objective of *"safe, sustainable agriculture worldwide"*. It has set voluntary standards for the certification of agricultural products around the world and has as its mission *"putting food safety, sustainability and the environment in the hands of growers"*. The GLOBALG.A.P. community is comprised of 400 voluntary members spread across 124 countries. GLOBALG.A.P. has 40 standards and programs to which it certifies and has more than 170,000 certifications worldwide.

Products are certified under GLOBALG.A.P.'s Integrated Farm Assurance Standard (IFA) Version 5 (IFA v5) which is comprised of its Crops Base (Fig. 4.33), Aquaculture Standard and Livestock Base, the latter being subdivided into pigs, poultry, turkey, cattle and sheep. Cattle is further divided into Dairy (DY) and Calf/Young Beef (CYB). Aquaculture covers finfish, crustaceans and molluscs while the Crops Base is subdivided into five (5) categories: fruits and vegetables, plant propagation materials, combinable crops, tea and flowers and ornamentals.

GLOBALG.A.P. (IFA) v5 has 218 Control Points in 4 categories under its Fruits and Vegetables standard, the one most germane to most producers from developing countries. The number and type of control points are as follows:

- *Traceability (22)*
- *Workers Occupational Health & Safety (28)*
- *Environment (69)*
- *Food Safety (99)*

FIGURE 4.33 Fields of crops prepared and grown in compliace with good agricultural practices. *Source: A. Gordon (2018).*

GLOBALG.A.P supports single producer certification as well as producer group certification.

The requirements of GLOBALG.A.P's IFA Version 5 standard cover several categories, including:

- Food Safety
- Traceability
- Quality Assurance
- Workers' Occupational Health & Safety
- Site Management
- Soil Management
- Fertilizer Application Management
- Integrated Pest Management
- Plant Protection Products Management
- Water Management

Because the standard undergoes review and updating periodically (at least once every three years as for all other GFSI-benchmarked standards), it is best for interested persons to ensure that the requirements being used are the most current. These are currently as indicated IFA v5 but will change over time. Details can be had at the GLOBALG.A.P website https://www.globalgap.org/uk_en/.

Organic certification and organic agriculture

Introduction

Certification of agricultural production and the foods produced by the certified practice as organic is among the areas of growth in the food industry globally as the demand for these products continues to expand. Attainment of certification requires adherence to, and the ability to demonstrate compliance with the principles of organic agriculture. Organic agriculture is the practice of agriculture in a manner that avoids synthetic pesticides, easily soluble mineral fertilizers and genetically modified organisms (GMO), protects the environment and promotes biodiversity, produces healthy food, recycles nutrients and uses locally adapted (traditional) methods that comply with its requirements. Because of the restrictions involved in the range of options and practices available to the producer/farmer (Fig. 4.34), organic farming requires more thought, planning and an ecosystem of support around the farming operation in order to achieve compliance. Consequently, while the demand for organically produced food is and will remain strong for the foreseeable future, compliance with organic standards in developing countries will require the kind of collaboration that is best achieved through viable, sustainable value chains in which the buyer is actively engaged with the primary producer. This section discusses the requirements for achieving organic certification, including a limited discussion on some of the options available to attain certification and the costs involved.

FIGURE 4.34 Use of environmentally "unfriendly" practices in field preparation is not allowed in organic farming although acceptable under GAPs. *Source: A. Gordon (2018).*

Organic agriculture

Effective practice of organic agriculture necessitates that practitioners of organic farming are trained in its requirements, including those for certification. This training should, at a minimum, cover modules on:

- The principles of organic farming,
- Organic seedling production,
- The nature and properties of soil,
- Soil fertility management,
- Organic pest and disease management,
- Water management, and
- Organic farm certification

Each of these areas of organic farming must be compliant with the standard to which the particular crop is being grown if it is to be certifiable as organic. An important point of departure between this kind or certification and others is that compliance with the requirements is not enough to attain the commercial objective of being able to sell the product as organic. The farm and the product *must be certified* organic in order to legitimately be sold as organic. This is different from some types of food safety or quality-based systems where compliance may be enough to allow at least initial access to specific markets. With organic labeling, the product is either organic or not and hence certification, not simply compliance, is required for market access. While there are many routes to organic certification and many providers, this section of the chapter will focus on three main certification program owner (CPO) programs: CERES[5], one of the largest organic certification body globally, the United States of America National Organic Program (NOP) and the Japanese Organic Agricultural Standard (JAS).

---

[5] CERES is the largest organic certification body in the world. They are EU-based and have operations in 45 countries

Organic certification

For organic certification, it is important to understand the requirements and be able to effectively implement them. These requirements fall into several categories, viz.

- Soil Fertility and Plant Nutrition
- Crop Protection
- Approved & Certified Inputs
- Seeds, Seedlings and Vegetative Planting Materials
- Conversion Period (from Conventional Farming)
- Cross Contamination
- Biodiversity
- Records
- Knowledge

To be effective at implementing and maintaining certified organic cultivation, the farmer and his/her team must know and understand the various considerations for each of the areas that are required to be implemented and managed. These are outlined below.

Considerations for soil fertility and plant nutrition

Among the specific considerations for soil fertility and plant nutrition are the following:

- Soil fertility has to be conserved or improved;
- Hydroponic cultivation is not allowed;
- Soil erosion must be avoided;
- For annual crops, a wide crop rotation has to be used, including legumes to assure biological nitrogen fixation;
- For perennial crops, wherever possible, legumes have to be planted in interrow spaces
- Organic manuring should be used to maintain soil fertility;
- Nitrogen fertilisers and superphosphate are not allowed, potassium chloride (from mined sources) is allowed only by Japan;
- Rock phosphate, potassium sulfate, and single trace element fertilisers can be used where analyses show nutrient deficiencies;
- Organic and inorganic fertilization must not exceed crop requirements;
- Lime ($CaCO_3$) can and should be applied, when necessary;

Considerations for crop protection

Effective crop protection without the use of a range of pesticides typically used to target fungi, weeds, mites, flies and other pests requires a combinations method approach, based on pest prevention, routine pest scouting, early identification of infestation (Fig. 4.35) and other techniques that are used in integrated pest management (IPM). Among the considerations are:

- Some systems (e.g. the US National Organic Certification Program - NOP) restrict not only the active substance, but also the inert ingredients of natural pesticides;

FIGURE 4.35　Pest identification and pest scouting: integral aspects of integrated pest management as a part of organic production. *Source: A. Gordon (2017).*

- Synthetical herbicides, insecticides, and fungicides are not allowed;
- Pests and plant diseases must be prevented, using adapted species and resistant varieties, adequate crop rotations, and promoting natural enemies;
- Only natural or mineral substances mentioned in Annex II to Reg. (EC) 889/08, Annex 2 to Japanese Organic Agricultural Standard Notification 1605, and the US National Organic Program (NOP) National List can be used if certification to any of these standards is sought;
- Weeds must be controlled by mechanical or thermal means, through adequate soil tillage and crop rotation. The NOP allows botanical herbicides, as long as they comply with the National List.

Considerations for certification of farm inputs

Not all organic certification programs require that inputs used in organic farming come only from among those on an approved list. Among certification programs, the use of certified inputs is not for certification to the EU organic standard or those private certification program owner (COP) systems that target EU market compliance for its certified farming operations. It is, however, required for

- the US National Organic Certification Program (NOP). Certification of inputs must be by NOP or an approved third Party;
- the Japanese Organic Agriculture Standard (JAS). Certification must be by JAS or an approved third Party;

Considerations for planting materials

Among the considerations for planting materials are:

- seeds, seedlings and vegetative planting material *must* be organic **or**
- The farmer *must prove that no organic planting material was available* **and** get a derogation from the EU *before planting* (a CERES requirement)
- NOP and JAS have no such additional requirement

Considerations for conversion period

The term "conversion period" refers to the time required for a conventional farming operation to successfully undergo transformation to an organic farm with effective observation of all organic farming principles and compliance with all of the requirements, including for the soil, chemical usage and pest management. All conventional farms must undergo a conversion period before it can sell its products as organic. During this period, *all rules of organic production must be followed*. The conversion period for EU Organic Certification are:

- For *annual crops*, 24 months
- For *perennial crops*, 36 months

Considerations for cross contamination

Cross contamination between conventional and organic agricultural production is a major concern for CPOs as it is for consumers who want to ensure that their organic foods are not, in any way, contaminated by production or production practices which are non-organic. As such there are requirements by all certification programs to ensure adequate protection against cross contamination. These are:

- The NOP and JAS require specific *buffer zones* between organic and conventional crops
- The EU requires *"precautionary measure ... in order to reduce the risk of contamination."* between conventional and organic crop fields, including if there is a risk of pesticide drift.

Considerations for biodiversity

The protection and conservation of the natural biodiversity of the area under cultivation is an important requirement of organic certification systems. These are summarized as follows:

- Organic standards require soil fertility conservation and promotion of natural enemies;
- Biodiversity is a key element of organic farming (as per EU regulation EC 834/2007 for EU organic certification);
- CERES believes that these conditions cannot be met on huge monocrop fields;
- CERES have therefore established a maximum field size of 20—40 ha for annual crops, according to erosion risk;
- Bigger fields must be subdivided by hedgerows/buffer zones

Considerations for records

Basic considerations include:

- The farm needs at least a simple system of *bookkeeping for sales of organic products*
- In addition, *JAS requires "grading" records*: before selling products with the JAS logo, the producer has to double-check and record fulfillment of JAS standards.
- A *farm diary must be kept*, recording the main activities on each plot on the farm
- Before the first inspection takes place, *the farm has to present an organic management plan (to be updated annually) to the certifier*
- *Invoices for purchases* of fertilisers, pesticides, seeds, etc., must be filed
- *Harvested quantities* must be recorded for each crop

Considerations for knowledge

- Records: The farmer has to have a copy of the respective standards and know them
- The farmer needs an adequate level of knowledge on organic farming rules and technologies.

Cost considerations for organic certification

One of the major issues often cited with organic certification is the additional costs and care required to implement a certifiable production program. Because of the care required, there are some additional cost incurred in producing organically grown products. This varies by product type. For example, for organically certified cocoa, these include:

- The need for additional labor;
- The cost of sourcing improved varieties;
- The requirement for the planting of shade trees;
- The need to purchase only the recommended chemicals (such as ridomil and actara) which may be more costly than other alternatives.

Benefits of organic certification

Despite the more stringent requirements for compliance with and certification to organic standards, it can be well worth the while of producers because the benefits can be significant. The benefits of organic agriculture include the sometimes substantially higher prices that are available to the vendor of organically produced foods as consumers, both in developed and developing countries seek to eat more natural, healthier foods that are produced in a sustainable manner with due care for the environment. There are, however, several other benefits that are often overlooked. These include significant environmental benefits and benefits related to productivity and yields for some crops. More details on some of the benefits are described below.

Financial benefit for selected organically grown crops

Cocoa

For cocoa, there are some additional benefits gained in producing certified organic cocoa. These benefits were realized in both Nigeria and Ghana for the farmers for which published information is available (Victor et al., 2010; Oseni and Adams, 2013), including:

- Better yields per acre;
- Higher prices per metric tonne (MT);
- Increased net revenue/crop;
- Higher overall profitability (market dependent).

At the right pricing for bulk organic cocoa in Ghana, the returns were much better for the farmers,

- Farmers benefited from a 25% yield increase following training as a part of the certification process;
- The gain in yields exceeded the costs of certification;
- The gains were dependent on the variety of cocoa grown and the relative pricing received for the organic product;

For Nigeria, the significant additional benefits include:

- Increased yields of crops leading to increased output (for example, up to 20,000 MT more for organic cocoa in Nigeria);
- 18% higher price per MT;
- Greater than 50% higher net revenue per crop
- 145% higher net profit for organic cocoa in Nigeria vs. conventional production

Spices

The results for the production of organic spices (black and chilli peppers) in Tanzania (Akyoo and Lazaro, 2008) were very different from those attained for organic cocoa in Ghana and Nigeria. The expected premium prices, higher net revenue and increased yields did not materialize.

It is important to note, therefore, that organic production and organic certification are not equally beneficial for all crop types and all markets. While there are financial benefits that can and have been derived in several cases, the outcome is heavily crop, market and situation dependent, including how the costs associated with the implementation and certification processes are met.

Environmental benefits of organic certification

Among the environmental benefits of organic certification are:

- Less or no use of chemical fertilizers and pesticides result in better soil health and significantly reduced environmental damage;
- No/reduced contamination of ground water resources;
- Greater natural biodiversity as less damage is done to natural plant and animal life;

- Reduced energy costs because of methods applied;
- Reduced production costs (in some cases) because of reduced need for high cost fertilizers and pesticides;
- The improved soil quality makes it more useful for other crops;
- Application of better crop rotation practices, pest management and other practices improves the overall health of the soil and productivity of the farm;
- Better air quality as well as better water quality long term and for future generations.

Halal certification

A description of Halal: what is it?

Halal foods are foods that are allowed under Islamic dietary guidelines. In Arabic, *Halal* means "permitted" or "lawful" (USA Halal Chamber of Commerce Inc., 2019). Consequently, in the food industry *Halal* products are those products, including meats, that are permitted for consumption by practicing Muslims under Islamic law as outlined in the Qur'an, the holy book of practicing Muslims. The adoption of *Halal* certification is therefore based on the concept of *Halal* which encompasses both the Sharia Islamic requirements, as well as the sustainability concepts of sanitation, hygiene and safety. The "wholesomeness" focus of *Halal* makes it particularly attractive to non-Muslim consumers concerned about food quality, health and safety, as well as social justice and animal welfare. The *Halal* industry is gaining more importance with the increasing Muslim population and the attention *Halal* products are receiving in the global market by non-Muslims attracted to its wholesomeness concept. In fact, the *Halal* product market has been getting more attention globally and is one that producers in developing countries should take note of. The industry was estimated to be valued at US$2.3 trillion in 2019, involving some 3.1 billion consumers around the world (Baharuddin and Ismail, 2018). This is only likely to continue to grow.

Based on the guidelines provided by the Qur'an, followers of Islam (Muslims) cannot consume a range of food products. These include the following:

- pork or pork by products;
- animals that were dead prior to slaughtering;
- animals not slaughtered properly or not slaughtered in the name of Allah;
- blood and blood by-products;
- alcohol;
- carnivorous animals;
- birds of prey;
- land animals without external ears.

These prohibited foods and food ingredients are called *haram*, meaning forbidden in Arabic.

In the food industry, *Halal* certification is recognized as a measure of quality standard (Muhammad et al., 2009; Baharuddin and Ismail, 2018). Halal certification, such as that encapsulated by the first *Halal* standard published in 2004, the Malaysian *Halal* Standard

MS1500:2004, is recognized internationally (Muhammad et al., 2009). *Halal* certification is seen by the Islamic community as the benchmark for food safety, quality assurance and provides many beneficial characteristics which are not only to be enjoyed by Muslim consumers, but also meant for non-Muslims consumers. Halal certification is equivalent to and incorporates the elements of many of the conventional quality standards, such as Good Hygienic Practice (GHP), Good Manufacturing Practices (GMP) (Muhammad et al., 2009), veterinary inspection, Codex Alimentarius and Hazard Analysis Critical Control Point (HACCP). The USA Halal Chamber of Commerce Inc. (ISWA Halal Certification Department) (2019) promotes unified standards of acceptance and certification. A unified standard bridges the gap between Muslim consumers and the industry. It establishes credibility and assures the Muslim consumer of strict compliance to the *Halal* process (USA Halal Chamber of Commerce Inc., 2019). Some authors claim that a food industry which implements the *Halal* requirements will produce better quality food products compared to those who implement the conventional standards (Talib and Ali, 2009).

Why seek Halal certification?

Halal products allow the national and international Muslim communities to comply with Islamic law (Sharia). Muslim consumers choose products that are in compliance with the *Halal* process and procedure as defined by Islamic Law (Sharia). While being discussed primarily in the context of food industry certification, it should be noted that *Halal* is not only for food but is also applicable to the production of cosmetics and pharmaceuticals and to tourism, logistics, finance and the entertainment industry (Muhammad et al., 2009). Industry and producers of goods and services are often not aware of *Halal* requirements and therefore often overlook the needs of this segment of the population. *Halal* certification requirements necessitate:

* Purity and cleanliness of the sources from which inputs are derived and the process by which the products are made.
* Safety of human lives by consuming only that which is wholesome and healthy
* Integrity and ethical way of life by avoiding cruelty to animals, harm to the environment and unfair business practices.

Halal is therefore seen by Islam and Muslims as a way of life that benefits an individual in their physical and spiritual well-being. *Halal* products can be consumed by everyone whether or not they are a Muslim. *Halal* is the ultimate stamp of purity and safety, guaranteeing superior quality products.

Applying for Halal certification

While the procedures may vary between certification bodies, the general procedure for Halal certification is similar.

Firms are required to formally apply for the certification through an authorized official requesting an audit of the production facility with a view to getting certification. An appointment is then made with a halal inspector to visit and assess the facility. The

Inspector will then prepare and submit his report for review. Once successful, the plant will be issued with a *Halal* certificate valid for a one-year period, renewable each year. This certifies that the plant has met the requirements. Subsequently, each load or shipment from the plant will be issued with an original *Halal* certificate for the product.

Kosher certification

What are Kosher foods?

Foods or premises in which food is sold, cooked or eaten, satisfying the requirements of Jewish dietary regulations according to *halakha* (Jewish law) are termed *"kosher"*. The word *kosher* is Hebrew for "fit" or "appropriate" and describes the food that is suitable for Jews to eat. In Jewish tradition, *Kosher* laws were given by God to the Children of Israel in the Sinai Desert. Moses, a leader, prophet and teacher of central importance in Judaism, taught these laws to the Jewish people and wrote the basics of these laws in Leviticus 11 and Deuteronomy 14. The details of the laws have been handed down through generations and are found in the Mishnah and Talmud[6]. The observance of *kosher* has been a hallmark of Jewish identity throughout their 4000-year history. *Kosher* laws are complex and comprehensive. This section is intended as an introduction to some of the fundamentals of *kosher* foods and as a guide to the general procedures for *kosher* certification.

The global demand for Kosher foods

The demand for *kosher* certified products has increased dramatically. According to Berry (2017), the Orthodox Union (O.U.) has estimated that almost 80% of all *kosher* food sales are actually outside the traditional Jewish market. There are more than 12 million American consumers of *kosher* food products and more than 40% of the United States' new packaged food and beverage products are labeled as kosher (Kosher Industry Facts, 2015), with the market being valued at about $17 billion. *Kosher* foods are particularly attractive to non-Jewish consumers for reasons related to health, food safety, lactose intolerance, vegetarianism and other dietary restrictions. The *kosher* market is estimated to be experiencing annual growth rates of approximately 15% (Kosher Industry Facts, 2015). In order to meet this demand and expand their current markets and sales, companies have been increasingly pursuing *kosher* certification.

Categories of Kosher foods

There are three categories of Kosher foods. These are:
Dairy - All foods derived from, or containing, milk are classified as dairy. These include foods such as milk, butter, ice cream, yogurt and all cheese (hard, soft and cream - often

---

[6] The Talmud is the central text in Rabbinic Judaism and is the body of Jewish civil and ceremonial law and tradition.

described with the Yiddish word *milchig*). Even trace amounts of dairy products in foods cause a food to be designated dairy.

Dairy products must meet the following criteria in order to be certified *kosher*:

- They must come from a *kosher* animal;
- All ingredients must be *kosher* and free of meat derivatives (conventional rennet, gelatin, etc., are of animal origin and may not be used in *kosher* dairy products);
- They must be produced, processed and packaged on *kosher* equipment;

Meat - This includes all kosher animals and fowl slaughtered in the prescribed manner and their derivative products (often referred to with the Yiddish word *fleischig*). All meat and fowl and their by-products, such as bones, soup or gravy are classified as meat, including products that contain meat or fowl derivatives such as liver pills.

To be considered *kosher*, items designated "meat" must conform to the following requirements:

- *Kosher* meat must come from an animal that chews its cud and has split hooves. (e.g., cows, sheep and goats);
- *Kosher* fowl are identified by a universally accepted tradition and include the domesticated varieties of chickens, Cornish hens, ducks, geese and turkeys. The Torah[7] names the forbidden species of fowl, including all predatory and scavenger birds.
- Animal and fowl must be slaughtered with precision and examined by a skilled shochet, an individual extensively trained in the rituals kosher slaughtering.
- Permissible portions of the animal and fowl must be properly prepared (soaked and to remove any trace of blood) before cooking.
- All utensils used in slaughtering, cleaning, preparing and packaging must be *kosher*.

Pareve - These are foods that are neither dairy nor meat, such as eggs and fish, tofu, nuts, seeds, fruits and vegetables, as long as they are not prepared with any milk or meat products. *Pareve* is a Yiddish word meaning "neutral". *Pareve* foods presents fewer kosher complexities than meat or dairy. Nevertheless, certain important points about these foods should be considered:

- *Pareve* status is lost if the food is processed on equipment also used to process meat or dairy products or when additives are used.
- Certain fruits, vegetables and grains must be checked for the presence of small insects and larvae, which are not kosher.
- Eggs must be checked for the presence of blood spots, which are not *kosher*.

According to *kosher* laws, all dairy and meat foods must kept completely separate from other food. This requires separate sets of dishes and cooking utensils. *Pareve* foods, on the other hand may be mixed in and served with either dairy or meat. There is a specific rule that governs the production of wine. Wine is *kosher* only if its production is done

---

[7] Torah means "the law" in Hebrew and refers to God's guidance for his people as encapsulated in the first five (5) books of the Bible, also called the Pentateuch (Genesis, Exodus, Leviticus, Numbers, and Deuteronomy). The word also is used to mean all of the teachings, guidance and doctrine in Jewish law and tradition, whether written or oral.

exclusively by Torah-observant Jews, even if all the ingredients in wine are of *kosher* origin. Importantly also, the eight-day Jewish holiday of *Passover* involves a unique set of *kosher* laws. During Passover, no leavened products (including grains that can ferment and become leavened[8] such as wheat, barley, spelt, oats and rye) or their derivatives may be consumed, even if they are *kosher* the rest of the year. During *Passover*, Jews can eat only unleavened grains. Wheat flour is permitted only if it is baked into Matzah i.e. unleavened bread.

The Kosher certification process

Kosher certification agencies comprise individual Rabbis, rabbinical organizations, food technologists and field supervisors with the expertise and resources to execute the *Laws of Kashrut*[9]. *Kosher* certification agencies are responsible for ensuring that food products and ingredients meet all *kosher* requirements. These organizations enter into agreements with food producers, manufacturers/processors and food businesses. Once *Kosher* requirements have been met, a letter of certification is issued to the manufacturer. The manufacturer is then authorized to advertise its *kosher* certification status, display the *kosher* symbol and promote the product as *kosher*. The United States dominates global *kosher* sales and also has the largest *kosher* certification agencies, known as the "Big Five". These five agencies are: the Orthodox Union (OU), OK Kosher Certification (OK), KOF-K Kosher Certification Agency (KOF-K), Star-K Kosher Certification (Star-K), and Chicago Rabbinical Council (CRC). Together, these five agencies certify more than 80% of the *kosher* food sold in the US. Whereas the procedures used by these main certification agencies may vary, the procedure are similar to those used for *Halal* certification.

The *kosher* certification process employed by Star-K Kosher Certification (2019) is summarized below as an example of what the process is like:

a. Application - submitted by the company requesting kosher certification;
b. Star-K contacts the applicant to begin the review process;
c. The applicant is required to provide a list of all ingredients and names of suppliers;
d. Star-K reviews data and estimated fees;
e. Star-K makes an initial inspection which includes a thorough evaluation of the company's ingredients, products, all plant equipment, and manufacturing processes;
f. Star-K submits a contract proposal detailing all the requirements, obligations and agreed terms for kosher certification. All kosher requirements, as well as a list of all ingredients and products, will be enumerated;
g. Star-K Rabbinic Field Representative (RFR) visits the company to monitor compliance with the terms of the agreement;
h. After all terms and issues have been satisfied, Star-K issues a letter of certification, authorizing use of the Star-K symbol on the products approved.

---

[8] Any substance that causes a dough or batter to rise or produces fermentation such as yeast or baking powder. This also applies to wine that is produced by fermentation.

[9] Jewish religious dietary laws

The steps required for Orthodox Union (OU) Kosher (kashrut) certification (OU, 2019) can be summarized as follows:

a. Application - submitted by the company requesting kosher certification;
b. OU Plant Consultation/Facility visit: A Rabbinic Field Representative (RFR) visits the plant to observe the operation and the feasibility of certifying the products (including estimation of fees)
c. OU field representative tours plant then files written inspection report to the OU headquarters
d. OU Rabbinic Coordinator reviews the company's application and inspection report and advises whether OU can grant certification.
e. OU drafts a contract to include all OU requirements for kosher certification. If the contract is acceptable to the company, a letter of certification is issued by OU to the company
f. OU provides final approval of the company's product labels carrying the OU symbol.

References

Akyoo, A., Lazaro, E., 2008. An accounting method-based cost-benefit analysis of conformity to certified organic standards for specis in Tanzania (No. 2008: 30). DIIS Working Paper.

Baharuddin, S.A., Ismail, R.M., 2018. Halal compliance impact on organizational performance. the role of religiosity. Int. J. Sup. Chain. Mgt. 7 (5), 455–460.

Berry, D., 2017. The trends fueling kosher certification. Food Business News, 03.07.2017. Accessed at: <https://www.foodbusinessnews.net/articles/9015-the-trends-fueling-kosher-certification>.

DeJong-Hughes, J., John F. Moncrief, J.F., Voorhees, W. B., and Swan, J.B. "Soil Compaction: Causes, effects and control." University of Minnesota Extension, University of Minnesota. Accessed 3 April 2018. https://www.extension.umn.edu/agriculture/soils/tillage/soil-compaction/.

Food and Agricultural Organization, 2003. Development of a Framework for Good Agricultural Practices. FAO Committee on Agriculture, seventeenth Session, Rome, 31 March-4 April 2003.

Kosher Industry Facts, 2015. Kosher Fest: The Business of Kosher Food & Beverage. http://www.kosherfest.com/about-kosher.

Muhammad, N.M.N., Isa, F.M., Kifli, B.C., 2009. Positioning Malaysia as halal-hub: integration role of supply chain strategy and halal assurance system. Asian Social Sci. 5 (7), 44–52.

OK Kosher Certification, 2019. What is Kosher?. Accessed on 29 July 2019 from: <http://www.ok.org/companies/what-is-kosher/meat-dairy-pareve-setting-boundaries/>.

OU Kosher Certification, 2019. Steps in Kosher certification. Accessed on 29 July 2019 from: < https://oukosher.org/get-certified-application/>.

Oseni, J.O., Adams, A.Q., 2013. Cost benefit analysis of Certified Cocoa production in Ondo state, Nigeria (No. 309-2016-5179), 4th International Conference of the African Association of Agricultural Economists, September 22–25, 2013, Hammamet, Tunisia.

Spink, J., Moyer, D.C., 2011. Defining the public health threat of food fraud. J. Food Sci. 76 (9), R157–R163.

Spink, J., 2019. Food counterfeiting: a growing concern. *Encyclopedia of Food Chemistry*, 2019. Elsevier, NY, pp. 648–651.

Spink, J., Moyer, D.C., Speier-Pero, C., 2016. Introducing the food fraud initial screening model (FFIS). Food Control 69, 306–314.

Smith, G.C., 2016. What is Food Fraud? FSNS News, Food Safety Net Services. December 13, 2016. Accessed on 3 September 2019 from: <http://fsns.com/news/what-is-food-fraud>.

STAR-K Kosher Certification, 2019. The Global Demand for Kosher. Accessed on 29 July 2019 from: < https://www.star-k.org/articles/category/advantage-kosher-certification/?sec=certified>.

Talib, H.A., Ali, K.A.M., 2009. An overview of Malaysian food industry: the opportunity and quality aspects. Pakistan Journal of Nutrition 8 (5), 507–517.

USA Halal Chamber of Commerce Inc., 2019. Halal Overview. Accessed on 29 July 2019 from: < https://www.ushalalcertification.com/halal-overview.html>.

USA Halal Chamber of Commerce, Inc. (ISWA Halal Certification Department), 2019. Become Halal Certified Today. Accessed on 22 July 2019 from: <https://www.ushalalcertification.com/>.

Victor, A.S., Gockowski, J., Agyeman, N.F., Dziwornu, A.K., 2010. Economic cost-benefit analysis of certified sustainable cocoa production in Ghana (No. 308-2016-5113).

Further reading

Ismaeel, M., Blaim, K., 2012. Toward applied Islamic business ethics: responsible halal business. J. Manage. Dev. 31, 1090–1100.

Religious Council of Brunei Darussalam, 2007. Guideline for Halal Certification. BCG Halal 1. First Edition, 2007. Accessed on 22 July 2019 at: <http://www.halalrc.org/images/Research%20Material/Report/GUIDELINE%20FOR%20HALAL%20CERTIFICATION.pdf>.

Star-K, 2019. Certification Process. Accessed at: <https://www.star-k.org/articles/category/certification-process/?sec=certified>.

CHAPTER

5

Microbiological considerations in food safety and quality systems implementation

Aubrey Mendonca[1], Emalie Thomas-Popo[2] and André Gordon[3]

[1]Department of Food Science and Human Nutrition, Iowa State University, Ames, IA, United States [2]Interdepartmental Microbiology Graduate Program Iowa State University, Ames, IA, United States [3]Chairman & CEO, Technological Solutions Limited, Kingston, Jamaica, West Indies

OUTLINE

Introduction

There is a growing global awareness of food safety and quality issues as food products from both developed and developing countries enter the rapidly expanding international food trade. Food safety issues are a major concern in developed countries due to extensive media coverage of foodborne disease outbreaks and increasing public awareness of the relationship between wholesome, nutritious foods and health. In this regard, food safety and quality standards, including stringent microbial specifications for various food products, have become the norm in developed countries. More importantly, these standards have extended into the international food trade and must be met by developing countries to assure safety of their food exports and exploit the economic gains from acceptance of their food products in the international food trade.

Several barriers exist with regard to entrance of food products from developing countries into the global food trade arena. A major barrier is the scarcity of information on the microbial ecology of foods that hinder the development of microbial standards, specifications, and microbial control measures. Epidemiological information that links contaminated foods to cases of foodborne illnesses is almost non-existent in many instances. There is little or no research on improving production methods, product testing and certification, and compliance with sanitary and phytosanitary standards. The vast array of fresh products from developing countries is more likely to be associated with food safety risks and face barriers to global market access from non-compliance with required sanitary and phytosanitary standards.

While consumers in developed countries are more aware of microbial food safety risks and demand assurance of safe handling of local and imported foods, many developing countries are now implementing or modifying existing food safety regulations to focus more on process control and risk prevention from farm to consumer. Certainly, future growth in developing country food exports will occur within the context of improved food safety standards of developing countries. Overall, overcoming these barriers is a crucial prerequisite for food product exports to enter the international food trade. In this respect, developing countries must implement approaches involving transparent science-based standards to produce safe food products for export.

This chapter aims to provide an overview of microorganisms of concern for food safety and product quality in a range of categories of food products from developing countries. It focuses on the incidence of pathogenic bacteria associated with these foods, as well as aspects of the microbial ecology of emerging market foods that are potential candidates for export. Suggested microbiological criteria are presented for broad groups of foods that are part of the diet of consumers in developing countries. Additionally, specific challenges that create barriers to selected foods from developing countries from entering the global food trade are discussed.

Sources of pathogenic bacteria in foods: a brief overview

All foods can be potential vehicles for transmission of enteric pathogens to humans. For foods of animal or plant origin, their populations of natural microflora originate from many

FIGURE 5.1   Vegetables grown in contact with or close proximity to soil.

sources in the natural environment where both pathogenic and non-pathogenic saprophytic microorganisms are likely to occur. For example, vegetable crops grown in fields and fruits in orchards are inevitably exposed to numerous sources of microbial contamination. These include soil, water, windblown dust, insects, and feces of wild animals, reptiles and birds. Vegetables in contact with or close proximity to soil (Fig. 5.1) and/or decaying vegetation are at high risk of contamination with pathogens such as *Listeria monocytogenes*, *Bacillus cereus*, *Clostridium botulinum* and *Clostridium perfringens* [International Commission on Microbiological Specifications for Foods (ICMSF), 2000]. Agricultural practices involving use of poor microbial quality irrigation water and fertilization of vegetable crops with non-composted animal waste can also add several pathogenic microorganisms to those crops. These include *Salmonella* spp., pathogenic *Escherichia coli* and *L. monocytogenes*, along with viruses and parasites. Although washing with potable water or approved chemical solutions may reduce the microbial populations on fresh vegetables and fruits, it is unlikely to eliminate all microorganisms from these products. Therefore, hygienic handling and proper temperature control for fresh produce are crucial for the microbial safety and quality of these products (ICMSF, 2000).

Animals harbor numerous populations of microorganisms on their skin or hide, mouth, nostrils, ears and in their digestive tract. Seafood does as well. During slaughter and dressing, meat from food animals becomes contaminated from various sources including soil, dust and fecal material from the animal's skin or hide, spilled animal intestinal contents, workers' hands, tools, processing equipment and the slaughter facilities (Roberts, 1980). In their intestinal tract, animals can harbor pathogens such as *Salmonella* spp., pathogenic *Escherichia coli*, *Campylobacter jejuni*, *Yersinia enterocolitica*, *Listeria monocytogenes*, *Staphylococcus aureus*, *Clostridium botulinum*, and *Clostridium perfringens* (International Commission on Microbiological Specifications for Foods (ICMSF), 2000). These pathogens contaminate fresh meat during the slaughter and carcass dressing processes including skinning, evisceration, trimming and washing. Additional spread of pathogens occurs via cross-contamination during fabrication of meats and seafood

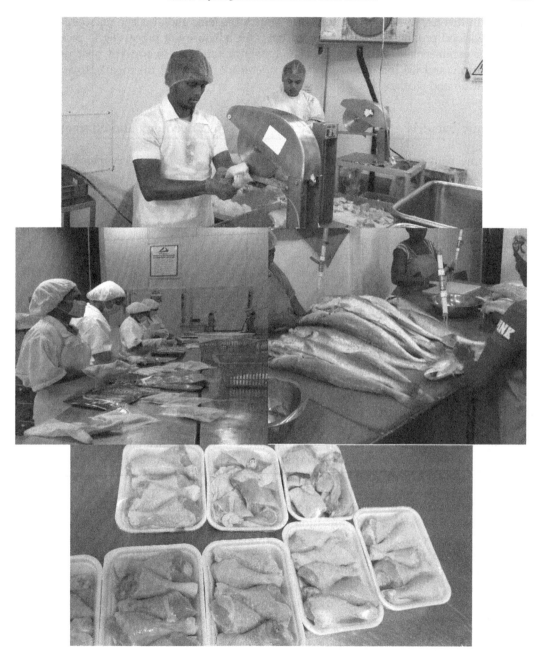

FIGURE 5.2   Processing of meat, poultry and seafood. *Source: André Gordon (2019) and* Rickey *Ong-a-Kwie (2017).*

into primal, sub-primal, and retail cuts, as well as during grinding, packaging and other activities (Fig. 5.2). Further meat processing involving addition of salt, phosphate, spices, nitrite, smoking, cooking, and post-processing handling can significantly alter the distribution, numbers and types of both pathogenic and spoilage microorganisms in the final ready-to-eat products. In this respect, the concentration of both pathogenic and spoilage microorganisms in the finished food product will be influenced by: i) the adequacy of processing to produce pasteurized or commercially sterile product, ii) hygienic handling to prevent post processing contamination, and iii) the use of proper temperature to control microbial growth during transportation and storage.

Pathogenic bacteria of major concern in food safety: ecology and characteristics

Foodborne pathogens such as *Salmonella, Campylobacter*, and *Enterohaemorrhagic Escherichia coli* and *Vibrio cholerae* are among the most common microorganisms, in addition to certain enteric viruses, that cause illness in millions of people globally every year. Since foods are a vehicle for transmitting these pathogens to humans, it is crucial for food manufacturers in both developed and developing countries to understand the ecology of pathogens and their characteristics that are pertinent to food safety. Such understanding is very important before embarking on the design of interventions for microbial inactivation or control of microbial growth in foods.

Salmonella spp.

Salmonellae are widely distributed in the natural environment due to numerous reservoirs for this pathogen. In fact, many wild and domestic animals, birds, reptiles and rodents are primary sources of *Salmonella* due to harborage of this pathogen in the intestinal tract. Infected humans can carry this pathogen in their intestinal tract and continue to shed the pathogen in fecal material long after the illness has subsided. *Salmonella* is continuously shed into the environment via animal feces and can survive in dried fecal material, soil and vegetation for an extended period of time (Bell and Kyriakides, 2002; Spector and Kenyon, 2012; Waldner et al., 2012). Prolonged survival of this pathogen increases its chances of infecting animal hosts via ingestion of contaminated vegetation. Several *Salmonella* serotypes have been isolated from vegetables and outbreaks of salmonellosis in humans were attributed to consumption of contaminated bean sprouts (O'Mahony et al., 1990) and lettuce, cabbage, celery, watercress and endive (Beuchat, 1996).

The extent of *Salmonella* contamination on the meat from food animals is influenced by its concentration in the intestinal tract of the animals as well as the level of contamination of the animal's skin, fleece or hide (Roberts, 1980). More importantly, carcass contamination is also influenced by the amount of care taken to prevent cross-contamination during the slaughter and dressing processes and by the sanitary condition of workers' tools, hands, and food contact surfaces of equipment (Sheridan, 1998). Generally, relatively low numbers of *Salmonella* are found on carcasses; however, those numbers increase following insufficient chilling, storage, or transport at temperatures above 7 °C which can allow multiplication of the pathogen (International Commission on Microbiological Specifications for Foods (ICMSF), 2000). While much emphasis has been placed on control of *Salmonella* in meat and egg products, a variety of other foods including milk, dairy products, vegetables, fruits and seafood can be sources of that pathogen (Todd, 1997).

Salmonella is a facultative anaerobe; therefore, it can grow with or without oxygen. It grows optimally from 35 to 37 °C; however, it can grow within a temperature range of 5–46 °C. It is destroyed by pasteurization temperature and time and is sensitive to low pH (4.5 or less). Generally, it will not grow at a water activity (a_W) of 0.94 or lower; however, this characteristic can vary among serovars and is affected by certain environmental factors including nutrient availability, pH, temperature, gaseous atmosphere, and presence of antibacterial substances. Salmonellae can survive in dry environments for very long times. They can also proliferate to high concentrations in foods without substantially altering the acceptance qualities of the foods (Bell and Kyriakides, 2002). Some serovars such as *S. Typhimurium* and *S. Enteriditis* infect a wide spectrum of animal hosts; however, others such as *S. typhi* are restricted to specific host animals (Stevens et al., 2009).

Shiga-toxin-producing *Escherichia coli*

Shiga-toxin-producing *Escherichia coli* (STEC) are a pathogenic group of major concern worldwide. Ruminants are a major reservoir of these human enteric pathogens. Ruminant animals that carry this pathogen in their intestinal tract include cattle, sheep, goats, and deer with cattle as the major source of infection in humans. The STEC are sometimes isolated from non-ruminant animals; however, the pathogens might be transient in the intestinal tract of those animals. Non-ruminant animals most likely acquired them from dietary sources contaminated by feces from ruminants (Caprioli et al., 2005). The STEC bacteria typically do not cause disease in animals except for diarrhea in calves (Kang et al., 2004). Bacteria in this group can cause severe illness including hemolytic uremic syndrome (HUS) and death in humans (Scallan et al., 2011; WHO—World Health Organization, 2015). The STEC O157:H7 has traditionally been associated with foodborne illness in humans (Robertson et al., 2016); however, more than 400 STEC serotypes cause foodborne illness globally (EFSA—European Food Safety Authority, 2013). Among the most prevalent non-O157 serotypes are O26, O45, O103, O111, O121, and O145, otherwise known as the "big six" which are among the most prevalent non-O157 serotypes that cause foodborne disease in the United States (Hoang Minh et al., 2015; USDA—United States Department of Agriculture, FSIS—Food Safety and Inspection Service, 2012) and other countries (EFSA—European Food Safety Authority, 2013). Ground beef, vegetables and fruits, and dairy products, are most frequently associated with foodborne disease in humans caused by STEC (Hoffmann et al., 2017).

The STEC organisms are facultative anaerobes, acid tolerant and can resist acidic pH (near pH 2.5). This acid tolerance aids their survival in acid foods, helps them to withstand the acidity of the animal host stomach, and facilitates their growth and colonization of the intestinal tract (Doyle et al., 2007; Cutrim et al., 2016; Salim et al., 2017). An important characteristic of the STEC group is the ability of the strains to survive at freezing temperatures for several months (De Schrijver et al., 2008; Strawn and Danyluk, 2010). They can survive in foods at −20 °C (Doyle et al., 2007). As observed with many vegetative bacterial cells, the STEC organisms are destroyed by the heat pasteurization process. Therefore, survival of these organisms in heat-treated foods is indicative of either under processing or post-processing contamination.

Campylobacter jejuni

Campylobacter jejuni inhabits the intestinal tract of a wide range of infected warm-blooded animals. This pathogen has been isolated from the intestinal tract of sheep, pigs, and cattle, and from the cecum of poultry. Due to the higher body temperature of poultry and other birds, poultry is the most common animal that harbors *Campylobacter* spp., and serves as a major source of *Campylobacter* infection for humans (Silva et al., 2011). Of the three *Campylobacter* species namely, *C. jejuni, C. coli*, and *C. lari*, which can all colonize the intestinal tract of poultry and cause illness in humans, *C. jejuni* is most frequently associated with outbreaks of human campylobacteriosis (Cean et al., 2015; Ugarte-Ruiz et al., 2018). *Campylobacter jejuni* is microaerophilic and requires a gaseous environment consisting of 2–5% oxygen and 5–10% carbon dioxide for optimal growth because 21% oxygen is toxic to this pathogen (Alterkruse et al., 1999). The pathogen can grow optimally at 37–42 °C; however, it is very sensitive to dry conditions and requires a water activity above 0.912 to survive. It is also sensitive to sodium chloride and its growth rate decreases at sodium chloride concentrations higher than 0.5% (Lund et al., 2000; Doyle and Roman, 1982). *C. jejuni* is a highly virulent pathogen requiring only a few hundred cells to cause foodborne disease in humans (Bhaduri and Cottrell, 2004). Most cases of campylobacter-related illness in humans result from ingestion of contaminated poultry (Skarp et al., 2016; Lee et al., 2017). Interestingly, Lynch et al. (2011) reported a higher incidence of *Campylobacter* in beef (36%) compared to pork (22%) and chicken (16%); however, in most countries about 50–80% of raw poultry at retail outlets tests positive for *C. jejuni* (Shane, 1992; Haba, 1993; Moran et al., 2009).

Yersinia enterocolitica

Yersinia enterocolitica is widely distributed in the natural environment and has been isolated from several sources including soil, milk, undercooked pork, ground pork and chitterlings (pork intestine). The pathogen has also been isolated from a variety of warm-blooded animals including humans, poultry and pigs, which are a major reservoir for this pathogen (Lund et al., 2000). Zdolec et al. (2015) reported a high incidence of *Y. enterocolitica* in the tonsils and mandibular lymph nodes of pigs.

Y. enterocolitica is a facultative anaerobe, which can grow over a broad temperature range from about −1.3 °C to 42 °C. Its optimum growth temperature is between 20 and 30 °C. The organism is psychrotrophic and therefore can grow at refrigeration temperatures making it a health hazard in contaminated foods stored under refrigeration for an extended period of time. Although many strains can grow at 1 °C, they grow relatively slowly and are usually out competed by fast growing spoilage microflora in foods (Mossel et al., 1995). In fact, growth of *Y. enterocolitica* in refrigerated ground meat was inhibited by background microflora up to a temperature of at least 15 °C (Fukushima and Gomyoda, 1986). A major food safety concern is the use of head meat in production of ground pork because remnants of tonsils and oral mucosa are highly to likely occur in the final product (International Commission on Microbiological Specifications for Foods (ICMSF), 2000).

Vibrio parahaemolyticus

Vibrio parahaemolyticus strains are halophilic and their natural habitat is the sea. Although *V. parahaemolyticus* is widely distributed in oceanic and coastal waters, it is generally not common in open oceans and is intolerant of the hydrostatic pressures found in ocean depths (Sakazaki, 1983). Some strains of *V. parahaemolyticus* are not pathogenic, however, foodborne gastroenteritis caused by pathogenic strains of *V. parahaemolyticus* is contracted mainly from seafood. *V. parahaemolyticus* has been commonly associated more with shellfish than other fish from the ocean. Oysters, crabs, clams, shrimps, lobsters and other shellfish are common vehicles for outbreaks. Fish such as mackerels, sardines and codfish were also linked to cases of gastroenteritis caused by *V. parahaemolyticus*. When gastroenteritis is caused by other foods, cross-contamination from seafood products is most times involved. Outbreaks have been linked to the consumption of raw, improperly cooked, temperature-abused or post-heat-contaminated seafoods.

Pathogenic strains of *V. parahaemolyticus* produce an important toxin called thermostable direct haemolysin (TDH) which is not destroyed by heating in a food. Other virulence factors include TDH-related haemolysin (TRH), siderophores (vibrioferrin), Type 3 secretion systems, capsules and biofilm production. *V. parahaemolyticus* cells are known to be heat (pasteurization) sensitive. $D_{47\,°C}$ values from 0.8 to 65.1 minutes have been reported (Beuchat, 1996). *V. parahaemolyticus* cells are also sensitive to drying, refrigeration and freezing (Ray and Bhunia, 2013). Because cells are sensitive to the low pH present in the stomach, large amounts of cells (10^{5-7} CFU) need to be consumed for symptoms to develop. Symptoms develop 10–24 hours after consumption of contaminated food and include diarrhea, cramps, weakness, nausea, chills, fever, headache and vomiting.

Under optimal conditions, *V. parahaemolyticus* has a generation time (9–13 minutes) that is shorter than that of *Escherichia coli* (20 minutes). The growth of *V. parahaemolyticus* is dependent on several factors. These include NaCl content, pH and temperature of the water. *V. parahaemolyticus* are halophiles and grow best in 2–4% NaCl but can also grow in 1–8% NaCl. They are sensitive to 10% salt and will not grow in distilled water. *V. parahaemolyticus* grows optimally in pH range of 7.6–8.6; however, it can grow within a pH range of 4.8–11. Its minimum growth pH is dependent on the temperature and NaCl content of the water. Although *V. parahaemolyticus* will not grow in a cold-water temperature of 4 °C, its growth at 5 °C has been shown at pH 7.6% and 7% NaCl, or pH 7.3% and 3% NaCl (Beuchat, 1996). It grows optimally from 30 °C to 37 °C; however, it can grow within a temperature range of 5 °C to 42 °C (Ray and Bhunia, 2013). Although the minimum temperature for growth in open waters has been reported as 10 °C, *V. parahaemolyticus* has grown at 9.5–10 °C in foods (Kaneko and Colwell, 1973). In a study of the ecology of *V. parahaemolyticus*, water temperature was shown to be correlated to the incidence of this microorganism in the Rhode River area of Chesapeake Bay (Kaneko and Colwell, 1973). The study showed that during the cold winter months, *V. parahaemolyticus* survived in the sediment. However, from spring to early summer, with water temperatures 14 +/− °C, the microorganisms are released into the water column where they attach to zooplankton. As the water temperature rose, they proliferated. *V. parahaemolyticus* therefore shows a seasonal variation with the highest microbial counts during the summer months. Hence, high microbial counts of *V. parahaemolyticus* have been isolated from seafood mostly in the summer months.

Shigella spp.

The only known habitats for *Shigella* spp. are the intestines of humans and primates. Human and some primates can harbor the pathogen in their intestines and shed it through fecal material without showing symptoms. Some infected humans who have recovered from the disease can remain a carrier for a long time. Humans and some primates which serve as their hosts can transmit the organisms directly through fecal-oral routes. Poor personal hygiene increases the risk of developing foodborne shigellosis via direct transmission. Transmission can also occur indirectly through fecal-contaminated food and water. In many developing countries with poor sanitation such as some countries in South America and Mexico, fecal-contaminated drinking water poses a serious risk for transmission of shigellosis (Ray and Bhunia, 2013). Indirect contamination can occur if fecal-contaminated water is used to wash foods that will not be heat processed. Prominent vehicle foods of *Shigella* spp. include shellfish, chicken, fruits and vegetables, and salads (Jay et al., 2005). Cross-contamination of ready-to eat foods can also lead to shigellosis.

The genus *Shigella* is a member of the *Enterobacteriaceae* family, along with salmonellae and escherichiae. There are four species in the genus Shigella: *S. dysenteriae, S. flexneri, S. boydii* and *S. sonnei. Shigella* spp. are endemic in countries such as South Asia, Sub-Saharan Africa and tropical and sub-tropical regions of the world (Kotloff et al., 2013; Liang et al., 2007). *S. dysenteriae* is the main pathogen implicated in bacillary dysentery (Jay et al., 2005). *Shigella* spp. have a very low infective dose and as little as 10 colony forming units (CFU) can cause an infection in susceptible individuals. Once ingested with contaminated food or water, the pathogen passes through the stomach, then small intestine before reaching the large intestine. Shigellae cells can invade the epithelial mucosa of the small and large intestines. Once inside the epithelial cells, they produce an exotoxin known as the Shiga toxin. Symptoms of the disease such as abdominal pain, often bloody diarrhea, fever, chills and headache are due to the Shiga toxin and the invasion of the epithelia mucosa.

Shigella spp. are Gram-negative facultative anerobic rods. Generally, strains of *Shigella* grow between 7 °C and 46 °C but grow optimally at 37 °C. They are capable of growing in several foods once the storage temperature is within the temperature growth range, although growth in food may not be very important for shigellosis as the infective dose is low. *Shigella* spp. grow best at pH 6 to pH 8 but growth has been recorded at pH 5. Although cells of *Shigella* spp. can survive physical and chemical stresses such as freezing, refrigeration, pH 4.5% and 5% NaCl, they are destroyed by pasteurization (Ray and Bhunia, 2013).

Staphylococcus aureus

Staphylococcus aureus is widespread in nature, inhabiting the skin and mucous membranes or the nostrils and throat of most warm-blooded animals including all food animals and humans. The pathogen is opportunistic and while it can exist as a harmless commensal in the animal host, it can cause infections of the skin via open wounds such as cuts and abrasions. It is resistant to desiccation and can survive in dry environments on skin and surfaces of inanimate objects (Chaibenjawong and Foster, 2011). Outside of the animal host, *S. aureus* can survive for a long time in a dry state and has been isolated from dust, air, water and sewage (Doyle et al., 1997).

S. aureus can grow within a temperature range of 7–48.5 °C with optimum growth at 30–37 °C. The pH range for its growth is pH 4.2–9.3 with optimum growth at 7–7.5. The pathogen is a facultative anaerobe which is salt tolerant and can grow in up to 15% sodium chloride. If conditions in foods allow its growth to approximately 10^6 colony-forming units (CFU) per gram or milliliter, it can produce heat stable enterotoxin in sufficient concentration to cause food poisoning. It is a poor competitor and easily outgrown by competing microflora; therefore, outbreaks of *S. aureus* foodborne intoxication from raw or unprocessed foods are non-existent. In contrast, contaminated food products such as pre-cooked ham, cream-filled pastry, cooked turkey meat and canned mushrooms have caused outbreaks of staphylococcal foodborne poisoning (CDC Centers for Disease Control, 1996).

Listeria monocytogenes

Listeria monocytogenes is widely distributed in the natural environment. It has been isolated from numerous sources including water, soil, mud, pasture grasses, silage, fecal material from domestic and feral animals, and decaying vegetation (Swaminathan, 2001). Primary habitats of this pathogen are soil and decaying vegetation where it lives as a saprophyte. Growth of *L. monocytogenes* can take place in improperly fermented or moldy silage with a pH > 5.0–5.5 (Fenlon, 1985; Ryser et al., 1997). Such silage presents a source of infection in farm animals such as calves (Fenlon, 1986), goats (Gitter et al., 1986), and sheep (Fensterbank et al., 1984). *L. monocytogenes* is a hardy organism, which can survive for years in food processing environments and serve as a constant source of post-processing contamination of foods (Gravani, 1999). The pathogen's formation of biofilms facilitates its persistence in food processing and retail environments (Fenlon, 1999).

Listeria monocytogenes is a facultative anaerobe, which can tolerate several antimicrobial hurdles in foods that would otherwise suppress growth or result in inactivation of other foodborne vegetative pathogens. For example, *L. monocytogenes* is tolerant to salt and nitrite (McClure et al., 1997) and can grow at a water activity as low as 0.92 (Ingham et al., 2004). A wide array of food products, including dairy products such as raw milk, pasteurized milk, chocolate milk, butter, soft cheeses, and processed meats such as turkey frankfurters and various types of sliced deli meats have been associated with outbreaks of human listeriosis (Dalton et al., 1997; Headrick and Tollefson, 1998; Maijala et al., 2001; Ryser, 1999). In accordance with good hygienic and sanitation practices in food processing plants, it is crucial to check for and recover nonpathogenic *Listeria* spp. since these can serve as indicator organisms for the likely presence of *L. monocytogenes* (Lakicevic et al., 2010).

Bacillus cereus

Bacillus cereus cells and spores are widely distributed in nature and are commonly found in dust and soil. Small numbers have also been isolated from various raw and processed food products. During grazing, *B. cereus* spores or cells can contaminate udders of cows resulting in the common contamination of milk (Andersson et al., 1995; Lin et al., 1998). Spores or cells can enter the dairy farm via feed and bedding material (Magnusson et al., 2007). Cells or spores of *B. cereus* can be introduced into other food production areas when accompanying plant material and can contaminate food-processing equipment (Stenfors Arnesen et al., 2008). *B. cereus* persistence in food processing environments can be attributed to their ability to form biofilms which protects their cells and spores from the antimicrobial action of sanitizers (Ryu and Beuchat, 2005). The highly adhesive characteristic of their endospores and their ability to survive heat treatments are also very advantageous (Stenfors Arnesen et al., 2008). Another habitat of *B. cereus* is the intestinal tracts of humans. Under normal conditions, 10% of the intestinal tracts of healthy adult humans harbor *B. cereus* (Ray and Bhunia, 2013). *B. cereus* has also been isolated from the guts of soil-dwelling arthropod species (Margulis et al., 1998).

B. cereus cells are toxin producers. They produce toxins during the exponential or stationary phase of *B. cereus'* growth cycle or in the intestinal tract of humans upon lysis of *B. cereus* cells (Ray and Bhunia, 2013). *B. cereus* toxins can be found in the intestines of animals as well as in foods. They produce several toxins such as cytotoxin K, enterotoxin T, enterotoxin FM, emetic (cereulide) and diarrheagenic toxins. The emetic and diarrheagenic toxins are the most important. *B. cereus* emetic toxin (cereulide) is produced in foods before ingestion, so therefore this emetic form of the disease is classified as an intoxication (Stenfors Arnesen et al., 2008). This toxin is extremely heat stable and therefore once the toxin is formed in the *B. cereus* cell in foods, it will not be destroyed by heating (Ray and Bhunia, 2013). The symptom specifically associated with emetic toxin is severe vomiting. It causes severe symptoms (nausea and vomiting) to occur just 1−5 hours after consuming food containing the toxin and viable cells. Diarrhea and abdominal pain may also occur. Another important characteristic of the emetic toxin is that it is active at a wide pH range of pH 2−11. On the other hand, the diarrheagenic toxin is heat-labile and the symptom specifically associated with it is profuse watery diarrhea (Ray and Bhunia, 2013). Other symptoms include abdominal pain and nausea but generally no vomiting or fever. Symptoms occur 6−12 hours after consuming foods containing viable *B. cereus* cells. The diarrheal form of the disease is classified as a toxicoinfection because viable vegetative cells or spores are ingested but the toxin is produced in the intestines (Stenfors Arnesen et al., 2008).

Consumption of foods with small numbers of *B. cereus* cells and spores does not result in the disease. However, when these foods are held at improper temperatures, the spores germinate and cells multiply causing them to reach dangerous levels that can result in the disease. In addition, inadequate cooking, improper cooling, contaminated equipment and poor personal hygiene are other contributing factors that are associated with *B. cereus* gastroenteritis (Ray and Bhunia, 2013). Spores or cells of *B. cereus* can also contaminate other foods by cross-contamination. Foods such as salads, vegetables, pudding, meats, sauces, casserole and soups are commonly implicated in diarrheal outbreaks. In emetic outbreaks however, mostly starchy foods such as fried or boiled rice and pasta are involved. In fact, fried rice served at Chinese restaurants seem to be the most predominant food associated with the emetic form of the disease (Ray and Bhunia, 2013). It has been reported that holding cooked rice at room temperature for a long period of time could lead to the germination of surviving or post-contamination spores and multiplication of cells. Once cell numbers reach a very high level, consumption of the contaminated rice results in the emetic form of the disease. Other foods associated with the emetic form of the disease include mashed potatoes, vegetable sprouts and pasteurized cream.

B. cereus are Gram-positive, endospore formers. Their endospores are resistant to heat, dehydration and other physical stresses (Stenfors Arnesen et al., 2008). They are aerobic but can also grow under some anaerobic conditions. Optimally, they grow within a temperature range of 35–40 °C, however, they can multiply in a wide temperature range of 4–50 °C. Although their spores can survive high cooking temperatures, they may be destroyed by pressure-cooking (Ray and Bhunia, 2013). Their cells are however destroyed by pasteurization (Ray and Bhunia, 2013). Other factors necessary for growth include a water activity (a^w) $\geq$ 0.95, pH of 4.0–9.3 and a NaCl concentration <10%.

Clostridium perfringens

Clostridium perfringens cells and spores are widely distributed in the environment. They are found in soil, dust, water, and sewage. They are also found as part of the microbiota of animals and humans. *C. perfringens* is found in the intestinal tracts of humans and other animals such as birds. Raw foods such as fruits, vegetables and spices can become contaminated from soil and dust while raw meat can be contaminated with spores and cells from the intestinal contents of some animals. During slaughter, *C. perfringens* can contaminate meat directly or meat can become contaminated subsequently from containers, dust or handlers (Jay et al., 2005).

C. perfringens causes gastroenteritis in humans when a large number ($\geq 5 \times 10^5$/g) of viable cells is ingested with food (Ray and Bhunia, 2013). Consuming inadequately cooked and improperly held foods are generally associated with food poisoning by *C. perfringens* (Hailegebreal, 2017). Foods that encounter contaminated equipment can also contribute to food poisoning. In addition, slowly cooling foods provides the ideal temperature for spore germination, outgrowth and cell multiplication and cooking large volumes of food creates an anerobic environment which can result in food poisoning by *C. perfringens*. Following the consumption of large number of vegetative *C. perfringens* cells, the cells enter the gut. In the small intestines, vegetative cells undergo sporulation with the vegetative cells lysing and producing the toxin which is released in the intestine (Kalinowski et al., 2003; Ray and Bhunia, 2013). Protein-rich foods such as meats, meat products, meat stews (beef and poultry), casseroles, bean dishes, meat pies, roasts, some Mexican foods such as enchiladas and tacos, and gravies and sauces have been associated with *C. perfringens* food poisoning (Hailegebreal, 2017; Ray and Bhunia, 2013). *C. perfringens* has also been isolated from vegetable products such as herbs and spices as well as in other raw and processed foods (Hailegebreal, 2017).

C. perfringens are Gram-positive, spore forming bacteria. The bacterium is classified into 5 types (A, B, C, D, and E) according to their production of 4 major types of extracellular toxins (alpha, beta, epsilon and iota). Type A strains are predominantly associated with food poisoning cases. The *C. perfringens* enterotoxin that is associated with the foodborne disease is heat labile. *C. perfringens* is anerobic but is not as strict an anaerobe as other clostridia (Jay et al., 2005). A few strains are aerotolerant (Hailegebreal, 2017). Their spores are very heat resistant however their vegetative cells are sensitive to low-heat temperatures (pasteurizations). In addition to heat, spores of *C. perfringens* are also resistant to drying and toxic compounds (Jay et al., 2005). *C. perfringens* cells require numerous amino acids for growth and thus grow best in protein-rich foods. They are mesophilic, therefore the optimum growth temperature for the growth of vegetative cells and germination of spores and outgrowth is 45 °C, however, their growth temperature ranges from 10 °C to 52 °C. Cell multiplication occurs in as little as about 9 minutes under optimum conditions (Ray and Bhunia, 2013). Factors that inhibit growth include a water activity (a^w) < 0.93, pH < 5.0, 500 ppm nitrite and a NaCl concentration >5%.

Clostridium botulinum

Clostridium botulinum spores are ubiquitous in nature. They are widely distributed in soil, mud, sewage, sediments of lakes, coastal waters and lakes and intestines of fish and animals such as birds and mammals. Some organisms such as plants, algae and invertebrates may harbor *C. botulinum* but are unaffected by its toxins (Espelund and Klaveness, 2014). Spores can be released into the environment and be transported large distances via wind (spores present in dust) or by surface waters in heavy rain (Long and Tauscher, 2006). Another source of *C. botulinum* is household solid waste. Flies can spread the cells or spores to foods or other areas (Böhnel, 2002). Several foods can become contaminated with spores of *C. botulinum*. Spores from the soil can contaminate fruits and vegetables. Fish can become contaminated from spores from water and sediments. Spores can survive in minimally or incorrectly processed foods (Long and Tauscher, 2006).

Foodborne botulism requires the consumption of food containing the highly toxic botulinum toxin. However, cell growth is necessary for toxin production. For toxin production to occur, the food product must first be contaminated with *C. botulinum* spores. Then once the intrinsic characteristics of the food support the germination and outgrowth of spores as well as the subsequent multiplication of vegetative cells, botulinum toxin is produced. Outbreaks are mainly associated with low-acid vegetables such as corn, spinach, green beans, peppers, asparagus, mushrooms and carrots (Ray and Bhunia, 2013). Fruits such as figs and peaches have also been involved in outbreaks. Improper home canning of contaminated foods such as vegetables has been the main cause of botulism (Ray and Bhunia, 2013). Fin fish as well as improperly cooked, fermented and smoked fish and fish eggs have also been involved in foodborne botulism outbreaks (Ray and Bhunia, 2013). Foods such as honey and corn syrup have been known to be sources of *C. botulinum* spores that causes infant botulism. Because dairy, poultry and meats are often heated and consumed immediately, they are seldomly involved in botulism outbreaks. However, unlikely foods such as baked potatoes and sautéed onions have resulted in outbreaks because they were temperature abused (Ray and Bhunia, 2013).

Consumption of foods containing the potent botulinum neurotoxin of *C. botulinum* results in a deadly foodborne disease, foodborne botulism. Only a small amount (1 ng/kg body weight) of the neurotoxin is required for severe symptoms and death (Ray and Bhunia, 2013). Following consumption of food containing botulism toxin, the toxin is absorbed through the intestinal tract and enters the circulatory system. It then spreads to peripheral nerves where it disrupts release of the neurotransmitter, acetylcholine, as well as blocks nerve impulses (Ray and Bhunia, 2013). This results in paralysis. Thus, botulism causes neurological symptoms such as blurred or double vision, dryness of mouth, difficulty breathing and swallowing, and paralysis. It also causes gastroenteritis (vomiting, nausea, diarrhea and constipation). In infant botulism, infants (less than 1 year old) consume *C. botulinum* spores in foods, such as honey, which germinate, grow and produce the toxin in the intestines.

C. botulinum cells are Gram-positive rods. They are obligate anaerobes and form single terminal spores. Spores of *C. botulinum* are unable to germinate in the presence of nitrite (250 ppm). Spores are also very heat resistant. They are known to withstand temperatures greater than 100 °C (Böhnel, 2002). Temperatures as high as 115 °C are required to destroy spores (Ray and Bhunia, 2013). On the contrary, cells of *C. botulinum* are very sensitive to several conditions. They are sensitive to low a_w (0.93), low pH (<4.6), moderately high salt (5.5%), and moderate heat (pasteurization) (Ray and Bhunia, 2013).

C. botulinum produces toxins during growth and optimum toxin production is facilitated by optimum growth of *C. botulinum* cells. There are 7 types (A, B, C D, E, F, and G) of *C. botulinum* strains based on the type of toxin produced. Of these, Types A, B, E and F cause food intoxication in humans. Type A strain is proteolytic, Type E strain is nonproteolytic, however, Types B and F strains can be either proteolytic or nonproteolytic. Generally, the proteolytic strains grow between 10 °C and 48 °C but grow optimally at 35 °C. The nonproteolytic strains grow between 3.3 °C and 45 °C but grow optimally at 30 °C. *C. botulinum* toxin is heat labile and can therefore be destroyed in food by high, uniform heating (boiling for 5 minutes or 90 °C for 15 minutes) as well as by radiation (5–7 mrad) (Ray and Bhunia, 2013).

Establishing microbiological criteria for food products: use of indicator organisms

A microbiological criterion is a risk management measurement for the acceptability of a food or the effectiveness of a food safety control system. In this respect, acceptability is based on results of sampling and testing for microorganisms, their toxins/metabolites or indicators of pathogens or food spoilage at a specific stage of the flow of food from farm to consumer. Microbiological criteria are applicable to assess the acceptance of samples of raw material or finished product. More importantly, in the context of the international food trade, they are used by developed and some developing countries to assess the acceptability of imported food in addition to evidence of an implemented food safety program such as a Hazard Analysis Critical Control Points (HACCP) system. The three types of microbiological criteria are microbiological standards, microbiological guidelines and microbiological specifications. For participation in the international food trade, it is crucial for exported foods from developing countries to meet microbial standards, guidelines or specifications of international trading partners.

Indicator organisms

Indicator organisms or groups are used to evaluate general sanitation or environmental conditions that may indicate the potential presence of pathogens of food safety concern. These organisms are not necessarily pathogenic; however, their use is attractive to food manufacturers because tests for pathogens can be expensive, time-consuming, and inefficient especially for testing large amounts of food and/or environmental samples. Also, indicator microorganisms are useful for developing microbiological criteria to define the acceptability of food products. In this respect, acceptability relates to the absence or presence of certain microorganisms or quantifiable limits of those or other microorganisms per unit of mass, volume, area or lot.

The following section provides basic information to help food manufacturers in developing countries to make science-based selection of indicator organisms as components of microbiological criteria for their food products destined for export markets.

Aerobic plate count

The aerobic plate count (APC) is also known as standard plate count, aerobic mesophilic count, total plate count or aerobic colony count. The APC is used to estimate the bacterial population in a food sample. It is not an evaluation of the entire bacterial population nor does it indicate differences among bacterial types in a food product. It provides an estimate of the numbers of microorganisms that can grow aerobically at mesophilic temperatures. The APC may be used to judge sanitary quality, sensory acceptability, and conformance with good manufacturing practices (GMPs). Results of the APC can provide a food processor with information on the quality or handling history of raw materials, food processing and storage conditions, and handling of the finished product. Additionally, it can be used to determine the shelf-life or forthcoming sensory change in a food product (Silliker, 1963). Detectable changes in food quality characteristics due to microbial growth and enzyme production generally occur when the APC increases reaches about $10^6 - 10^7$ per g or ml.

The APC is an unreliable index of microbial food safety because it has no direct correlation to the occurrence of pathogens or toxins. A low APC does not signal the absence of pathogens in a product or its ingredients. However, if an unusually high APC is observed in a food product or ingredient, it can be reasonably assumed that a public health hazard exists pending results of pathogen testing. In interpreting APC results one must consider the type of food product and whether or not a high APC is typical of that product. For example, a fermented food such as yogurt is expected to have a high APC.

Psychrotrophic and thermoduric organisms

The APC can be modified to estimate populations of specific microbial groups such as psychrotrophic or thermoduric bacteria. The microbial quality and shelf-life of refrigerated foods are negatively affected by the growth of psychrotrophic organisms. Although psychrotrophs do not grow at their optimal rate at refrigeration temperatures they can cause food quality defects (off-odors, unpleasant tastes, or sliminess) when their populations (CFU per g or ml) reach about 10^7 (off-odors) and 10^8 (sliminess) (Jay et al., 2005). The time taken for those organisms to grow to high numbers and spoil foods depends on the initial level of contamination and temperature. There is an inverse relationship between initial psychrotrophic count and shelf-life of refrigerated foods with a higher initial count resulting in shorter shelf life.

Many psychrotrophic bacteria excrete heat-resistant enzymes (proteolytic and lipolytic) that adversely affect the quality of foods during extended storage following heat processing (Griffiths et al., 1981). Some psychrotrophic bacteria such as *Pseudomonas* spp. and *Acinetobacter* spp. are strongly proteolytic and/or lipolytic and cause serious defects in meat, poultry, dairy and seafood products when high counts (10^6 per g or ml or higher) are attained during refrigerated storage. The proteolytic count is a component of a microbiological criterion for butter. Both the proteolytic count and lipolytic count for butter can serve as an index of poor manufacturing practices.

Thermoduric organisms survive heat pasteurization of milk and constitute a substantial part of the APC in pasteurized milk. Several non-endospore-forming bacteria such as *Streptococcus* spp., *Enterococcus* spp., and *Lactobacillus* spp. have a higher tolerance to heat than expected, surviving heating of foods at 60−80 °C. Although thermoduric organisms will normally grow in the mesophilic temperature range (15−37 °C), some of them can grow at refrigeration temperatures (Johnston and Bruce, 1982; Washam et al., 1977). High numbers of thermoduric organisms are usually associated with inadequate cleaning and sanitizing of equipment on the dairy farm or at the milk processing facility. The dairy industry uses the "Laboratory Pasteurization Count" also known as the thermoduric count (APHA, 1978) to determine which batches of milk have excessive numbers of those microorganisms. Counts of thermoduric organisms greater than 1,000 CFU/mL of producer's milk are indicative of a sanitation problem (Johnston and Bruce, 1982). Similarly, a bacterial concentration of greater than 1,000 CFU/mL in pasteurized whole egg is unacceptable (Shafi et al., 1970).

Enterobacteriaceae and coliforms

The most significant application of *Enterobacteriaceae* and coliform tests is in assessment of overall quality of a food product and hygienic conditions maintained during processing. These indicator organisms can be used to determine the adequacy of a thermal process for destroying vegetative bacteria in a food or beverage product. Also, they can be used to determine post-pasteurization contamination of a food produce because *Enterobacteriaceae* and coliforms are both destroyed if effective pasteurization temperature and time are utilized. Although both microbial groups serve the same purpose, the *Enterobacteriaceae* is not used much in the United States but it is widely used in Europe.

Fecal coliforms

The term "fecal coliforms" is defined as coliforms which ferment lactose to produce acid and gas in EC broth at 44.5−45.5 °C within 48 h (Hitchins et al., 1998). The fecal coliform test may recover strains of *E. coli*, *Enterobacter* spp., *Klebsiella pneumoniae* and *Citrobacter freundii* based on the type of food product and temperature of incubation (Splittstoesser et al., 1980). Populations of fecal coliforms can build up over time on improperly cleaned and sanitized processing equipment and utensils and become a constant source of contamination to processed foods. Those organisms are readily killed by heat and may become sub-lethally injured or may die off during frozen storage of food products. Therefore, they are not a reliable indicator group for frozen foods. Fecal coliforms are mainly utilized in microbiological standards to assess the wholesomeness of shellfish and the microbial quality of waters in which shellfish is grown ((USDHEW) U.S. Department of Health, Education and Welfare, 1965). The intended purpose of the fecal coliform test is to decrease the risk of harvesting shellfish from feces-polluted waters. Since *E. coli* is the best indicator of fecal contamination, the availability of fast, direct plating methods for *E. coli* (Anderson and Baird-Parker, 1975; Holbrook et al., 1980), including methods for resuscitation of sub-lethally injured cells, makes it advantageous to use *E. coli* as opposed to fecal coliforms as a component of microbiological criteria for foods.

Escherichia coli

The occurrence of *E. coli* in a food product suggests the possibility of fecal contamination and that other microorganisms of fecal origin, including pathogenic microorganisms, might be present. Compared to commonly used fecal-indicator microorganisms, *E. coli* is currently the best indicator of fecal contamination. For safety evaluation of food products, the presence of *E. coli* indicates a greater likelihood of hazard than the presence of other coliforms. However, a negative *E. coli* test result for a food is no assurance that enteric pathogens are absent (Mossel, 1967; Mossel, 1978; Silliker and Gabis, 1976). In raw meats the occurrence of small numbers of *E. coli* is not surprising because of the likelihood of carcasses contamination from animal hide and feces during the slaughter and dressing operations. Since *E. coli* is readily killed by heat processing of foods, its presence in heat-treated foods is suggestive of under processing or post-processing contamination from poorly cleaned and sanitized equipment, employees, or from contaminated raw foods. If a food processor's aim is to determine post-processing contamination, then the target organisms should be coliforms. High populations of *E. coli* in certain foods such as soft cheese may be a consequence of microbial growth in the cheese or accumulation of the organism on unclean food contact surfaces of equipment. *E. coli* serves an important role in microbiological criteria particularly in instances where it is necessary to determine if fecal contamination of a food occurred.

Enterococci

Generally, the natural habitat for the enterococci is the intestinal tract of warm-blooded and cold-blooded animals and insects (Mundt, 1982). Insect infestation of foods can therefore contribute substantial amounts of enterococci to those foods. Some enterococci are associated with growing vegetation. The enterococci have some characteristics that make them unique as indicators. They are salt-tolerant (grow in 6.5% NaCl), aerotolerant, relatively resistant to freezing, unlike *E. coli*, and can grow at 45 °C (113°F). The entero-cocci can also grow at 7–10°C (44.6–50°F). *Enterococcus faecalis* and *Enterococcus faecium*, both of which are common enterococci in foods, are thermoduric and may survive heat pasteurization of milk (ICMSF, 2001). Psychrotrophic, thermoduric enterococci can spoil marginally heated perishable canned cured meat products including hams, pate, bacon, and emulsion-style sausages (Mendonca, 2010). These are meat products that are heated to about 65–75 °C (147–167 °F) in hermetically sealed, high gas barrier films and sold as "cook in the bag" products.

The occurrence of substantial numbers of enterococci in dairy products signals inade-quate level of sanitation in the production and processing of milk (Girafa et al., 1997). Their resistance to pasteurization and their adaptability to various environmental stress conditions may explain their occurrence in food products made from raw materials or in heat-pasteurized foods. Although enterococci are used as a fecal indicator in the environ-ment, their importance as hygiene indicators in food processing is questionable (Birollo et al., 2001).

Generally, many foods may have small to large amounts of *E. faecalis* and *E. faecium*. For example, raw meat and poultry products, and raw vegetables can have levels of enterococci at approximately 10 –1,000 CFU per gram. In contrast, certain cheeses and fer-mented sausages may carry >10^6 CFU per gram. Therefore, enterococci counts in foods are not a reliable indication of fecal contamination. A good understanding of the signifi-cance of enterococci in specific foods is necessary before any meaning is ascribed to their occurrence and the quantity present in food products. Counts of enterococci are of little value to microbiological criteria for foods. Their use to determine poor manufacturing practices would require ascertaining the normal enterococci counts at various stages of processing and handling using a standardized testing method.

Staphylococci

The proliferation of *Staphylococcus aureus* in foods is a potential food safety hazard because many strains of this organism produce a heat-stable enterotoxin that causes food-borne intoxication. Staphylococci colonize the nasal passages, mouth, skin, and lesions of man and other mammals. The occurrence of *S. aureus* in foods that have undergone pro-cesses that kill the pathogen is commonly indicative of contamination by food handlers. To a lesser extent, contact of the processed food with contaminated equipment or air con-tributes to foodborne contamination with that pathogen. Justifications for testing foods for *S. aureus* are to: i) ascertain that the organism is the cause of foodborne intoxication, ii) assess whether a food or ingredient is the source of pathogenic staphylococci, and iii) pro-vide evidence of post-process contamination. Processed foods contaminated with enterotoxin-producing strains of *S. aureus* are a significant food safety hazard because competing microorganisms that inhibit pathogen growth and toxin production are absent. Toxin production by *S. aureus* is detectable in foods when populations of the pathogen reach about 10^6 CFU/g; therefore, it is crucial to prevent contamination and temperature abuse that boost the growth of the pathogen in foods. *S. aureus* is applicable as part of microbiological criteria for cooked foods, foods (such as sandwiches) that are touched extensively during preparation, and other foods that endure post-heat handling. The *S. aureus* counts are useful as a component of microbiological criteria to indicate the potential presence of enterotoxin and lack of sanitary handling of sensitive foods.

Pseudomonas aeruginosa

Pseudomonas aeruginosa is ubiquitous in the natural environment and can be isolated from fresh waters, soil, and vegetation. Viable counts of this pathogen up to about 10^3 CFU per gram of vegetables including cucumbers, lettuce, tomatoes, onions, and radishes have been reported (Stiles, 1989). *P. aeruginosa* is constantly recovered from wastewater (Leclerc and Oger, 1974, 1976) and in surface waters contaminated with polluted effluents. Growth of this pathogen in water may not be directly attributed to organic matter because it can grow in pure water (Leclerc et al., 2002). It has an oligocarbo-tolerant characteristic, which allows it to grow in low nutrient water after a period of adaptation to that environment.

Since the late nineteenth century, several publications revealed that some *P. aeruginosa* strains produce enterotoxins and have caused diarrhea in infants and young children (Williams and Cameron, 1897; Hunter and Ensign, 1947; Florman and Schifrin, 1950; Henderson et al., 1969). In a later report, the occurrence of *P. aeruginosa* in drinking water was deemed a risk to public health (Hoadley, 1977). However, due to the very widespread distribution of this organism and several other *Pseudomonas* spp., most likely, we inevitably eat substantial numbers of *P. aeruginosa* with raw vegetables. Therefore, one may assume that the health risk related to this organism in water is insignificant. Hardalo and Edberg (1997) are of the opinion that establishing microbiological criteria related to *P. aeruginosa* in drinking water would not provide any benefits for protecting public health. Since *P. aeruginosa* is an opportunistic pathogen, it is of concern for infants who may become ill when contaminated water or contaminated processing equipment is used in the manufacture of infant formulas. In Europe, *P. aeruginosa* has been used as an indicator organism in bottled mineral water. Currently, in the United States, this organism is not used in microbiological criteria for water.

Thermophilic spore count

Many types of bacterial endospores, including thermophilic spores, enter food processing facilities in soil and dust on raw foods and in food ingredients, e.g. sugar, starch, flour and spices (ICMSF, 2001). Where suitable conditions exist in the processing plant, the endospores can germinate, grow and produce vegetative cells, which in turn multiply and subsequently sporulate to amplify the number of endospores. Counts of thermophilic spore-formers are utilized by the canning industry as a component (specification) of the industry's microbiological criteria. These counts are important for monitoring the quality of ingredients such as starch, sugar, flour, non-fat dry milk, spices, whey protein powder, and cereals obtained from suppliers and destined for addition to low-acid heat processed foods. Thermophilic spores are of concern in food ingredients because of their very high resistance to heating and their spoilage of foods held at elevated temperatures, especially during insufficient cooling and/or storage of heat-processed foods at high temperatures (Olson and Sorrells, 2001). Also, they are difficult to inactivate in a food product or on equipment due to their exceptional resistance to heat and chemical agents. Therefore, for processors to meet stringent specifications regarding thermophilic spore counts in food products, implementation of ways to minimize spore concentrations in incoming raw foods and ingredients and effective plant sanitation practices are crucial.

Yeast and mold counts

Yeast and molds (fungi) are widespread in the environment and can enter foods via contaminated air or inadequately sanitized food contact surfaces. These organisms thrive on foods under conditions that are not conducive to bacterial growth such as low pH, low water activity, or high sugar or salt concentration. Therefore, several fungi eventually become the predominant spoilage microflora in foods such as fruits and fruit beverages, fermented products, dairy products, pickled/marinated products, dried foods, intermediate moisture foods, soft drinks, and alcoholic beverages. While fungi can cause food spoilage in the form of discoloration, musty odors, off-flavors, gas or sediment, several molds can also produce mycotoxins which can cause toxic and carcinogenic outcomes in consumers (Wu et al., 2014). Satisfactory methods are available for yeast and mold counts that are applicable for purposes of microbiological criteria. Yeast and mold counts are used as microbiological standards for sugar (National Soft Drink Association, 1975) and several dairy products.

Incidence of pathogenic microorganisms in food products from developing countries

Knowledge of the incidence of pathogens in specific foods or food groups is important for identifying foods that pose the greatest risk to human health. Such knowledge also helps food manufacturers and regulatory agencies in making science-based decisions in allocating resources to improve the microbial safety of foods for local consumption and for export markets. Generally, any food, if mishandled, can serve as a vehicle for infectious doses of human enteric pathogens and pose a health risk to consumers. Some foods that are mostly associated with human foodborne illnesses are raw foods of animal origin, especially raw or undercooked meat and poultry, raw or lightly cooked eggs, raw milk, fresh produce (vegetables and fruits) and raw seafood [Centers for Disease Control and Prevention (CDC), 2018]. Incidences resulting in outbreaks associated with foods from developing countries and the identification of causative pathogenic bacteria are less than 20% of foodborne illness outbreaks in developed country markets (CDC, 2018). Additionally, while increasing in number in recent years, the data on pathogenic bacteria in food products from developing countries as well as any associated foodborne illness outbreaks have been less available in mainstream sources. Nevertheless, several authors have reported their findings in this regard and Gordon (2016) discussed selected major outbreaks in developed countries involving fenugreek sprouts from Egypt, papaya and mangos from Mexico and raspberries from Guatemala. This section therefore seeks to fill this gap and provides a summary of reports and findings involving pathogenic microorganisms in developing country foods.

Incidence of pathogenic microorganisms and foodborne illness outbreaks in developing countries

Pathogenic bacteria in fruits, vegetables, meats, prepared meals and other foods in the Caribbean

Information on the microbiological status and safety of foods from the Caribbean region, the world's premier vacation travel destination, is limited, as is information from many developing countries and regions. In seeking to address this, Gordon et al., (2017) undertook a study over eleven years (2004–2014) in which 28,527 food and environmental samples from the Caribbean were evaluated for pathogens. They were also assessed for compliance with the International Commission on the Microbiological Specifications for Foods (ICMSF) microbiological standards for those foods (Skovgaard, 2012). Of the samples examined, 95.9% of the samples were compliant for pathogens and indicator organisms, with 4.1% exceeding the recommended limits. The pathogens that were evaluated were *Listeria* spp., *E. coli*, *Clostridium perfringens*, *Staphylococcus aureus* and *Salmonella*. The foods and food environments assessed included prepared foods, seafood, meat and meat processing, poultry, fruits, vegetables and beverages. For *E. coli*, the foods examined were dairy products, fish and other seafood, raw and cooked meats, poultry, and processed meat products (ham, sausages, bacon). No *E. coli* were found in dairy products, with the incidence overall in all food products being very low, when present, and the organism being absent from samples examined in the last three years, 2012–2014 (Fig. 5.3). Of those instances where *E. coli* was found (Fig. 5.4), less than 1% (0.73%) exceeded the ICMSF-proposed limit (Gordon et al., 2017). *E. coli* was found in meats (5.8%), in 0.6% of fruit and vegetables (cucumbers), prepared meals (1.9%) and 0.3% of environmental samples (swabs).

The incidence of *Staphylococcus aureus* in Caribbean food samples was low (Fig. 5.5) while the occurrence of both *S. aureus* and *Clostridium perfringens* were reduced over time in dairy and environmental samples such that none were present in 279 samples examined in the period 2010–2014. Very low incidences of positive samples for *Listeria* spp. (2.31%) and *Salmonella* spp. (0.91%), respectively, were found in the foods examined (Gordon et al., 2017). Specimen were taken from meats, dairy products, prepared meals, vegetables and environmental swabs. Although the number, range and type of samples evaluated for *Listeria* and *Salmonella* increased over the years, particularly with the passage of the US FDA's Food Safety Modernization Act, the incidents of positive findings remained low (Fig. 5.6). In general, Gordon et al. (2017) found that the number of pathogen positive samples, including those resulting from environmental monitoring, declined over the period under study, a finding attributed to greater vigilance and a response to the introduction of HACCP-based food safety systems and increasingly stringent local standards.

Studies from the region on the incidences of *Listeria* in beef, pork and goat meat are discussed below and included in Table 5.2 (Adesiyun, 1993), Gordon and Ahmad (1991, 1993) having examined the incidence and types of organisms in the hatchery environment and

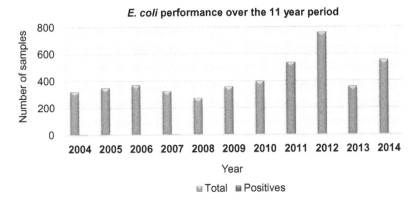

FIGURE 5.3 *E. coli* counts (CFU) vs. total bacterial numbers in food samples from the Caribbean examined over the period 2004−2014.

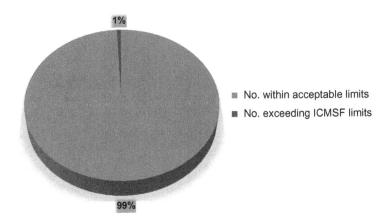

FIGURE 5.4 *E. coli* counts (CFU) in food samples from the Caribbean examined over the period 2004−2014. *Source: Adapted from Gordon et al. (2017).*

the impact of this on the poultry production system, as well as the nature of thermoduric enterococci, including *Streptococcus faecium* (now *Enterococcus faecium*) found on processed meats. Mendonca et al. (2014) and Mendonca et al. (2015) undertook the assessment of pathogen survival in traditional seasonings, finding that the conditions were inimical to the survival of *Listeria, E. coli, Clostridium perfringens, Salmonella* and *S. aureus*. Other studies have included investigation of the microorganisms involved in cocoa and chocolate production (Camu et al. 2007, 2008), characterization of the microflora of ackees, a traditional Jamaica fruit (Gordon and Jackson, 2013) and the role of microorganisms in the quality and safety of coconut water (Jackson et al., 2004). A study reported the finding and successful development of effective intervention strategies to reduce the prevalence of *Alicyclobacillus* in tropical fruit beverages in a Caribbean beverage manufacturing plant (Knight and Gordon, 2004). While not of importance as a pathogen, *Alicyclobacillus* is of major economic significance in the beverage industry as it can wreak havoc on fruit juice and beverage production operations. This organism, its impact and control are examined in detail in Gordon (2016).

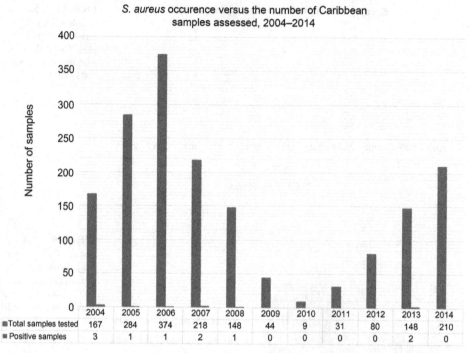

FIGURE 5.5 Incidence of *S. aureus* in selected Caribbean over the period 2004–2014. *Source: Adapted from Gordon et al. (2017).*

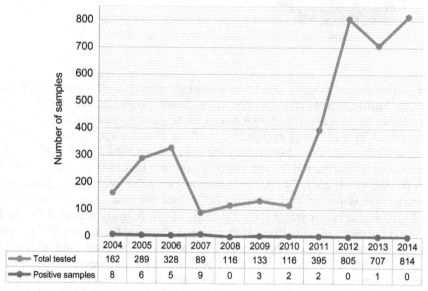

FIGURE 5.6 Total number of food samples and positive samples from the Caribbean examined for *Listeria* spp. *and Salmonella* spp. over the period 2004–2014.

Incidence of pathogenic microorganisms and foodborne illness outbreaks in developing country regions

As the global food chain becomes more diverse with a greater dependence on developing countries for a significant part of the supply, there is increased interest from developed countries in the production and food safety systems in these countries and, hence, in the safety of the foods produced. There has also been an increase in the research on developing country foods and publication of studies on their microbiology and, where relevant, foodborne illnesses associated with them. While the information remains relatively sparse, this section summarizes some of the more important work on pathogens and reports on foodborne illnesses in developing countries and is followed by detailed suggested limits that could help to assure the microbiological safety of developing country foods intended for local consumption or export.

A detailed review of foodborne illness outbreaks in 20 countries in Latin American and the Caribbean (LA&C) between 1993 and 2010 found that there were 6,313 bacterial outbreaks caused mainly by meat, dairy products, water and vegetables in the 1990s, and eggs, vegetables, and grains and beans in the 2000s (Pires et al., 2012). While the situation in the Central and Southern American countries of the LA&C region vary, foodborne illness outbreaks in the Caribbean have been relatively infrequent as reported by the Caribbean Public Health Agency (CARPHA), a Pan-American Health Organization (PAHO) affiliate. Where they have been reported, however, they have had significant impact, particularly those which have occurred at hotels and involve some of the large number of visitors who transit the region. Outbreaks have involved *Vibrio cholera*, *Salmonella* Enteritidis, norovirus, *Staphylococcus aureus*, and *Campylobacter* and have affected Barbados, Bermuda, the Dominican Republic, Jamaica, Trinidad and Tobago and the Turks and Caicos Islands, among others (Brown et al., 2001; Doménech-Sánchez et al., 2011; CAREC, 2015 (unpublished); CBS News, 2019). The outcome has included hotel closures and attendant lawsuits, in some instances, with the properties affected never recovering financially from the incidents. The consequence has been a notable reduction in reported incidences as a result of significantly increased intervention and public education by regional and local health authorities and food safety professionals over the last 20 years.

Foodborne illness outbreaks in China from 2000 to 2014 were reviewed by Luo et al. (2017) and outbreaks in South Africa over the period 2013–2017 were discussed by Shonhiwa et al. (2019), while extensive reports on outbreak in various African countries have also been reviewed by Dewaal et al. (2010), Paudyal et al. (2017) and Richter et al. (2019). *Vibrio cholera* is one of the more prominent causative agents of foodborne illnesses in developing countries, having implicated in major outbreaks in Africa and Western Pacific (Dewaal et al., 2010). It resulted in a multinational outbreak following a wedding in the Dominican Republic in 2011 as a result of the contamination of the seafood served (Jimenéz et al., 2011). There was a cholera outbreak in Laos People's Democratic Republic Sekong Province which extended from the end 2007–2008 (Lenglet et al., 2010). *Vibrio cholera* also caused outbreaks in Kenya and Zambia, among other Sub-Saharan countries (Ajayi and Smith, 2019). *Salmonella* has been the cause of several cases of foodborne illness in several African and Asian countries, having been implicated in typhoid fever outbreaks in Uganda and Zimbabwe (Kabwama et al., 2017; N'cho, 2019). *Salmonella* Enteritidis has also been identified as the causative agent in an outbreak at a major hotel in the Caribbean in 2011 (KDHE, 2011) and Mexico. Salmonella on Mexican papayas was also the causative agents of a multistate illness outbreak in 2017 (CDC, 2017; Larsen, 2019). In 2017, there was an outbreak caused by Shigella sonnei in India (Srinivasa et al., 2009) and cases of botulism were reported in Thailand in March 2006 (Kongsaengdao et al., 2006).

Virus mediated foodborne illness outbreaks have also been reported in several developing countries. Norovirus outbreaks have been reported in Brazil (Gallimore et al., 2004), China (Lu et al., 2015), the Dominican Republic (Doménech-Sánchez et al., 2011) and on a cruise ship headed to Jamaica (CBS News, 2019). In a systematic review of shellfish-borne viral outbreaks from 1980 to July 2012, Bellou et al. (2013) reported that the majority were located in East Asia, followed by Europe, America, Oceania, Australia and Africa with 63.6% being in Japan where the consumption of raw seafood is common. There have also been studies on the microbiology and pathogen prevalence in street foods in Nigeria, Ethiopia and several developing countries (Eromo et al., 2016; Obaji et al., 2018) and we reviewed the microbiology of cocoa in Chapter 3. These, along with other studies have helped to inform the recommendations for practical microbiological limits for an extensive list of developing country foods discussed below.

Prevalence of pathogenic microorganisms on selected foods from developing countries

Painter et al. (2009) proposed a scheme for categorizing foods incriminated in foodborne disease outbreaks reported to the CDC over a 33-year period (1973–2006). Those same authors defined the following three broad groups from which almost all foods are derived: land animals, land plants and aquatic animals. For the three groups the authors defined six food commodities (beef, game, poultry, pork, dairy, eggs) in the land animal group, nine commodities (leafy vegetables, sprouts, fruits-nuts, root, fungi, vine-stalk, root, grains-beans, oils-sugars) in the plant group and three commodities (fish, crustaceans, mollusks) in the aquatic group. More than 90% of human exposure to foodborne pathogens is largely through consumption of contaminated meat and dairy products, fish and shellfish (WHO, 2014). This should also be evident from the nature of the reports on foodborne illness outbreaks in developing countries discussed previously. The information in the following section will therefore generally address the incidence of pathogenic bacteria in developing country foods of animal origin namely, various red meats, poultry and seafood that pose a relatively high risk for causing foodborne illness.

Red meat

In several developing countries in various global regions, the less traditional meats from animals such as small ruminants (sheep and goat), deer, camel, springbok, buffalo, horses, llamas and yaks are an important part of the diet of consumers. For example, in several African and Asian countries and in the Caribbean region, small ruminants such as goats and sheep are important animals for both domestic meat and a source of income (Casey et al., 2003; Vokaty and Torres 1997). Table 5.1 shows the variety of meats

TABLE 5.1 Some meat and meat products that are part of the diets in developing countries in four regions.

Region	Meat and Meat Products
Africa	Beef, pork, lamb, veal, mutton, goat, springbok, camel, buffalo
Caribbean	Beef, pork, lamb, veal, mutton, goat, rabbit, turtle
Central & South America	Beef, pork, lamb, veal, mutton, goat, deer, horse, llama, guinea pig, rabbit, agouti, paca, turtle
South & Southeast Asia	Beef, pork, lamb, veal, mutton, goat, guinea pig, rabbit

consumed in some developing countries in four global regions. While production in many parts of those countries is linked to subsistence agriculture, commercial scale production for metropolitan markets is carried out in parts of Asia, Africa, and Central and South America (Casey et al., 2003). In Ethiopia, while there is a high local demand for mutton and goat especially during religious festivals, the growing demand for those meats by neighboring Middle Eastern countries and their taste preference for meat from Ethiopia's indigenous animals, are advantageous for Ethiopia's meat export market (Sebsibe, 2008; Worku and Dereje, 2017). Therefore, depending on culture, religion, and dietary patterns of persons globally, there could be international trade of those non-traditional or exotic meats among countries irrespective of their level of economic development. In this regard, microbial safety and product quality standards should be established and harmonized to facilitate international trade.

The percent incidence of four major foodborne pathogenic bacteria (*Salmonella, Campylobacter* spp., STEC, *Listeria monocytogenes*) in selected meat and meat products from some developing countries is shown in Table 5.2. For each pathogen, the data reveal wide variations in pathogen incidence within and among meat types. This situation is not surprising in view of signification variations in hygienic meat handling practices at meat processing facilities within and among developing countries. Variations in hygienic practices can occur from the animal slaughter stage through carcass dressing, fabrication of primal and sub-primal meat cuts at abattoirs and handling at points of retail sales at meat markets and butcher shops.

Salmonella

Based on published reports from several developing countries, the incidence of *Salmonella* on beef, goat meat and pork ranged from none detected to 7.07%, 3.3–60%, and none detected to 43.8%, respectively. While buffalo meat, camel meat, and mutton are an important part of the diet of consumers in several developing countries, there is a scarcity of published reports on the incidence of pathogens on those meats. Therefore, with relatively few data on pathogen incidence on meats, it is challenging to provide a reliable assessment of the food safety risk posed by certain meat products. However, the recommendations of the authors of the few existing studies frequently include improved sanitation and hygienic practices in meat slaughter to reduce the incidence of pathogens such as *Salmonella* on raw meats.

TABLE 5.2 Percent incidence of four major foodborne pathogenic bacteria in selected meat and meat products from some developing countries.

Pathogen incidence[a]

Country	Product	SAL	CAM	STEC	LIS	Reference[b]
Argentina	Beef	–	–	12.1	–	14
Argentina	Beef (ground)	–	–	40.7	–	14
Argentina	Beef	–	–	2.6	–	15
Brazil	Beef (ground)	–	–	3.5	–	16
Brazil	Beef	–	–	0.3	–	27
Chile	Beef	–	–	10.0	–	13
Egypt	Beef (ground)	–	–	–	32	28
Ethiopia	Beef	–	–	–	47.5	26
Ethiopia	Beef	7.07	–	–	–	3
Ethiopia	Beef	–	6.2	–	–	10
Ethiopia	Beef	–	–	–	51.3	24
Ethiopia	Beef	–	–	8.0	–	32
India	Beef	–	–	41.6	–	19
Iran	Beef	–	–	8.2	–	29
Iran	Beef	–	2.4	–	–	20
Malaysia	Beef	–	7.5	–	–	18
Mexico	Beef	6.0	–	0.4	–	30
Nigeria	Beef	–	43.0	–	–	23
Nigeria	Beef	–	12.25	–	25	
Trinidad	Beef	nd	nd	–	6.6	21
Trinidad	Beef (ground)	nd	nd	–	11.4	21
India	Buffalo meat	–	–	16.9	–	19
Nepal	Buffalo meat	3.5	–	–	–	1
Nepal	Buffalo meat	20	–	–	–	22
Iran	Camel meat	–	0.9	–	–	20
Iran	Camel meat	–	–	2.0	–	29
Nigeria	Camel meat	30.1	–	–	–	4
DR Congo	Goat meat	–	41.2	–	–	11

(Continued)

TABLE 5.2   (Continued)

Ethiopia	Goat meat	—	9.4	—	—	9
Ethiopia	Goat meat	—	7.6	—	—	10
Ethiopia	Goat meat	—	—	2.0	—	28
Ethiopia	Goat meat	9.01	—	—	—	3
India	Goat meat	3.57	—	—	—	2
India	Goat meat	—	—	28.8	—	19
Iran	Goat meat	—	9.4	—	—	20
Iran	Goat meat	—	—	1.7	—	29
Nepal	Goat meat		—	—	—	1
Nepal	Goat meat	33.3	—	—	—	22
Nigeria	Goat meat	23.3	—	—	—	4
Tanzania	Goat meat	60.0	—	—	—	58
Trinidad	Goat meat	2.9	—	—	7.0	21
Ethiopia	Mutton	—	10.5	—	—	10
Ethiopia	Mutton	—		2.5	—	28
Ethiopia	Mutton	8.4	—	—	—	3
Ethiopia	Mutton	—	10.6	—	—	9
India	Mutton	—	64.0	—	—	12
India	Mutton	—	—	40.0	—	31
Iran	Mutton	—	12.0	—	—	20
Iran	Mutton	—	—	4.8	—	29
Nigeria	Mutton	50.1	—	—	—	4
Argentina	Pork	6.3	—	—	—	17
Colombia	Pork (carcass)	—	—	—	10.0	8
Colombia	Pork (deboned)	—	—	—	76.0	8
Ethiopia	Pork	—	—	—	69.8	26
Ethiopia	Pork	—	8.5	—	—	10
Ethiopia	Pork	43.8	—	—	3	
Nepal	Pork	10.0	—	—	—	22
Nigeria	Pork	8.0	—	—	—	7
Trinidad	Pork	nd	nd	-—	1.4	21

(Continued)

TABLE 5.2   (Continued)

Colombia	Ham	–	–	–	5.0	8
Colombia	Chorizo	–	–	–	3.0	8
Colombia	Sausage	–	–	–	6.0	8
Argentina	Morcilla	3.0	–	–	–	6

[a]Values expressed as percentage (%) of samples tested; nd, none detected SAL, Salmonella spp.
[b]1, Maharajan et al. (2006); 2, Makwana et al. (2015); 3, Tadesse and Tessema, (2014); 4, Musa et al. (2017); 5, Mwanyika et al. (2016); 6, Oteiza et al. (2006); 7, Yemisi et al. (2011); 8, Gamboa-Marin et al. (2012); 9, Woldemariam et al. (2009); 10, Dadi and Asrat (2008); 11, Mpalang et al. (2014); 12, Sharma et al. (2016); 13, Toro et al. (2017); 14, Etcheverria et al. (2010); 15, Masana et al. (2010); 16, Morato Bergamini et al. (2007); 17, Colello et al. (2018); 18, Premarathne et al. (2017); 19, Sethuleksmi et al. (2018); 20, Rahimi et al. (2010); 21, Adesiyun (1993); 22, Bantawa et al. (2018); 23, Adeleye et al. (2018); 24, Gebretsadik et al. (2014); 25, Braide et al. (2017); 26, Molla et al. (2004); 27, Brasileiro et al. (2018); 28, El-Malek et al. (2010); 29, Rahimi et al .(2012); 30, Navarez-Bravo et al. (2013); 31, Kiranmayi et al. (2011); 32, Hiko et al. (2008).
CAM, Campylobacter spp.; STEC, shiga-toxin-producing E. coli; LIS, Listeria monocytogenes and other Listeria spp.

Maharajan et al. (2006) tested meat samples collected from meat markets in Nepal and reported 13.5% and 3.3% incidence of *Salmonella* on buffalo and goat meat, respectively. More than 80% of the meat samples tested were positive for coliform bacteria, a microbial group which indicates the potential presence of enteric pathogens. In Anand City (Gujarat) India, Makwana et al. (2015) reported a similar incidence (3.57%) of *Salmonella* in samples of goat meat obtained from retail shops while the incidence of the pathogen on mutton samples was 6.25%. In that same study, 84.6% of the *Salmonella* serotypes isolated were *Salmonella* Typhimurium. In an Ethiopian study, the incidence of *Salmonella* on dressed carcasses of cattle, sheep, goats and pigs were 7.07% 8.4%, 9.01% and 43.8%, respectively (Tadesse and Tessema, 2014). No significant difference in incidence of *Salmonella* was found among meat from cattle, sheep and goats (P > 0.05); however, the *Salmonella* incidence on pork samples was significantly higher (P < 0.05). While S. Mishmarhaemek, S. Infantis, and S. Hadar were the predominant serotypes on cattle, small ruminant (sheep and goat) and pig carcasses, respectively, S. Typhimurium was isolated from all of those carcasses. Based on those findings, S. Typhimurium seems to have a broader animal host range compared to other *Salmonella* serotypes isolated in that study.

In Nigeria and Tanzania higher incidences of *Salmonella* on meat samples have been reported for meat samples tested. For example, Musa et al (2017) tested meat samples procured from meat retailers in abattoirs, markets and shops in Maiduguri and surrounding areas in Nigeria. Those same researchers reported 23.3%, 50.1%, and 30.1% incidence of *Salmonella* in goat meat, mutton, and camel meat, respectively. Percentages of antibiotic (ampicillin) resistant *Salmonella* isolated from meats were 33% (mutton) and 15% (camel meat). Average incidences of the pathogen on goat meat from slaughter facilities in Arusha, Tanzania were 63% and 60% for small and large facilities, respectively (Mwanyika et al., 2016). Those same researchers reported a low incidence of resistance (2–4%; n = 219 isolates) to ampicillin, amoxicillin, streptomycin, sulphamethoxazole and trimethoprim with complete sensitivity to ciprofloxacin, ceftazidime and cefotaxime. In both of those previously mentioned reports the researchers recommended improved hygienic practices in meat slaughter and sales outlets to reduce the incidence of meat samples testing positive for *Salmonella* and ensure the microbial safety of fresh meat.

In Argentina, Colello et al. (2018) conducted a study to assess the incidence of *Salmonella* in the pork production chain and to characterize the *Salmonella* isolates. Samples for testing were taken from pig farms (environmental samples), slaughterhouses (rectal swabs, carcasses, environment), boning rooms (carcasses, ground meat, environment), and retail markets (meat cuts, ground meat, environment). From 764 samples tested, 35 (4.6%) were positive for *Salmonella* based on biochemical tests and presence of the *inv*A gene. The highest incidence of *Salmonella* was observed in samples taken from boning rooms (6.3% from pig carcasses; 17.5% from environment) and from retail markets (46% from ground pork). *S.* Typhimurium was the most prevalent serotype in the entire pork production chain. All 35 *Salmonella* isolates were resistant to at least one antibiotic and 30 isolates (85.7%) exhibited multidrug resistance to different classes of antibiotics.

In both developed and developing countries control of *Salmonella* in foods, especially foods of animal origin, continues to be a major undertaking considering the numerous reservoirs for this pathogen. There is therefore need for more pathogen testing to gather data on the incidence of *Salmonella* in meat and meat products produced in developing countries.

Campylobacter *spp.*

Adeleye et al. (2017) tested meats (including beef) from five markets within the Owerri zone of Imo state in Nigeria for the prevalence of Campylobacter species. The incidence of *Campylobacter* species (specifically *C. jejuni* and/or *C. coli*) on meats ranged from none detected to 43% (beef), 7.6−41.2% (goat meat), 10.5−64% (mutton). They reported a 12.25% incidence of the pathogen with C. jejuni being the predominant species in all meat samples tested. In a subsequent study (Adeleye et al., 2018), the prevalence of *C. jejuni* in beef samples from two markets (Ihiagwa market and Relief market) in the same Owerri zone was reported as 56% (Relief market) and 43% (Ihiagwa market). Considering the low infectious dose [500 colony forming units (CFU)] of *Campylobacter* for humans (Keener et al., 2004), the relatively high prevalence of that pathogen in beef is of major concern and emphasizes the need for proper cooking of beef and other meats along with stringent measures to prevent post-cook contamination.

Woldemarian et al. (2009) evaluated the prevalence of *Campylobacter* species in goat and sheep carcasses at a private export slaughtering facility in Debre-Zeit, Ethiopia. Of the 398 carcasses sampled, the pathogen was isolated from 10.6% and 9.4% of sheep and goat carcasses, respectively. Of all the *Campylobacter* isolates recovered from the carcasses *C. jejuni* and *C. coli* represented 72.5% and 27.5%, respectively. Dadi and Asrat (2008) isolated *Campylobacter* from 10.5%, 7.6% and 6.2% of samples of mutton, goat meat, and beef respectively, from Addis Ababa and Debre-Zeit, Ethiopia. Those authors also reported *C. jejuni* as the predominant species of the pathogen recovered from the carcasses tested. In this study, 78% of the isolates were *C. jejuni*, 18% were *C. coli* and 4% were *C. lari*. These findings suggest that *C. jejuni* is the major species of *Campylobacter* on sheep and goat carcasses in Ethiopia. Jayasekara et al. (2017) assessed the incidence of *Campylobacter* species on beef procured from wet market and supermarket in Malaysia. The overall incidence of the pathogen was 14.2% and 7.5% in raw beef from wet market and supermarket, respectively. Based on the multiplex-polymerase chain reaction, the percent occurrence of *Campylobacter* species among all isolates of that pathogen was 55% (*C. jejuni*), 26% (*C. coli*) and 19% (other *Campylobacter* species). Rahimi et al. (2010) reported that the incidence of *Campylobacter* species in various meats samples including lamb (12.1%), goat meat (9.4%), beef (2.4%) and camel (0.9%) in two provinces in Iran. In that same study, the most prevalent species isolated was *C. jejuni* (84%) followed by *C. coli* (16%).

Shiga-toxin-producing E. coli

The incidence of shiga-toxin-producing *E. coli* (STEC) in meat products from meat processing facilities and retail outlets in some developing countries has been addressed by some researchers in Ethiopia, India, Iran, Argentina, Chile, Brazil and Mexico. Overall, the reported incidence of STEC on meats in these developing countries ranges from 0.4% to 41.6% (beef), 1.7% to 28.8% (goat meat), and 2.5% to 64% (mutton). Of two hundred and ninety five raw meat samples representing beef (n = 85), water buffalo (n = 38), sheep (n = 62), goat (n = 60), and camel (n = 50) tested for STEC, fourteen (4.7%) were positive for *E. coli* O157. In studies done in Iran, the highest incidence of *E. coli* O157 was 8.2% (beef), followed by 5.3% (water buffalo), 4.8% (sheep), 2.0% (camel), and 1.7% (goat) (Rahimi et al., 2012). The scarcity of published studies on the incidence of pathogens on the types of meats consumed in developing countries precludes the reporting of this information with a reasonably degree of scientific certainty.

In Ethiopia, of 125 swab samples taken from beef carcasses held at retail shops, only one sample (0.8%) tested positive for STEC (*E. coli* O157:H7). That isolate was resistant to the antibiotic amoxicillin and moderately resistant to cefoxitine and nitrofurantoin but was susceptible to seven other antibiotics used in that study (Abdissa et al., 2017). Those same authors concluded that there is a low incidence *E. coli* O157: H7 in beef cattle in Ethiopia and that current sanitary carcass dressing methods in the processing plants, as well as storage conditions used at retail shops visited were effective against that pathogen.

Bergamini et al. (2007) evaluated 250 samples of raw ground beef obtained from grocery stores in Ribeirão Preto (114 samples) and Campinas (136 samples), São Paulo State, Brazil for the prevalence of STEC. While samples from Campinas tested negative, STEC isolates of serovars O93:H19, ONT:HNT, ONT:H7, and O174:HNT were recovered samples from Ribeirão Preto, representing a 3.5% incidence of the pathogen in ground beef. In Argentina, the prevalence of STEC on beef carcasses and cuts of meat in the market chain of Buenos Aires Province was evaluated by Etcheverria et al. (2010). In that study, 12.34% and 18.64% of the beef carcasses sampled at slaughter and at sanitary control cabins, respectively, were positive for STEC. The sanitary control cabin is the facility where carcasses arrived from abattoirs from outside the city. The incidence of STEC on carcasses increased at butcheries (24.52%) and about 25% of retail beef cuts tested positive for STEC. In this respect, there were significant differences among meat cuts. For example, the incidence of STEC was 12.12% on beef chuck or rump roast, and 40.74% in ground beef. More recently, Toro et al. (2018) reported a 10% incidence of STEC in 430 ground beef samples obtained from retail centers in Santiago, Chile. In another study conducted in Argentina, testing of carcasses from nine beef exporting abattoirs for STEC revealed a 2.6% incidence of the pathogen (Masana et al., 2010).

Due to several foodborne outbreaks, the life-threatening nature of illness, and deaths linked to *E. coli* O157:H7 in ground beef, this pathogen is deemed an adulterant in that product in the United States. In this context, an adulterant is a contaminant that renders the ground beef unfit for human consumption. Therefore, the United States has a "zero tolerance" policy for *E. coli* O157:H7 in ground beef (ICMSF, 2002; Doyle et al., 2006). Based on this stringent policy, US imports of ground beef or other red meats are likely to be held to those same microbiological standards described for *E. coli* O157:H7. To the authors' knowledge, to date, no other developed country has such a stringent standard for *E. coli* O157:H7 in ground beef.

Listeria *spp. and* L. monocytogenes

Relatively few reported studies are available on the incidence of *Listeria* species or *L. monocytogenes* in meats produced in developing countries. The incidence of that pathogen ranges from 11.4% to 90.9% (in raw pork) and from 1.4% to 76% (in raw beef), respectively. From meat samples obtained from retail supermarkets in Assuit Egypt, *Listeria* spp. were detected in 32% of frozen ground beef tested (El-Malek et al., 2010). Based on a study conducted by Adesiyun (1993) in Trinidad, the incidence of *Listeria* spp. in beef and ground beef, respectively, was 6.6% and 11.4%. The mixing of beef trim and subsequent grinding of the meat most likely aided distribution of the *Listeria* to increase its incidence in samples of the final ground product. In that same study, a higher incidence of *Listeria* spp. was reported in goat meat (10%) compared to pork (1.4%). Since *Listeria* spp., including *L. monocytogenes*, are widely distributed in the natural environment, they can easily be ingested with vegetation by grazing animals, including goats. Information on whether the pigs tested were free range was not provided.

Gebretsadik et al. (2017) demonstrated widespread occurrence of *L. monocytogenes* and other *Listeria* species in foods of animal origin in Addis Ababa, Ethiopia. Those authors reported 51.3% and 2.6% incidence of *Listeria* spp. and *L. monocytogenes*, respectively, in raw beef. In another study conducted in Ethiopia, Molla et al. (2004) reported the incidence of *Listeria* spp. in raw beef (47.5%) and pork (69.8%) obtained from retail supermarkets and shops in Addis Ababa.

An investigation to determine the source of *L. monocytogenes* contamination of ready-to-eat (RTE) meats in a meat processing plant in Trinidad was undertaken by Gibbons et al. (2006). These researchers tested both raw and processed meats, air in the plant, and meat contact surfaces for *Listeria* spp., *Salmonella* spp., and *Campylobacter* spp. Results revealed that of eleven raw meat (pork) samples tested, ten (90.9%) were positive for *Listeria* spp. with two of those positive samples yielding *L. monocytogenes*. Based on all the data collected, inconsistent sanitary practices, insufficient heating of meat products, and the presence of *Listeria* in biofilms on meat contact surfaces were cited as likely contributory factors for pathogen contamination of the RTE meat products.

Poultry meat

The types of poultry meat consumed in developing countries include chicken, duck, goose, guinea fowl, turkey, ostrich, pigeon and emu (Table 5.3). Of all those types of poultry meat, the meat of chicken (*Gallus gallus*) is the most widely consumed in developing countries as well as countries with developed economies. Chicken meat represents about

TABLE 5.3   Some poultry meats that are part of the diets in developing countries within four regions.

Region	Poultry and poultry products
Africa	Chicken, turkey, duck, goose, guinea fowl, ostrich
Caribbean	Chicken, turkey, duck, goose, guinea fowl
Central & South America	Chicken, turkey, duck, goose, emu
South & Southeast Asia	Chicken, turkey, duck, goose, pigeon

88% of global poultry meat production and offers an affordable protein source for consumers (Skarp et al., 2016). The global production of broiler meat increased from 84.3 million tons in 2013 to 90.3 million tons in 2017 (USDA, 2017). The United States contributed about 21% of that amount followed by Brazil (16%), China (13%), European Union (12%), India (5%), Russia (4%) and Mexico (4%) (USDA, 2017). The United States of America continues to be the leading producer of broiler meat globally, producing approximately 19.7 million tons of broiler meat followed by Brazil with an output of about 13.8 million tons (Statistica, 2019). In developing countries, there is high demand for poultry meat that signals the need for larger scale production of this popular meat. In this respect, larger scale production without infrastructure for proper sanitation and hygienic handling of poultry meat can result in higher incidences of enteric pathogens in the finished products. This in turn can increase the risk of transmitting pathogens from animals to consumers (Soomro et al., 2011).

The percent incidence of four major foodborne pathogenic bacteria (*Salmonella, Campylobacter* spp., *Staphylococcus aureus*, and *Listeria monocytogenes*) in poultry meat and poultry products from some developing countries is shown in Table 5.4. The published data on the incidence of pathogens on poultry meat presented are derived from selected studies conducted in the following countries: Barbados, Bangladesh, Brazil, Egypt, Ethiopia, Ghana, India, Iran, Iraq, Malaysia, Pakistan, Nigeria, Nepal, Sri Lanka, Thailand, Trinidad & Tobago, and Vietnam. While there are other published studies on microorganisms found in poultry meat from developing countries, the major focus was on numbers of spoilage and pathogenic organisms and not on prevalence of pathogens. The vast majority of published reports on the incidence of pathogens in poultry meat relate to studies performed on chicken meat. As observed in the pathogen incidence data on red meats, the data for poultry meat reveal wide variations in pathogen incidence in chicken meat. Significant variations in hygienic meat handling practices at poultry processing facilities and retail markets in developing countries are a major contributory factor to those observations. The following information will provide a summary of overall incidence of pathogens in poultry meat in developing countries along with brief descriptions of conditions that contribute to increased pathogen incidence in poultry meat.

Salmonella and *Campylobacter*

Globally, millions of people are affected by enteric diseases caused by *Salmonella* and *Campylobacter*, which are among the most common pathogens that cause gastroenteritis in

TABLE 5.4 Percent incidence of foodborne enteric bacterial pathogens in poultry meat (chicken, duck, goose)/poultry products from some developing countries.

Pathogen Incidence[a]

Country	Product	SAL	CAM	SA	LIS	Reference[b]
Barbados	Chicken	—	79.8	—	—	47
Bangladesh	Chicken	—	—	—	8.33	37
Brazil	Chicken	—	53.3	—	—	20
Brazil	Chicken	14.1	—	—	—	3
Brazil	Chicken	—	19.6	—	—	21
Brazil	Chicken	—	91.7	—	—	38
Brazil	Chicken	—	14.2	—	—	19
Brazil	Chicken	—	—	—	40	36
Brazil	Chicken	—	10.8	—	—	22
Brazil	Chicken	—	40	—	—	23
Brazil	Chicken	—	16.8	—	—	24
Brazil	Chicken	—	36.3	—	—	25
Brazil	Chicken	—	37.1	—	—	26
Brazil	Chicken	—	45	—	—	27
Egypt	Chicken	44	—	—	—	4
Egypt	Chicken giblets	—	23.5	—	—	29
Ethiopia	Chicken	—	—	—	15.4	9
Ghana	Chicken	10.8	—	—	—	7
India	Chicken	52.9	—	—	—	18
India	Chicken	—	—	46.6	—	1
India	Chicken	22.8	—	82.86	—	48
India	Chicken	10.8	—	—	—	7
Malaysia	Chicken	100	—	—	—	32
Malaysia	Chicken	—	—	—	20	35
Malaysia	Chicken	—	92.5	—	—	42
Malaysia	Chicken	—	20	—	—	44
Malaysia	Chicken	—	66.67	—	—	43
Malaysia	Chicken	72.7	—	—	—	34
Malaysia	Giblets	—	—	—	26.4	17

(*Continued*)

220

5. Microbiological considerations in food safety and quality systems implementation

TABLE 5.4 (Continued)

Country	Product	SAL	CAM	SA	LIS	Ref
Pakistan	Chicken	–	29	–	–	45
Pakistan	Chicken	48.75	–	–	–	45
Pakistan	Chicken	–	–	–	12.5	16
Nepal	Chicken	60	–	53.3	–	12
Nigeria	Chicken	–	12.7	–	–	10
Nigeria	Chicken	–	96	–	–	11
Nigeria	Chicken	–	–	80	–	49
Sri Lanka	Chicken	59	–	–	–	5
Sri Lanka	Chicken	–	–	–	34	31
Thailand	Chicken (RTE)	–	–	–	0.2	39
Trinidad	Chicken (RTE)	12.5	–	–	–	6
Trinidad	Chicken	92.3	–	–	–	8
Trinidad	Chicken	–	48.7	–	–	13
Trinidad	Chicken#	20.5	–	–	–	14
Trinidad	Chicken##	8.3	–	–	–	14
Trinidad	Chicken	60.7	60.7	–	–	41
Iraq	Chicken	–	–	–	13	15
Iran	Duck meat	–	35.5	–	–	28
Egypt	Duck meat	–	19	–	–	29
Malaysia	Duck meat	10	–	–	–	30
Malaysia	Duck meat	28.4	–	–	–	33
Vietnam	Duck meat	22.3	–	–	–	40
Iran	Goose meat	–	26.5	–	–	28
Egypt	Turkey giblets	–	14.5	–	–	29

[a]Values expressed as percentage (%) of samples tested.

[b]1, Herve and Kumar (2017); 2, Das and Mazumber (2016); 3, Volz Lopes et al. (2017); 4, Abd El-Aziz (2013); 5, Kottawatta et al. (2017); 6, Syne et al. (2013); 7, Pesewu et al. (2018); 8, Rodrigo et al. (2006); 9, Molla et al. (2004); 10, Adeleye et al. (2017); 11, Adeleye et al. (2018); 12, Bantawa et al. (2018); 13, Aesiyun (1993); 14, Khan et al. (2018); 15, Ahmed et al. (2017); 16, Mahmood et al. (2003); 17, Kuan et al. (2013); 18, Gautam et al. (2017); 19, Silva et al. (2018); 20, Maziero et al. (2010); 21, DeMoura et al. (2013); 22, Carvalho et al. (2013); 23, Ferro et al. (2015); 24, Hungaro et al. (2015); 25, Brizio et al. (2015); 26, Perdoncini et al. (2015); 27, Panzenhagen et al. (2016); 28, Rahini et al. (2011); 29, Khalafalla (1990); 30, Adzitey et al. (2012); 31, Gunasena et al. (1995); 32, Nidaullah et al. (2016); 33, Adzitey et al. (2012); 34, Modarressi and Thong (2010); 35, Goh et al. (2012); 36, Loura et al. (2005); 37, Islam et al. (2016); 38, Wurfel et al. (2019); 39, Kanarat et al. (2011); 40, Phan et al. (2005); 41,Thomas and Badre (2006); 42, Usha et al. (2010); 43, Nur Llida and Faridah (2012); 44, Ibrahim et al. (2018); 45, Nisara et al. (2018); 46, Shah and Korejo (2012); 47, Workman et al. (2005); 48, Ruban et al. (2012); 49, Yemisi et al. (2011).

#Chicken from cottage processors; ##chicken from supermarkets

SAL, Salmonella spp.; CAM, Campylobacter spp.; SA, Staphylococcus aureus; LIS, Listeria monocytogenes and other Listeria spp.

humans (WHO (World Health Organization), 2015). In this regard, poultry products are the most common vehicles for transmission of those pathogens to humans. The reported incidence of those two pathogens in chicken meat produced in developing countries ranges from 8.3% to 100% (*Salmonella*) and 8.3–96% (*Campylobacter*) (Table 5.4). Several reports on pathogen incidence on poultry meat in developing countries indicate that *Salmonella* or *Campylobacter* are consistently detected in most meat and environmental samples taken from small poultry slaughter facilities and processing units, pluck shops, cottage poultry processors, local retail markets and shops, or from wet markets (Abd El-Aziz, 2013; Adeleye et al., 2018; Gautam et al., 2017; Khan et al., 2018; Kottawata et al., 2017; Nidaullah et al., 2016; Rodrigo et al., 2006; Thomas et al., 2006; Wurfel et al., 2018).

Those previously mentioned small-scale poultry processing operations and retail sales markets are prevalent in developing countries in several regions of the world where freshly slaughtered poultry is highly preferred by consumers. A good example is sale of raw poultry meat in wet markets in Asian countries (Nidaullah et al., 2016). At wet markets, vendors sell poultry meat non-packaged and uncovered. Under those conditions, the poultry meat is exposed to numerous sources of microbial contamination, including but not limited to aerosols, windblown dust, flies and other insects, and human hands. Although at wet markets the slaughter of poultry is not permitted by local food and health authorities, this practice is still evident as buyers prefer freshly slaughtered poultry (Nidaullah et al., 2016). Also, the poultry meat is held at ambient temperature with occasional spraying with water to prevent surface drying. Ice is seldom used for chilling fresh food products, except for seafood, which is covered with crushed ice to main product freshness (Nidaullah et al., 2016). The floors are frequently sprayed with water to maintain the humidity of the atmosphere, which also helps to prevent dehydration and quality loss of fresh products. Unfortunately spraying the floors creates highly contaminated aerosols that can deposit pathogens on the exposed surfaces of food products. These conditions ultimately increase the incidence of meat-borne pathogens and the health risk posed by consuming poultry meat.

There are two major sources of raw poultry meat to consumers in developing countries. One source is more traditional and involves meat vendors selling non-packaged and uncovered poultry meat in retail shops or wet markets (discussed above). The second meat supply source, which is steadily emerging, involves supermarkets and department outlets where packaged poultry meat is held for sale in refrigerated display cases (Vaskas et al., 2012). With changes in consumer lifestyle with increased urbanization, fresh chilled ready-to-cook (RTC) poultry products are becoming readily available in supermarkets in major cities in countries with developing economies (Gautam et al., 2017). In this regard, the incidence of pathogens in poultry meat sold in supermarkets is likely to be influenced by the extent of sanitation and hygienic practices during animal slaughter and carcass dressing operations, as well as maintenance of proper temperature control during storage and distribution of the product.

With regard to ready-to-eat poultry meat products, 12.5% of pre-cooked chicken frankfurters from a meat processing facility in Trinidad tested positive for *Salmonella* (Syne et al., 2013). The cooking temperature for frankfurters should destroy vegetative cells of pathogens including *Salmonella*. The survival of that pathogen was therefore likely due to inadequate cooking or post-processing contamination. The researchers concluded that there was a relatively high risk for microbial contamination of RTE meats at the processing plant and emphasized the need for improvement of the quality assurance program to improve the microbial safety of the meat products.

Staphylococcus aureus

The incidence of *Staphylococcus aureus* on raw chicken meat in developing countries ranges from 46.6% to 82.86%. In Nigeria, of 75 samples of raw chicken meat tested, 60 (80%) were positive for *S. aureus* (Yemisi et al., 2011). Herve and Kumar (2017) reported a 46.61% incidence of *S. aureus* in fresh raw chicken samples obtained from retail shops in Jalanghar, Punjab, India. Several of those isolates exhibited multiple drug resistance. An almost similar incidence of staphylococci (48.57%) was reported for raw chicken meat from Southern Assam, India (Das and Mazumder, 2016). In that same study, 80% of all the isolates (from poultry and goat meat) exhibited maximum resistance to penicillin (73.33%) followed by erythromycin (36.66%) and tetracycline (26.66%). Also, the percentage of isolates that exhibited resistance to oxacillin, ciprofloxacin, chloramphenicol and vancomycin were 23.33%, 16.66%, 10.0% and 3.33%, respectively. The high incidence of antibiotic resistant staphylococci in raw meat samples is a major health concern for animal slaughter personnel as well as consumers.

In Brazil, Ribeiro et al. (2018) used meta-analytical methods (combining the results of multiple scientific studies) to evaluate the incidence of methicillin-resistant *Staphylococcus aureus* (MRSA) in chickens, turkeys, chicken meat, and turkey meat. The pooled MRSA prevalence for turkey meat and chicken meat was 13% (1−28%) and 5% (2−9%), respectively. Poultry from South America had the highest incidence of MRSA (27%; 17−37%) whereas poultry from North America had the lowest (1%; 0−2%). The MRSA typically associated with livestock was recovered from poultry and poultry meat, suggesting that the pathogen can spread from farm to consumer (Ribeiro et al., 2018). The occurrence of MRSA in poultry and poultry meat is a serious health risk because that pathogen causes infections in communities, hospital-acquired infections (Deurenberg et al., 2007) and livestock-associated infections (Neela et al., 2009); therefore, there is an urgent need for interventions to reduce the contamination and incidence of MRSA in the poultry production continuum from farm to consumer.

Listeria spp. and L. monocytogenes

The incidence of *Listeria* spp. on raw chicken meat in developing countries ranges from 0.2% to 40%. Of the samples of raw chicken obtained from retail supermarkets and shops in Addis Ababa,15.4% were positive for *Listeria* spp. Similar studies performed on raw chicken in Sri Lanka (Gunasena et al., 1995) and Malaysia (Goh et al., 2012) reported 34% and 20% incidence, respectively, of *L. monocytogenes*. Ahmed et al (2017) investigated the incidence of *L. monocytogenes* in various foods including chicken meat taken from various supermarkets, restaurants and veterinary quarantine sites in Duhok province, Iran. Using real time polymerase chain reaction (PCR), they reported an 8% incidence of *L. monocytogenes* in samples of frozen chicken meat. Those same authors highly recommended maintenance hygienic conditions to reduce the incidence of *L. monocytogenes* in foods. With respect to ready-to-eat (RTE) chicken products, Kanarat et al. (2011) reported a 0.2% incidence of *L. monocytogenes*. Considering that adequate cooking of chicken to safe internal temperatures (i.e. 72 °C) will destroy vegetative foodborne pathogens, the presence of *L. monocytogenes* in RTE chicken products suggests those samples were undercooked or contaminated with the pathogen after cooking. The United States considers *L. monocytogenes* as an adulterant in RTE foods and therefore has a zero-tolerance policy for that pathogen in those food products. In countries such as Canada, Germany and other European countries, regulations for *L. monocytogenes* in RTE foods, although stringent, do not include a zero-tolerance policy. In this regard, developing countries and others that intend to export RTE foods must ensure that those foods are in compliance with microbiological standards of the importing country.

Seafood

Over the past few decades, there has been a steady rise in seafood consumption in both developed and developing countries. The global per capita supply of seafood increased from 9.9 kg (1960s) to 14.4 kg (1990s), 19.7 kg (2013) and to more than 20 kg in 2014 (Food and Agriculture Organization of the United Nations (FAO), 2016). In 2013, the per capita consumption of seafood in East Asian countries increased from 10.8kg to 39.2 kg and, on average, consumption in the Africa continental increased to 10 kg (Food and Agriculture Organization of the United Nations (FAO), 2016). This trend is partly attributed to growing consumer awareness of significant health benefits from consuming seafood. Those benefits include improved neural, visual and cognitive development during pregnancy and infancy (Emmett et al., 2013) and reduced risk of cardiovascular disease (Zarrazquin et al., 2014). Population growth and greater per capita seafood consumption in the United States of America has seen imports of seafood increase substantially from below 50% of the overall seafood consumption in 1980 to greater than 91% in recent years to compensate for the deficit in domestic production [National Marine Fisheries Service (NMFS), 2013].

Almost 50% of the U.S. seafood imports are from aquaculture sources mostly in developing countries, and frozen seafood constitutes about 75% of gross imports [Wang et al., 2011; National Marine Fisheries Service (NMFS), 2010]. Since fisheries and aquaculture are a major source of income for about 56.6 million people globally (Food and Agriculture Organization of the United Nations (FAO), 2016), the international trade in seafood and seafood products is a major enterprise. In developing countries, persons involved in fisheries operations are engaged in wild capture of seafood or aquaculture farming. Those operations are either part-time, small, medium, or large-scale. About 84% of persons involved in seafood operations are from Asia, followed by Africa (10%), and Latin America and Caribbean (4%). The remaining 2% are spread throughout other areas of the world (Food and Agriculture Organization of the United Nations (FAO), 2016).

The previously mentioned increase in seafood consumption globally over the past few decades has raised concern about seafood safety [Centers for Science and Public Interest (CSPI), 2009]. In the United States, imported seafood is a major food product linked to foodborne disease outbreaks in which finfish and molluscan shellfish are mainly implicated. In 2007, finfish was associated with the largest number of foodborne disease outbreaks, which were more that those caused from consumption of beef or poultry meat (CSPI, 2009). According to the U.S. Centers for Disease Control and Prevention (CDC), a greater proportion of the seafood-related foodborne outbreaks were intoxications compared to infections (CDC, 2010).

Based on their incidence, the causal agents of seafood-related foodborne disease outbreaks are classified as pathogens of major or minor concern. Those of major concern include norovirus, certain *Vibrio* spp. and *Salmonella*. Agents of minor concern include hepatitis A virus, *Staphylococcus aureus* (toxins), *Shigella*, *Listeria monocytogenes*, and *Clostridium botulinum* (Elbashir et al., 2018). In this respect, published reports on the incidence of pathogens in seafood from developing countries are scarce and, when available, mostly focus on microbial populations in seafood relative to mere numbers of pathogens or spoilage microorganisms. The information in the present section will focus mainly on the overall incidence rates of selected seafood pathogens in various types of seafood harvested in some developing countries. Table 5.5 shows some examples of seafood that are part of the diets in developing countries within four global regions.

In Nigeria, Amusan and Sanni (2019) obtained a total of 108 fresh croaker samples from five retail centers (Badagry, Iyana Ipaja, Liverpool, Makoko and Mushin) in Lagos, Nigeria. Microbial testing on those fish samples revealed that 8 samples (7.4%) tested positive for *Listeria monocytogenes*. Genotype identification of the pathogen was carried out using PCR involving 16S rRNA and subsequent DNA sequence analysis.

In Vietnam, an investigation on the contamination levels of *Salmonella* spp. in raw food, including shellfish, chicken, beef and pork was carried out. Samples (n = 50) of shellfish were obtained from 14 markets and 4 supermarkets in Ho Chi Minh City and tested for *Salmonella* spp. Test results revealed that 18% of the samples were positive for the pathogen (Van et al., 2007). Wong et al. (1999), in a study of 686 samples of seafood imported from several Asian countries (Indonesia, Hong Kong, Vietnam and Thailand), found that *Vibrio parahaemolyticus* was detected in 45.9% of those samples. The incidence of the pathogen in seafood from Hong Kong and Thailand was significantly higher compared to seafood from Vietnam and Indonesia. Shrimp and crab had the highest incidence rate (75.8% and 73.3%, respectively) while incidence rates for other types of seafood were 44.3% (snail), 44.1% (lobster), 32.5% (sand crab), 29.3% (fish) and 21.1% (crawfish).

TABLE 5.5 Some seafood that are part of the diets in developing countries in four regions.

Global region	Seafood
Africa	Crayfish, tuna, mussels, oysters, mackerel, lobster
Caribbean	Conch, codfish/saltfish, crabs, lobster, shrimp, mackerel, flying fish, red snapper, tuna, dolphin fish (mahi-mahi), grouper, kingfish, octopus
Central & South America	Corvina (corvina drum), grouper, king crab, tuna, lobster, sardines, salmon, octopus, conch, red snapper, crawfish, shrimp
South & Southeast Asia	Crabs, shrimp, tuna, salmon, catfish, tilapia

Yucel and Balci (2010) examined 78 raw fish samples (30 freshwater fish and 48 marine fish) for *Listeria*, *Aeromonas*, and *Vibrio* species. Overall, the incidence of *Listeria* spp. in freshwater fish and marine fish, was 30% and 10.4%, respectively. Of all the *Listeria* spp. detected, *Listeria monocytogenes* (44.5%) was the most common in freshwater fish whereas *Listeria murrayi* (83.5%) was the most common in marine fish. Motile *Aeromonas* spp. were more prevalent in marine fish (93.7%) compared to freshwater fish (10%). Interestingly, only the marine fish were positive for *Vibrio alginolyticus* (40.9%), *Vibrio fluvialis* (38.6%), and *Vibrio damsela* (36.3%) indicating stronger adaptation of certain *Vibrio* spp. to the marine environment. For both freshwater and marine fish, their skins had higher incidences of *Listeria* and *Aeromonas* spp. whereas the highest incidence of *Vibrio* spp. was observed in gills of in marine fish. The incidence of *Listeria* spp. and *L. monocytogenes* was significantly different in regard to type of freshwater fish. For example, of all species of the freshwater fish examined, the brown trout (*Salmo trutta*) and horse mackerel (*Trachurus trachurus*) had the highest incidence of those microorganisms.

For seafood processors in developing countries, an understanding of the limiting growth conditions for pathogens is important for controlling their multiplication in seafood. This in turn contributes to improved microbial safety of seafood products for sale nationally and internationally. Table 5.6 shows the limiting conditions for growth of pathogens in seafood. Information is provided on limits of water activity (a_w), pH, salt concentration and temperature for growth of selected pathogenic bacteria. Similar to the meat of other animals, the flesh of fish and other seafood contain large amounts of water and nutrients to support microbial growth. Also, the pH of various fish and seafood range from 4.8 to 7.3 (FDA, Center for Food Safety and Applied Nutrition, 1992) which fall within the pH range for growth of all pathogens listed in Table 5.6. Therefore, seafood harvesters and processors have to depend largely on cold temperatures to prevent multiplication of pathogens and spoilage microorganisms. In this respect, the minimum and maximum temperatures for growth of important pathogenic bacteria in seafood are stated in Table 5.6 for ease of reference. In instances where fish is salted for preservation, consideration must be given to pathogens such as *Staphylococcus aureus*, *Listeria monocytogenes*, *Vibrio parahaemolyticus* and certain types of *Clostridium botulinum* (Table 5.6) that have a relatively high salt tolerance. Since seafood can support rapid microbial growth, the parameters of time and temperature are crucial in the control of the proliferation of both pathogens and spoilage microorganisms. In this regard, soon after harvesting of seafood and throughout processing, product handling, storage and distribution, every effort has to be made to avoid temperature abuse to prevent microbial growth and toxin production. Data on the effect of temperature and time on pathogen growth and toxin production in seafood are presented in Table 5.7.

Microbiological criteria for food products from developing countries

Microbiological criteria are standards, guidelines or specifications that are used to determine acceptability of raw or finished food products. They are based on the absence or presence of certain microorganisms or numbers of those or other organisms, per specified mass, volume, area or lot of food. Microbiological criteria are applied at any stage in the

TABLE 5.6 Limiting conditions for growth of selected pathogenic bacteria in seafood[a].

Pathogen	Min a_w	pH (min−max)	Max % Salt	Temperature (min−max)
Salmonella spp.	0.94	3.7−9.5	8	41.4 °F−115.2 °F
				5.2 °C−46.2 °C
Vibrio parahaemolyticus	0.94	4.8−11	10	41 °F−111 °F
				5.0 °C−44 °C
Vibrio cholera	0.97	5.0−10	6	50 °F−109.4 °F
				10 °C−43 °C
Vibrio vulnificus	0.96	5.0−10	5	46.4 °F− 109.4 °F
				8.0 °C−43 °C
	0.95	4.0−9.0	6.5	44.6 °F−120.9 °F
				7.0 °C−49.4 °C
Shigella spp.	0.96	4.8−9.3	5.2	43 °F−116.8 °F
				6.1 °C−47.1 °C
Campylobacter jejuni	0.987	4.9−9.5	1.5	86 °F−113 °F
				30 °C−45 °C
Yersinia enterocolitica	0.945	4.2−10	7	29.7 °F−107.6 °F
				−1.3 °C−42 °C
Clostridium botulinum type A & proteolytic B &F	0.935	4.6−9.0	10	50 °F−118.4 °F
				10 °C−48 °C
Clostridium botulinum type E & non-proteolytic B & F	0.97	5.0−9.0	5	37.9 °F−113 °F
				3.3 °C−45 °C
Listeria monocytogenes	0.92	4.4−9.4	10	31.3 °F−113 °F
				−0.4 °C−45 °C
Staphylococcus aureus (growth)	0.83	4.0−10	25	44.6 °F−122 °F
				7.0 °C−50 °C
Staphylococcus aureus (toxin production)	0.85	4.0−9.8	10	50 °F−118 °F
				10 °C−48 °C

[a]*Adapted from CFSAN (2001).*

flow of food from farm to consumer. Considering the growing international use of *hazard analysis* and *risk assessment* in managing food safety, some food manufacturers question whether microbial criteria are still necessary except for meeting food legislation requirements. Actually, in many instances, the numbers of certain microorganisms or microbial

TABLE 5.7 Pathogen growth and toxin production in seafood as influenced by product temperature and exposure time[a].

Pathogen	Growth/toxin production	Product temperature	Max exposure time (cumulative)
Salmonella spp.	Growth	41.4–50 °F (5.2–10 °C)	14 days
		51–70°F (11–21 °C)	6 hours
		Above 70°F (above 21 °C)	3 hours
Vibrio parahaemolyticus	Growth	41–50°F (5–10 °C)	21 days
		51–70°F (11–21 °C)	6 hours
		Above 70°F (above 21 °C)	2 hours
Escherichia coli (pathogenic strains)	Growth	44.6–50°F (7–10 °C)	14 days
		51–70°F (11–21 °C)	6 hours
		Above 70°F (above 21 °C)	3 hours
Shigella spp.	Growth	43–50°F (6.1–10 °C)	14 days
		51–70°F (11–21 °C)	6 hours
		Above 70°F (above 21 °C)	3 hours
Campylobacter jejuni	Growth	86–93°F (30–34 °C)	48 hours
		Above 93°F (above 34 °C)	12 hours
Yersinia enterocolitica	Growth	29.7–50°F (1.3–10 °C)	1 day
		51–70°F (11–21 °C)	6 hours
		Above 70°F (above 21 °C)	2.5 hours
Clostridium botulinum type A & proteolytic B &F	Growth and Toxin Production	50–70°F (10–21 °C)	12 hours
		Above 70°F (above 21 °C)	4 hours
Clostridium botulinum type E & non-proteolytic B & F	Growth and Toxin Production	37.9–50°F (3.3–10 °C)	24 hours
		51–70°F (11–21 °C)	12 hours
		Above 70°F (above 21 °C)	4 hours
Listeria monocytogenes	Growth	31.3–50°F (0.4–10 °C)	2 days
		51–70°F (11–21 °C)	12 hours
		Above 70°F (above 21 °C)	3 hours
Staphylococcus aureus	Growth and Toxin Production	44.6–50°F (7–10 °C)	14 days
		51–70°F (11–21 °C)	12 hours
		Above 70°F (above 21 °C)	3 hours

[a]Adapted from CFSAN (2001).

groups and their safe limits are still necessary for food safety. For example, microbiological criteria are important for establishing critical limits for HACCP-based food safety systems, verifying HACCP plans, assessing shelf life, and storage studies involving pathogen challenge tests. In microbial risk assessment, apart from human exposure, infective dose and pathogenicity of foodborne organisms, the occurrence, numbers and distribution of certain organisms in foods is important (Forsythe, 2002). In the use of mathematical predictive models for microbial survival, growth, or death, it is important to appreciate the levels of viable microorganisms that cause concern or signal the end of a food product's shelf life.

Irrespective of whether a country has a developed or developing economy, larger companies usually have the resources for microbial testing and development of microbiological criteria. These criteria invariably remain confidential and are seldom in the public domain for an obvious reason, namely, competitive advantage. In developing countries, small to medium-sized and very small food operations are common and have little or no access information on microbial standards. Such information, where available, are found in documents such as peer-reviewed scientific publications, food industry codes of practice, legislative publications, and International Commission for Microbiological Safety of Foods (ICMSF) publications. In fact, some of this latter information may be outdated and thus inadequate or irrelevant regarding current microbial issues, including the emergence of new pathogens, and new food products and processes.

This section of the chapter attempts to rectify this problem by providing guidance for the establishment and use of microbiological criteria for foods for food manufacturers in developing countries. It includes suggested microbiological limits (for groups of food products) adapted from several food microbiology publications (APHA, 2001; Doyle et al., 2007; FSANZ, 2018; International Commission on Microbiological Specifications for Foods ICMSF, 2000; Stannard, 1997). The authors caution that all microbial tests for each food group are not always necessary and that additional tests are sometimes needed. However, a substantial amount of information is provided on what can be attained by Good Manufacturing Practices (GMP) and microbial numbers representing limits of shelf-life and safety.

Types of microbiological criteria

Microbial tests are used to determine if food products conform to microbial criteria. However, such testing can never offer absolute assurance of the safety or quality of food products mainly because of the non-homogeneous distribution of microorganisms in foods and challenges associated with methods of sampling. There are three types of microbiological criteria namely, standards, guidelines, and specifications.

Standards

Standards are microbiological criteria stated in laws or regulations that require mandatory compliance by food manufacturers. They are established by regulatory authorities or governments and food manufacturers' compliance with these standards are monitored by enforcement agencies. Food products that are non-compliant with the standards are rejected and deemed unfit for the intended use. Food manufacturers may not be permitted to take delivery of non-compliant imported foods or face prosecution for manufacturing foods that fail to meet microbial criteria and pose a serious health risk to consumers. Several developing countries are in the process of developing or updating their food safety legislation.

Guidelines

Guidelines are microbiological criteria applied at any step in the processing and retailing of food. They refer to the microbiological status of the food sample and facilitate identifying situations that need attention for improvement of food safety and quality. Results of microbiological testing, when compared with guidelines, help in trend analysis. Test results that are inconsistent with a normal trend usually signal a situation that is out of control and needs correction. In most instances, guidelines established by food manufacturers are for in-house purposes; however, they may at times be included in legislation. Commodity groups and food industry associations sometime publish guidelines on microbial types and levels pertinent to certain foods produced under good manufacturing practice (GMP) for use by their members.

Specifications

Specifications are microbiological criteria applied to raw materials, ingredients, in-process or finished products. They are used in purchase agreements between supplier and buyer and can include limits for pathogens, indicator or spoilage microorganisms, and toxins. Raw materials and ingredients that do not comply with specifications warrant investigation to determine the cause so that corrective interventions can be applied. Noncompliant in-process material are reworked, if possible, or discarded and finished products can be rejected even if they are not hazardous at the time of testing. It is critical that both parties in an agreement regarding specifications fully concur that the components of the specifications are realistic, consistently achievable and relevant.

Components of microbiological criteria

The four parts of a microbiological criterion are statements of: i) the food product and sample type, ii) the microbial group, specific microorganism, or toxin, iii) the sampling plan, and iv) the microbiological limit(s). It is important for food manufacturers in developing countries to monitor product conformance to microbiological criteria via validated test methods for the target microorganisms or toxin in foods. Food safety auditors who represent the interests of potential international trading partners may ask for written evidence of such actions.

Food product and sample type

Microbiological criteria must describe the food product, its stage of processing, and conditions under which the product should be stored and utilized. Such descriptive information is important because the content and reliability of a microbiological criterion are influenced by food processing and storage conditions. For example, a specification may indicate a higher aerobic plate count (APC) for a dried food product at the point of production compared to the same product (non-dried) stored refrigerated. This is because bacterial growth will not occur in the dried food but may occur in the non-dried food during refrigerated storage. Additionally, a microbiological criterion must clarify the point in the shelf life of a food product to which it is applicable. In this regard, differences in the APC of perishable food products at the beginning of their shelf life compared to the end of the shelf life have to be considered. For multi-component food products such as various types of sandwiches including chevita, bauru, cemita, bake and shark, doubles, chivito, banh mi, barros luco, and choripan, microbiologically sensitive components should be tested separately.

Microorganism(s) or toxin

Microorganisms described in a criterion may be foodborne human enteric pathogens, indicator organisms, food spoilage organisms or microbial toxins (ICMSF, 2002). Relevance of the organism to the food product and/or applied food process is important. In this respect, knowledge of microbial ecology of foods and microbial responses to food processes is most useful. Testing for pathogens is in the interest of public health especially in instances where evidence indicates that a specific organism poses a serious risk to human health. Tests for indicator and spoilage organisms are performed for several reasons including cost, ease of performance, rapidity and trend analysis (Stannard, 1997). For food manufacturers the cost of microbial testing is an important consideration for economic reasons. For example, relatively higher costs are incurred in the direct testing for pathogens. Far more tests can be performed for spoilage organisms and microbial indicators of improper (unhygienic) food handling than for specific foodborne pathogens. Testing for indicator organisms is much simpler to perform compared to testing for pathogens. Additionally, results from testing for indicator organisms are obtained relatively fast compared to results of pathogen testing thus permitting more rapid corrective action. Microbial indicators and spoilage organisms usually outnumber pathogens and signal unsatisfactory conditions when there are significant increases in microbial populations. Monitoring microbial numbers over time facilitates establishment of trends which, when analyzed, can reveal potential problems before they reach an uncontrollable level. In contrast, tests for pathogens or toxins frequent yield numerous negative results and any positive result necessitates immediate corrective action (Stannard, 1997).

Sampling plan

The sampling plan describes how samples of food products will be taken and assessed for acceptability based on microbiological criteria. There are two types of sampling plans namely, two-class sampling plan and three-class sampling plan. Both plans should describe the strategy for sampling the food products and state the action to be applied based on results of testing. The food industry usually applies the principles of 2-class and 3-class sampling plans as described by International Commission for the Microbiological Specifications for Foods (ICMSF, 1986). The following information provides a summary of 2-class and 3-class sampling plans.

Two-class sampling plan

    n = number of samples tested
    m = maximum acceptable level of target microorganism(s) or toxin(s)
    c = number of samples allowed to exceed m

Three-class sampling plan

n = number of samples tested

m = maximum level of target microorganism(s) or toxin(s) acceptable under GMP

M = level of target organism(s) or toxin(s) above which the food is defective (considered)

c = number of samples permitted to fall between m and M without the food lot being deemed unacceptable

In the food industry, the two-class sampling plan is applicable when testing is performed to detect (presence or absence) pathogens and c = 0. For example, in a two-class sampling plan for *L. monocytogenes* in a batch of ready-to-eat meat products, where five 25-g meat samples are pulled and tested and any sample testing positive is considered unacceptable, the following is applicable: n = 5, c = 0, m = not detected in 25 g.

The three-class sampling plan is applicable when using quantitative (enumeration) microbial testing. For example, if from five samples of a batch of refrigerated raw chicken breast meat, it is acceptable for the APC to exceed 5×10^6 CFU/g in three instances but not more than 10^7 CFU/g at all, the following is applicable: n = 5, c = 3, m = 5×10^6 CFU/g, M = 10^7 CFU/g.

Microbiological limit

The microbiological limit for a food product is a level which requires action if it is exceeded. Microbial limits should be realistic and established based on historical knowledge of the microbial ecology of raw materials. Knowledge of how processing, handling, storage and use of the product influence its microbial ecology is important. Food manufacturers in developing countries may be able to estimate microbiological limits for their products by examining the microbial levels in products with similar characteristics and for which processing and storage conditions are known. Microbial limits should account for levels that represent insufficient control or a risk to human health.

It is impractical to prove the absence of microorganisms in a food product; therefore, negative results of a detection test should be stated as "not detected" (ND) in the quantity of sample tested. For example, a negative result of a presence/absence test for a pathogen should be expressed as not detected in 50 g or not detected in 10 mL. A negative result of any quantitative test should be expressed as less than the limit of detection of the test. For example, if there are no bacterial colonies observed from analyzing a 1:10 dilution of ground beef (25 g beef in 225 mL of diluent) that result has to be stated as <10 CFU/gram, which is equal to ND in 0.1 gram.

Use of information on microbiological criteria

For ease of reference, suggested microbiological criteria for groups of food products that are part of the diets of consumers in developing countries are organized in tables below. The criteria in the tables are relevant to broad product categories and may not be fully applicable to some specific food products. For several multicomponent food products such as ifisashi (vegetables mixed with pounded groundnuts and beans), beef or chicken enchilada, banh xeo (crispy crepe filled pork, shrimp, and bean sprouts), Jamaican beef patties, kuih kosi (glutinous rice with coconut dessert filling), and tapado (coconut-milk-based soup with seafood), readers should consider checking the appropriate information in more than one table.

The numerical values for microbiological criteria are under two headings namely, "GMP" (good manufacturing practice) and "Maximum". The values under GMP are criteria expected immediately after manufacture of the food product under good manufacturing conditions. There are likely to be instances when samples may exceed the limits; however, the results may not warrant rejection of the food product. In this respect, due consideration should be given to all of the important factors in the entire sampling plan. The values under "Maximum" are those that are acceptable at any stage throughout shelf life of the food product. Numbers of pathogens higher that the maximum value suggest a possible food safety risk whereas higher numbers of an indicator organism or group suggest process failure or lapse in hygienic practices. In using the information in the tables, the following points should be considered:

1) Some developing countries might already have legislation for microbial criteria for some foods at specific points in production, processing, and storage. It is important for food manufacturers to ensure that their products comply with pertinent legislation.
2) For some foods microbial growth will occur throughout their shelf life. In this regard, if microbial growth reaches near "Maximum" levels early in the predicted shelf life, the food will most likely be spoiled or become a health risk.
3) The type of food and processing conditions can cause variations in numbers of any indicator organism. Therefore, the stated "Maximum" levels related to process failure may differ because of those variations.
4) The values relate to representative samples food such as 25 g for pathogen detection tests. Food samples include those taken from the food surface or core and with regard to multicomponent foods, selected food components.
5) Ingredients incorporated in a multicomponent food product after the processing step described in the tables are not taken into consideration and therefore must be accounted for separately.
6) No sampling plan or method is in the tables and should be developed and applied in accordance with the considerations previously stated in this chapter.

The information presented here is not aimed at providing microbiological criteria for a comprehensive list of food products with consideration of typically applied processing technologies. Food products are in broad groups based on their intended use, for example, "to be cooked" and "ready-to-eat". The values in the following tables represent suggested microbiological criteria for use as a basis for developing appropriate criteria for foods in developing countries. The authors assume the use of standard methods of analysis or equivalent validated food industry method. Guidance is provided on limits for pathogens, important toxins and indicator microorganisms relevant to each food group.

The authors advise food manufacturers in developing countries to avoid indiscriminate application of microbial testing and criteria. Application of all the stated tests for any

TABLE 5.8 Suggested microbiological criteria for raw meat products from developing countries.

Product types (examples):	Beef, pork, lamb, veal, mutton, goat, springbok, camel, buffalo, llama, rabbit, paca, guinea pig, etc. (see Table 5.1)
Storage condition:	Refrigerated or frozen
Intended use:	To be cooked

Microorganism	GMP	Maximum
Bacterial pathogens	Criteria for absence are not usually applicable	

Microorganism	GMP	Maximum
APC	$<10^5$	10^7
E. coli	$<10^2$	10^4
Yeast (sausages & marinated meats)	$<10^4$	10^6

GMP: values expected just after production using good manufacturing practices.
Maximum: maximum values acceptable at any point in the shelf life of the food. In application of microbiological testing and criteria, it is important to specify all elements of the microbiological criterion including the sampling plan. Values for microbial counts are expressed as colony forming units (CFU) per gram.

group of food products described in the following tables may not be necessary. In some instances, it may be appropriate to test for additional foodborne microorganisms and/or toxins as new pertinent scientific information becomes available. In application of microbiological testing and criteria, it is important to specify all elements of the microbiological criterion including the sampling plan, which is not in any table. Food manufacturers are urged to consult ICMSF (1986), ICMSF (2000) and other pertinent documents for principles and specific applications in sampling, microbiological analysis, and microbial ecology of foods. The tables contain the suggested criteria, right below which are comments on the use of indicator organisms for that specific category of foods and the associated pathogens that may be considered (Table 5.8).

Indicators & spoilage microorganisms in meat

The APC should be used as an index of product quality. Generally, numbers of microorganisms are higher in ground meats than on whole meat cuts. *E. coli* may be used an indicator of hygienic slaughter practices. Coliforms or *Enterobacteriaceae* are useful for trend analysis.

Pathogenic microorganisms in meat

Salmonella, Campylobacter, pathogenic *Escherichia coli* and parasites might be present; for pork, *Yersinia enterocolitica* is also an important pathogen (Table 5.8).

Indicators & spoilage microorganisms in raw poultry

The APC should be used as an index of product quality. Generally, numbers of microorganisms are higher in ground/reformed poultry meats than on whole birds or cut-up parts. *Pseudomonas* spp. are usually used as an index of quality for raw poultry (Fig. 5.7 and Table 5.9).

FIGURE 5.7 Selected meat, vegetables and poultry from developing countries. *Source: A. Gordon (2019).*

TABLE 5.9   Suggested microbiological criteria for raw poultry meats from developing countries.

Product types (examples):	Chicken, turkey, duck, goose, guinea fowl, ostrich, pigeon, emu, (see Table 5.3)
Storage condition:	Refrigerated or frozen
Intended use:	To be cooked

Microorganism	GMP	Maximum
Bacterial pathogens	Criteria for absence are not usually applicable	

Microorganism	GMP	Maximum
APC	$<10^5$	10^7
Pseudomonas spp.	$<10^5$	10^7
Yeast (marinated meats)	$<10^4$	10^6

GMP: values expected just after production using good manufacturing practices.
Maximum: maximum values acceptable at any point in the shelf life of the food. In application of microbiological testing and criteria, it is important to specify all elements of the microbiological criterion including the sampling plan. Values for microbial counts are expressed as colony forming units (CFU) per gram.

Pathogenic microorganisms in raw poultry meats

Salmonella, Campylobacter and other pathogens may be present. Monitoring the prevalence of those pathogens is useful in trend analysis.

Indicators & spoilage microorganisms in raw vegetables

The APC, *Enterobacteriaceae* and coliforms are likely to occur in high numbers on vegetables due to contamination from soil, wind-blown dust, insects, feral animals and/or poor agricultural practices. Developing country vegetables are no different (Table 5.10). Numbers of *E. coli* can be used to monitor (irrigation) water quality and the extent of hygienic handling. Vegetables must be visually checked for bacterial soft rots and mold growth. Monitoring for bacterial endospores is recommended if the vegetables will be canned.

Pathogenic microorganisms in raw vegetables

Raw vegetables (Table 5.11) can carry pathogens and washing alone may not totally remove them. Low prevalence and numbers of pathogens levels are usually found on leafy vegetables and criteria for specific pathogens are typically applied (Table 5.10).

TABLE 5.10 Suggested microbiological criteria for raw vegetables from developing countries.

Product types (examples):	Mustard greens, cabbage, spinach, callaloo, water cress, fresh herbs, blanched vegetables, onions, garlic, lettuce, etc. (**see** Table 5.11)
Storage condition:	Ambient, chilled or frozen
Intended use:	To be washed and cooked, or ready-to-eat

Intended use	Microorganism	GMP	Maximum
To be washed or cooked	Pathogenic bacteria	Criteria for absence are not usually applicable	
Ready-to-eat	*Salmonella* spp.	ND in 25 g	ND in 25 g
	L. monocytogenes	ND in 25 g	ND in 25 g
	STEC	ND in 25 g	ND in 25 g

Intended use	Microorganism	GMP	Maximum
To be washed or cooked	APC, *Enterobacteriaceae*, or coliforms	Criteria for absence are not applicable	
Ready-to-eat	*Escherichia coli*	20 to $<10^2$	10^2

GMP: values expected just after production using good manufacturing practices.
Maximum: maximum values acceptable at any point in the shelf life of the food. In application of microbiological testing and criteria, it is important to specify all elements of the microbiological criterion including the sampling plan. Values for microbial counts are expressed as colony forming units (CFU) per gram.
STEC: shiga-toxin-producing *Escherichia coli*.; ND: not detected.; APC: aerobic plate count.

TABLE 5.11   Some vegetables and vegetable products that are part of the diets in developing countries in four regions.

Region	Vegetable and vegetable products
Africa	Okra, corn, tomatoes, cucumbers, mustard greens, squash, pumpkins, carrots, cabbage, spinach
Caribbean	Callaloo, avocado, tomatoes, peppers, yams, cabbage, carrot, yams, pumpkin, okra, sweet potatoes, cucumbers, watercress
Central & South America	Avocado, tomatoes, onions, garlic, asparagus, yucca (cassava), arrowroot, peppers, yams, potatoes, cabbage, eggplant, chayote, corn
South & Southeast Asia	Luffa, green papaya, winter melon, cabbage, pumpkins, eggplant

Indicators and spoilage microorganisms in seafood

wThe APC is usually high in raw seafood and monitoring APC is important in trend analysis. *Escherichia coli* is typically used as an index of fecal contamination in shellfish.

Pathogenic microorganisms and toxins in seafood

Salmonella spp. and pathogenic vibrios including *Vibrio parahaemolyticus* may be present. Fish and shellfish harvested from warm waters may harbor pathogenic vibrios. Monitoring the prevalence of pathogens in seafood is important for trend analysis. Bivalve mollusks such as oysters, clams, and mussels are filter feeders and concentrate microorganisms (bacteria and viruses) and algal toxins from the marine environment. It is advisable to monitor the prevalence of bacterial pathogens and algal toxins. The occurrence of scombrotoxin (histamine), which typically occurs in scombroid fish, should be taken into consideration. PSP: paralytic shellfish poisoning; DSP: diarrhetic shellfish poisoning. ND: not detected. APC: aerobic plate count (Table 5.12 and Table 5.13).

Indicators and spoilage microorganisms in fruits and fruit products

The APC is usually high in raw products. Yeast and molds are important spoilage organisms for fruits (Table 5.14) and visual monitoring is appropriate. Fruits with low numbers of yeast will yield juice with a longer shelf life. Populations of *E. coli* are an indication of poor harvesting and processing practices.

Pathogenic microorganisms in fruit and fruit products

Pathogens are usually not prevalent on fruits but may be present as a result of issues with pre- and post-harvest handling as has been found over the last several years with outbreaks from papaya, mangos and cantaloupes (Table 5.15), among others (Petashea, 2016; Marler & Clarke, 2017; CCC, 2013). The presence of pathogens such as *Listeria monocytogenes* and *Salmonella* on papayas and mangos and the protozoan parasite *Cyclospora*

TABLE 5.12 Suggested microbiological criteria for seafood from developing countries.

Product types (examples):	Red snapper, flying fish, mackerel, conch, codfish, crabs, lobster, shrimp, mackerel, tuna, dolphin fish (mahi-mahi), grouper, kingfish, octopus, etc. (see Table 5.5)
Storage condition:	Refrigerated or frozen
Intended use:	To be cooked or ready-to-eat

Intended use	Microorganism/toxin	GMP	Maximum
To be cooked	Pathogenic bacteria	Criteria for absence not usually applicable	
	Scombrotoxin (in fish)	<50 ppm	50ppm
	PSP (shellfish flesh)	ND in 100 g	<80 µg/100 g
	DSP (shellfish flesh)	ND (bioassay)	ND (bioassay)
Ready-to-eat	*Vibrio parahaemolyticus*	ND in 25 g	10^2
	Salmonella spp.	ND in 25 g	ND in 25 g
	L. monocytogenes	ND in 25 g	ND in 25 g
	S. aureus (cold smoked fish)	<10^2	10^3
	Histamine (fish)	<50 ppm	50 ppm
	PSP (shellfish flesh)	ND in 100 g	<80 µg/100 g
	DSP (shellfish flesh)	ND (bioassay)	ND (bioassay)

Microorganism	GMP	Maximum
APC	<10^6	10^7
Escherichia coli	<10	10^3
Yeast (pickled/marinated)	<10^4	10^6

GMP: values expected just after production using good manufacturing practices.
Maximum: maximum values acceptable at any point in the shelf life of the food. In application of microbiological testing and criteria, it is important to specify all elements of the microbiological criterion including the sampling plan. Values for microbial counts are expressed as colony forming units (CFU) per gram.

TABLE 5.13 Some seafood that are part of the diets in developing countries in four regions.

Region	Seafoods
Africa	Crayfish, tuna, mussels, oysters, mackerel, lobster
Caribbean	Conch, codfish/saltfish, crabs, lobster, shrimp, mackerel, flying fish, red snapper, tuna, dolphin fish (mahi-mahi), grouper, kingfish, octopus
Central & South America	Corvina (corvina drum), grouper, king crab, tuna, lobster, sardines, salmon, octopus, conch, red snapper, crawfish, shrimp
South & Southeast Asia	Crabs, shrimp, tuna, salmon, catfish, tilapia

TABLE 5.14   Suggested microbiological criteria for fruit and fruit products from developing countries.

Product types (examples):	Mango, oranges, bananas, sugar apple, coconut, guavas, pineapple, watermelon, soursop, passion fruits, papayas, mangosteen, figs, lychee, marang, tamarind, breadfruit, etc. (see Table 5.15)
Storage condition:	Ambient, chilled or frozen
Intended use:	Ready to eat

Microorganism	GMP	Maximum
Pathogenic bacteria	Criteria for absence are not usually applicable	
Salmonella spp.	ND in 25 g	ND in 25 g
L. monocytogenes	<20	<100
STEC	ND in 25 g	ND in 25 g

Intended use	Microorganism	GMP	Maximum
Fruit juice (raw)	Yeast	$<10^3$	10^6
Fruit juice (pasteurized)	Yeast	<10	10^6

GMP: values expected just after production using good manufacturing practices.
Maximum: maximum values acceptable at any point in the shelf life of the food. In application of microbiological testing and criteria, it is important to specify all elements of the microbiological criterion including the sampling plan. Values for microbial counts are expressed as colony forming units (CFU) per gram.
ND: not detected. APC: aerobic plate count. STEC: shiga-toxin-producing *Escherichia coli*.

TABLE 5.15   Selected fruits and fruit products that are part of the diets in developing countries in four regions.

Region	Fruits and fruit products
Africa	Mango, papaya, oranges, figs, bananas, plantains, coconut, citrus
Caribbean	Mango, oranges, bananas, sugar-apple, coconut, guavas, pineapple, watermelon, soursop, passion fruits, papayas, citrus (limes), breadfruit, tamarind, plantains, carambola or star fruit, grapefruit
Central & South America	Grapes, plums, apples, pears, coconut, breadfruit, bananas, plantains, tomatoes, watermelon, avocado, mango, citrus, pineapple
South & Southeast Asia	Bitter gourd (bitter melon), durian, jackfruit, marang, soursop, rose apple, tamarind, citrus, mangosteen, lychee, pomelo, dragon fruit, langsat, pineapples

cayetanensis on raspberries (Gordon, 2016) have led to a rethink of the risk of these pathogens on fruits and fruit products. While pathogen growth is unlikely in highly acidic fruits, they may become contaminated from water used for irrigation or washing, from the environment on the farm or in the pack house, or from poor hygiene or sanitation during handling and packing. Other pathogens of note in fruits and fruit products include molds and, as such, monitoring for mold toxins, for example, patulin in apple juice, is recommended.

Indicators & spoilage microorganisms in dried cereal products

Wide variations of APC and numbers of *Enterobacteriaceae* and spore forming bacteria will occur based on type of cereal, harvesting conditions, and post-harvest handling and storage. Monitoring those indicators as well can be informative with regard to trend analysis. The concentration of bacterial spores should be low for products destined to be canned.

Pathogenic microorganisms in dried cereal products

Pathogens (spore forming and non-spore forming) may be present (Table 5.16). Monitoring those organisms can be important in trend analysis. Spores can survive the cooking process, but vegetative cells are destroyed. Improper cooling can result in germination and outgrowth of spores (especially endospores of *Bacillus cereus*) and toxin production. Heat stable toxins of *Bacillus* spp. and *Staphylococcus aureus* are not destroyed by reheating or further cooking. The occurrence of mold toxins should be taken into consideration.

Indicators & spoilage microorganisms in herbs, spices and seasonings

Herbs and spices that are not fumigated or irradiated can have APC and numbers of *Enterobacteriaceae*, which include *E. coli*, as high as 10^7 per gram. Molds and yeasts may also be present. For individual products, monitoring can provide useful information for trend analysis.

TABLE 5.16 Suggested microbiological criteria for dried cereals from developing countries.

Product types (examples):	Corn, rice, wheat, oats, cornmeal, sorghum, noodles, quinoa, sorghum, bread, tortilla, millet, barley (see Table 5.17)	
Storage condition:	Ambient	
Intended use:	To be cooked	

Microorganism	GMP	Maximum
Bacillus cereus	$<10^2$	10^4
Staphylococcus aureus	$<10^2$	10^4

Microorganism	GMP	Maximum
Molds	10^3	10^5
Escherichia coli	10^2	10^3

GMP: values expected just after production using good manufacturing practices.
Maximum: maximum values acceptable at any point in the shelf life of the food. In application of microbiological testing and criteria, it is important to specify all elements of the microbiological criterion including the sampling. Values for microbial counts are expressed as colony forming units (CFU) per gram.

Pathogenic microorganisms

A range of pathogens can occur in herbs, spices and vegetable seasonings, among these being spores of bacteria and fungi (mold and yeast) which are usually found in herbs and spices (Table 5.18).

Pathogenic microorganisms in nut and nut products

Pathogens such as *Salmonella* may be present in nuts (Tables 5.20 and 5.21), especially tree nuts, which are likely to be contaminated by animals (birds, reptiles, insects). Spores

TABLE 5.17 Selected cereals and cereal products that are part of the diets in developing countries in four regions.

Region	Cereal and cereal products
Africa	Wheat, corn, sorghum, millet, barley
Caribbean	Bread, cornmeal, roti
Central & South America	Wheat, oats, maize, quinoa, sorghum, tortilla
South & Southeast Asia	Rice, wheat, maize, barley, sorghum, noodles

TABLE 5.18 Suggested microbiological criteria for spices, herbs, and vegetable seasonings from developing countries.

Product types (examples):	Nutmeg, cinnamon, thyme, clove, ginger, bay leaves, hot peppers, curry powder, black pepper, oregano, cumin, coriander, ginger garlic, cinnamon, achiote, cilantro, thyme, etc. (see Table 5.19)
Storage condition:	Ambient
Intended use:	Ready to eat

Microorganism	GMP	Maximum
Salmonella spp.	ND in 25 g	ND in 25 g
Bacillus cereus	$<10^2$	10^4
Clostridium perfringens	$<10^2$	10^3

Microorganism	GMP	Maximum
Escherichia coli	<10	10^3
Molds	$<10^3$	10^4
Yeast	$<10^3$	10^6

GMP: values expected just after production using good manufacturing practices.
Maximum: maximum values acceptable at any point in the shelf life of the food. In application of microbiological testing and criteria, it is important to specify all elements of the microbiological criterion including the sampling. Values for microbial counts are expressed as colony forming units (CFU) per gram.

TABLE 5.19 Selected spices, herbs and vegetable seasonings that are part of the diets in developing countries in four regions.

Region	Spices, herbs and seasonings
Africa	Hot peppers, chilies, not native but widely used: cumin, cinnamon, sesame seeds, curry, coriander, ginger, garlic, ogilie (African seasoning made from dried seeds which have been steamed, fermented, then ground into a paste)
Caribbean	Nutmeg, cinnamon, thyme, clove, ginger, bay leaves, jerk seasoning, oregano, allspice, scotch bonnet peppers, moruga scorpion pepper, scallion, cilantro (chandon beni), curry
Central & South America	Cilantro, oregano, achiote, cumin, coriander seeds, chilies, vanilla
South & Southeast Asia	Mint, coriander, basil, lemon grass, black pepper, cloves, cinnamon, curry, ginger, peppers, coriander, nutmeg

TABLE 5.20 Suggested Microbiological Criteria for Selected Nuts and Nut Products from Developing Countries.

Product types (examples):	Peanuts, cashew, kola nuts, macadamia, dhoum/doum nut, dika nut, coconut, almond, brazil nut, macadamia, etc. (see Table 5.21)
Storage condition:	Ambient
Intended use:	Ready to eat

Microorganism	GMP	Maximum
Salmonella spp.	ND in 25 g	ND in 25 g
Aflatoxin	<4 ppb	<4 ppb

Microorganism	GMP	Maximum
Escherichia coli	<10	10^3
Yeast and molds	<10^3	10^4

GMP: values expected just after production using good manufacturing practices
Maximum: maximum values acceptable at any point in the shelf life of the food. In application of microbiological testing and criteria, it is important to specify all elements of the microbiological criterion including the sampling. Values for microbial counts are expressed as colony forming units (CFU) per gram.

TABLE 5.21 Selected nuts and nut products that are part of the diets in developing countries in four regions.

Region	Nuts and nut products
Africa	Peanuts, cashew, kola nuts, macadamia, dhoum/doum nut, dika nut
Caribbean	Coconuts, coconut milk or cream, copra, coconut oil, almond, cashew, peanuts
Central & South America	Peanut, cashew, walnut, brazil nut, macadamia, palm chestnut, palm oil
South & Southeast Asia	Cashew oil, Illipe nuts, satay sauces, tempeh bongkrek, copra, coconut oil

of bacteria and molds are typically found in those products. Testing for aflatoxin is recommended.

Indicators & spoilage microorganisms in nut and nut products

The APC can be relatively low; however, higher bacterial counts are usually associated with damaged shells and contamination with soil and dust.

Indicators & spoilage microorganisms in milk and dairy products

For milk and dairy products from developing countries, the application of criteria depends on the extent of thermal treatment applied (Tables 5.22 and 5.23). Coliforms or *Enterobacteriaceae*, and *Escherichia coli* are used as indicators of quality and hygienic processing (Table 5.22). The APC can provide useful information for trend analysis but is of no value with regard to fermented products with live microorganisms

TABLE 5.22 Suggested microbiological criteria for some milk and dairy products from developing countries.

Product types (examples):	Cow's milk, fresh cheeses, goat, sheep and camel cheese as well, various fermented milks, tuttis (yogurt), gibna, ayib, dulce de leche, tres leches cake, queso blanco, queso fresco, chanco cheese, panquehue, ghee (clarified butter), yogurt, dadiah (fermented milk), paneer, dangke, etc. (see Table 5.23)
Storage condition:	Refrigerated
Intended use:	Ready to eat

Microorganism		GMP	Maximum
Salmonella spp.		ND in 25 g or mL	ND in 25 g or mL
Listeria monocytogenes		ND in 25 g or mL	ND in 25 g or mL
Staphylococcus aureus		<20	10^3
STEC		ND in 25 g or mL	ND in 25 g or mL

Items	Microorganism	GMP	Maximum
Soft Cheeses (raw milk)	*Escherichia coli*	$<10^2$	10^4
	Listeria spp.	<20	10^2
Other Cheeses	*Enterobacteriaceae*	$<10^2$	10^4
	Escherichia coli	<10	10^3
Milk & Cream (pasteurized)	*Enterobacteriaceae*	<10	10^2
Other Milk products (pasteurized)	*Enterobacteriaceae*	<10	$<10^4$

GMP: values expected just after production using good manufacturing practices.
Maximum: maximum values acceptable at any point in the shelf life of the food. In application of microbiological testing and criteria, it is important to specify all elements of the microbiological criterion including the sampling plan. Values for microbial counts are expressed as colony forming units (CFU) per gram or milliliter.

TABLE 5.23 Selected milk and dairy products that are part of the diets in developing countries in four regions.

Region	Milk and dairy products
Africa	Cow's milk, fresh cheeses (klila, warankasi, kariesh, ayib, gibna) local goat, sheep and camel cheese as well, fermented milk (Iben, zabady, rob, biruni, gariss, sussa/suusac, ergo, ititu, kule naoto, amabere amaruranu, kivuguto, mursik, amasi), butter, smen (fermented, clarified butter), tuttis (yogurt), cheese (wagasi)
Caribbean	Cow's milk, eggnog, queso blanco (Dominican Republic), custard, processed cheese, condensed milk, evaporated milk, yogurt, flan
Central & South America	Dulce de leche, tres leches cake, queso blanco, queso fresco, chanco cheese, panquehue
South & Southeast Asia	Ghee (clarified butter), yogurt, dadiah (fermented milk), paneer, dangke

Pathogenic microorganisms in milk and dairy products

Raw milk can be a vehicle for several pathogens including *Salmonella, Campylobacter, Listeria monocytogenes*, pathogenic *Escherichia coli* and *Staphylococcus aureus*. Unless there is under-processing (improper pasteurization) and/or post-processing contamination, vegetative pathogens will be absent in milk and cream. Testing of milk and milk products for *Listeria* and *Salmonella* is not usually routinely performed unless indicated by historical findings. Low levels of thermoduric microorganisms and bacterial spores usually are present.

Prepared meals are becoming a more important part of the food choices in some developing countries, particularly in more developed urban areas or those in which there is a wealthier population who have a similar lifestyle and expectations to their developed country counterparts. This is true in many parts of the Americas, Asia and the major retail supermarkets in African urban centers. Prepared meals such as those shown in Fig. 5.8 are prepared by caterers for various functions or events, are the norm in hotels and would be among those produced daily by airline catering firms throughout the developing world. Tables 5.24 and 5.25 present suggested microbiological criteria for and indicate some of the types of meals that fall in this category.

Indicators & spoilage microorganisms in prepared meals

The APC can vary substantially among prepared meals due to types of meal components, initial microbial quality, the extent of processing, post-process handling and stage of shelf life. *Enterobacteriaceae* are useful for monitoring hygienic practices. Gram negative organisms in high numbers ($>10^7$) usually cause spoilage. However, an APC of 10^8 due to lactic acid bacteria (i.e. in some prepared sliced fermented meats) may be typical, are not usually a food safety consideration and cause no objectionable change in food quality. Therefore, food industry practitioners should understand the nature of specific products and exercise caution in using and interpreting the APC for prepared meals (Table 5.24).

FIGURE 5.8 Meals prepared for catering in a developing country.

Pathogenic microorganisms in prepared meals

The number and types of pathogens in prepared meals (Table 5.24) will varying dependent on their composition, handling and temperature history. While there have been several instances of pathogens in prepared foods causing illnesses, this is not to be expected

TABLE 5.24 Suggested microbiological criteria for selected prepared meals from developing countries.

Product types (examples): Cooked meats, poultry and fish products, meat stews, rice dishes, soups, curried meats, jerk seasoned meats, pastries, meat patties, fish cakes and vegetables, airline meals etc. (see Table 5.25)

Storage condition: Refrigerated or frozen

Intended use: Ready to eat or reheated

Microorganism	GMP	Maximum
Salmonella spp.	ND in 25 g	ND in 25 g
Clostridium perfringens	$<10^2$	10^3
Listeria monocytogenes	ND in 25 g	ND in 25 g
Bacillus cereus	$<10^2$	10^4
Staphylococcus aureus	<20	10^3
Vibrio parahaemolyticus	ND in 25 g	10^2
Pathogenic *Escherichia coli*	ND in 25 g	ND in 25 g
Histamine (in scombroid fish)	<50ppm	50ppm
Paralytic shellfish poison	ND in 100 g	$<80\,\mu$g in 100 g
Diarrhetic shellfish poison	ND in bioassay	ND in bioassay
Microorganism	**GMP**	**Maximum**
APC (heat treated foods)	$<10^4$	Variable (product dependent)
Enterobacteriaceae	$<10^2$	10^4
Escherichia coli	<10	10^2

GMP: values expected just after production of the using good manufacturing practices.
Maximum: maximum values acceptable at any point in the shelf life of the food. In application of microbiological testing and criteria, it is important to specify all elements of the microbiological criterion including the sampling plan. Values for microbial counts are expressed as colony forming units (CFU) per gram or milliliter.

as proper preparation and handling, including refrigerated storage or keeping the food above 60 °C, should eliminate organisms of concern or prevent them from growing. Potential environmental contaminants such as *Listeria monocytogenes* must be excluded through proper hygienic zoning of the production facility and sanitation of the preparation areas. Temperature abuse (hot meals not being kept hot enough) potentiates the growth of *Clostridium perfringens*, increasing the risk of food poisoning. Similarly, there is increased risk of *Bacillus cereus* food poisoning from temperature abused rice dishes. Improper refrigerated storage can facilitate the growth of *Staphylococcus aureus*. It is also important to monitor for the prevalence of algal toxins in shellfish and scombrotoxin (histamine) in scombroid fish.

TABLE 5.25 Selected prepared meals that are part of the diets in developing countries in four regions.

Region	Prepared Meals
Africa	Fufu, jollof rice, ugali, gari, dukunoo, bunny chow, bobotie, kenkey, banku, akamu, waake, moi-moi, agidi, koko, koose, yam and plantain, gari, boiled rice, fried fish, chicken/beef stew, roasted sausage (beef/chicken), nkontmre, light soup, okra soup, groundnut soup, tomato stew, palm nut soup, nshima, buka fish, ifisashi (vegetables mixed with pounded groundnuts and beans), sazda, maize porridge (pap).
Caribbean	Roti and curry (chicken, beef, goat), soups, stews, akee or green bananas and salt fish, oxtail stew, bake and shark, jerk chicken, callaloo, peas and rice, "cook-up" rice, pepper pot, metemgee, paella, banana fritters, fried plantains, Jamaican beef patties, Johnny cakes, funjii, black pudding, souse, conkey, bread pudding, pholourie, fruit cakes, cassava pone
Central & South America	Corvina, ceviche, tamales, empanadas, humitas (pureed corn cooked in husks), stews (ajiaco and sancocho), tapado (coconut-milk-based soup with seafood), charque de llama, enchilada, chorizo, platanos fritos, baleada, arepas, churros, tacos, paella,
South & Southeast Asia	Curries, dahl and rice, satays, banh cuon (roll cake), pork/chicken adobo, sticky rice, boat noodles, pho (broth with rice noodles, chichen, beef and herbs), bun cha (grilled seasoned pork slices or patties), roti canai (flatbread), tandoori chicken, tempeh, adobo, nasi lemak (cooked rice in coconut milk), otak (fish cakes in banana leaf), kuih kosi (glutinous rice and coconut dessert filling), amok fish, banh xeo (crispy crepe filled with pork, shrimp, bean sprouts), pad thai, naan

Microbiological criteria for developing country beverages

The range and type of beverages consumed in developing countries varies substantially across and within regions of the world. Some of the beverages that are part of the diets in developing countries are shown in Table 5.26. The consumption of beverages made fresh at the point of sale is a very common practice in almost all developing countries. The wide array of hot and cold beverages including teas, coffee, beer and blends of fruit and vegetable juices are gaining much attention from investors who recognize the market potential of those beverages. This interest in commercializing beverages, especially tropical fruit juices, is largely due to the diversification of tastes and preferences for these products in developed countries, driven by tourism. Also, the perceived health benefits of some herbal teas and plant extracts from developing countries is a growing trend in the health food industry globally.

While there are increasingly more shelf stable product offerings in the Americas and parts of Asia, many commercially available beverages in developing countries are heat pasteurized or canned to ensure their safety for consumption and shelf stability. However, discerning consumers in countries with developed economies are demanding more fresh foods and beverages which are not rigorously processed (Jackson-Davis et al., 2018). This is because heat processing can result in loss of color, flavor, and heat labile vitamins in beverages such as fruit or vegetable juices (Elez-Martínez et al., 2006). For health conscious consumers, a viable alternative to heat pasteurization is the incorporation of certain plant essential oils or essential oil components into beverages such as juices to destroy human

TABLE 5.26 Selected beverages that are part of the diets in developing countries in four regions.

Region	Beverages
Africa	Beer, teas (mint, black, green, sweet milky chai, lemongrass, kenkiliba, ginger, hibiscus (karkadeh)), hibiscus juice (bissap), coffee, tamarind juice, baobab fruit juice (boeye), masala chai
Caribbean	Beer, rum, ginger beer, sorrel, mauby, irish moss/sea moss, coffee, sugar cane juice, herbal ("bush") tea, pina colada, tamarind juice, citrus juices (orange, lime, grapefruit), fruit juices (mango, pineapple, guava), malt beverages, coconut water, "soft" (carbonated) drinks (sodas), soursop juice, acerola, golden apple (June plum) juice, mango juice
Central & South America	Beers, rum, coffee, fruit smoothies (refrescos), fresh fruit juices, sodas, coconut water, sugar cane juice, tea, soursop (guanabana) juice, acerola, sugar cane juice, fruit juices (mango, pineapple, guava)
South & Southeast Asia	Tamarind juice, citrus juice (lime, orange), lychee drink, sugar cane juice, coconut water, beer, coffee, fruit shakes (pineapple, orange, mango, coconut, dragon fruit, melon, guava), teh tarik (black tea and condensed milk)

enteric pathogens (Mendonca et al., 2018; Thomas-Popo et al., 2019) to enhance microbial safety while allowing retention of heat labile nutrients in those juices.

Raw juices can be vehicles for human enteric pathogens and several foodborne disease outbreaks involved unpasteurized juice contaminated with pathogens including *Escherichia coli* O157:H7 and *Salmonella enterica* (Danyluk et al., 2012). Fresh fruits and vegetables are all susceptible to microbial contamination with enteric pathogens from environmental sources such as soil, dirt, fecal material from animals or insects and windblown dust (Doyle and Erickson, 2012). During pressing or blending of the fresh product, pathogens can contaminate the juices. While proper refrigeration can retard or in some instances stop microbial growth, enteric pathogens such as *Escherichia coli* O157:H7, *Salmonella enterica*, and *Listeria monocytogenes* have survived for extended time periods in refrigerated fruit juices (Duan and Zahao, 2009; Mosqueda-Melgar et al., 2008; Thomas-Popo et al., 2019). In this respect, all juices from developing countries (Table 5.26) have to undergo some validated antimicrobial process such as pasteurization, canning or other means of ensuring their safety to gain acceptance from international buyers. Criteria such as those in Table 5.27 are therefore useful.

Indicators & spoilage microorganisms in fruit drinks & alcoholic beverages

Routine checking of raw materials for defects, good manufacturing practices, monitoring process hygiene, and ensuring consistent product formulation are crucial for preventing product spoilage. The dominant spoilage microflora consists of yeasts, molds and lactic acid-producing bacteria (Table 5.27). Checks such as incubation of final products and visual examinations for signs of microbial spoilage including gas production, haze or sediment are important for quality control.

TABLE 5.27 Suggested microbiological criteria for fruit drinks and alcoholic beverages from developing countries.

Product types (examples):	Fruit drinks (e.g. mango, tamarind, pineapple, guava, orange), beers, cocktails, etc. (see Table 5.3)
Storage condition:	Refrigerated or ambient
Intended use:	Ready to drink

Microorganism	GMP	Maximum
Bacterial pathogens	Criteria for absence are not usually applicable	

Microorganism	GMP	Maximum
Lactic acid bacteria	<1 in 100 mL	Signs of growth
Yeast (filtered or heat treated)	<1 in 100 mL	Signs of growth
Lactic acid bacteria	<10	Signs of growth
Yeast (not filtered or heat treated)	<10	Signs of growth
Escherichia coli	<1 in 100 mL	10 in 100 mL

GMP: values expected just after production using good manufacturing practices.
Maximum: maximum values acceptable at any point in the shelf life of the food. In application of microbiological testing and criteria, it is important to specify all elements of the microbiological criterion including the sampling plan. Values for microbial counts are expressed as colony forming units (CFU) per mL.

Pathogenic microorganisms in fruit drinks & alcoholic beverages

Enteric pathogens do not survive well in non-alcoholic, highly acid beverages, such as most fruit juices. The exception to this has been *E. coli* which has been found in apple cider, leading to a foodborne illness outbreak (Peters, 1998). Distilled alcoholic beverages are free of pathogens which are unable to survive in the hostile alcoholic environment. However, fermented alcoholic beverages have been shown to facilitate the survival of *B. cereus*, *E.coli* O157:H7 and coliforms (Kim et al., 2014; Jeon et al., 2015), while non-alcoholic beer has supported the survival of *Salmonella* Typhimurium and *S. aureus* (Menz et al., 2011). This indicates that the need to reconsider the long-held traditional view that alcoholic beverages are free of pathogens.

Specific challenges for entry of developing countries products into the global food trade

As has been discussed in earlier chapters, the specific challenges that developing countries face to participating in the global food trade are mainly linked to satisfying food safety requirements for the production, processing, handling, storage and distribution of food products to meet the standards of potential international buyers. Even when such requirements are met, consistent efforts must be made to maintain the expected safety and quality of food exports. Failure to do so has resulted in severe economic consequences for some countries. For example, in Peru a cholera outbreak involving *Vibrio cholerae* in 1991 caused more than $700 million in lost exports of fish and fish products (CDC, 1991). In 1996 and 1997 an outbreak of *Cyclospora* in raspberries from Guatemala decimated the Guatemalan raspberry industry; foreign countries including the United States stopped importing Guatemalan raspberries and the number of raspberry growers in Guatemala decreased drastically from eighty-five to three from 1996 to 2002 (Calvin et al., 2003).

Generally, in developing countries the major areas that need improvement to meet global food standards are food production, processing and marketing systems and food control resources and infrastructure. The food systems in developing countries are often more complex than is expected. Depending on the product and the country, they may be characterized as very fragmented and driven by many small producers with food products passing through many food handlers and small distributors. Such extensive handling of food increases the risk of food contamination and adulteration. Many growers and food handlers lack knowledge and expertise in Good Agricultural Practice (GAP), food hygiene and safe food handling practices. This in turn increases the risk of contamination during harvest, post-harvest handling, processing and storage of food. This problem may be further exacerbated by inadequate facilities and infrastructure including a lack or shortage of potable water, power outages, and poor cold storage facilities.

There are wide variations in size of food processing facilities in developing countries. Processing facilities range from state-of-the-art to very small cottage scale operations that make traditional foods for the local market. Small and very small operations may sometimes lack the infrastructure and capability to implement and sustain food safety and quality assurance systems (FSQS). It is not uncommon, therefore, to find small operators or groups being supported by Governments or multilateral agencies because they create employment opportunities for residents in various regions and generate income for the owners. A major challenge for developing countries is to improve the capacity of these small operations to accommodate new scientific technology to assist them in meeting the standards of export markets. Many medium sized food business also find it challenging to comply with the food safety and quality system requirements of United States, Canadian and European Union (EU) importers regarding evidence of compliance with standards for labeling, residues and allergens. Successful exporters therefore often require assistance (at least initially) by food safety practitioners with a knowledge of the requirements and the ability to help firms modify their operations to implement sustainably compliant FSQS systems.

Many developing countries have inadequate food control infrastructure. This is often due to insufficient resources and inadequate management. This situation becomes worse when several agencies participate in food control with little or no interaction or sharing of information. As discussed in Chapter 1, many developing country food control laboratories are under-funded and in need of updated equipment and trained analytical staff. In several instances, there is no unified strategy or direction for food safety in some countries. This in turn results in inefficient utilization of limited resources. Additionally, compliance policies in food control systems may be poorly developed or inadequate. Henson and Loader (2001) and Jaffee et al. (2019) cited outdated laws, lack of knowledge sharing among agencies administering food safety issues, and lack of awareness of standards and quality as some of the main impediments to effectively participation of developing countries in international food trade.

For developing countries to participate more in the global food trade, their food control systems must be enhanced to involve decision-making processes based on science. To ensure the effectiveness of food control systems, there is need for involvement of knowledgeable and trained professionals in several disciplines. These include areas such as agricultural sciences, food science and technology, chemistry, biochemistry, microbiology, veterinary science, food law, quality assurance, auditing, and epidemiology. Food regulatory agencies need to consider and incorporate science and risk-based approaches to food safety. Not all developing countries will have adequate resources to develop and sustain efficient food control systems. In this regard, assistance from international partners is crucial and should include the following: i) development of national food safety policies and strategies, ii) updating food legislation, regulations, standards, and codes of hygienic practice, iii) implementation of food inspection programs, iv) application of HACCP-based food safety and quality assurance systems, v) establishment or improvement of food analysis capabilities, vi) development and delivery of food safety education and training programs, and vii) establishment and/or strengthening of foodborne disease surveillance activity.

References

Abdissa, R., Haile, W., Fite, A.T., Beyi, A.F., Agga, G.E., Edao, B.M., et al., 2017. Prevalence of *Escherichia coli* O157:H7 in beef cattle at slaughter and beef carcasses at retail shops in Ethiopia. BMC Infect. Dis. 17, 277. Available from: https://doi.org/10.1186/s12879-017-2372-2.

Adeleye, S.A., Braide, W., Chinakwe, E.C., Esonu, C.E., Uzoh, C.V., 2017. Chicken meat, beef and vegetables: potential sources of *Campylobacter jejuni* contamination in Imo State, Nigeria. Sust. Food Prod. 3, 63–71.

Adesiyun, A.A., 1993. Prevalence of Listeria spp., *Campylobacter* spp. *Salmonella* spp. *Yersinia* spp., toxigenic *Escherichia coli* on meat and seafoods in Trinidad. Food Microbiol. 10, 395–403.

Ahmed, S.S.T.S., Tayeb, B.H., Ameen, A.M., Merza, S.M., Sharif, Y.H.M., 2017. Isolation and molecular detection of Listeria monocytogenes in minced meat, frozen chicken and cheese in Duhok Province, Kurdistan Region of Iraq. J. Food Microbiol. Saf. Hyg. 2, 118. Available from: https://doi.org/10.4172/2476-2059.1000118.

Ajayi, A., Smith, S.I., 2019. Recurrent cholera epidemics in Africa: which way forward? A literature review. Infection 47 (3), 341–349.

Altekruse, S.F., Stern, N.J., Fields, P.I., Swerdlow, D.L., 1999. *Campylobacter jejuni*-an emerging food borne pathogen. Emerg. Infect. Dis. 5, 28–35.

Amusan E.E., Sanni A.I., Isolation and identification of *Listeria monocytogenes* in fresh croaker (*Pseudotolithus senegalensis*). IOP Conf. Series: Earth and Environmental Science 210: 012004, 2019. doi:10.1088/1755-1315/210/1/012004

Anderson, J.M., Baird-Parker, A.C., 1975. A rapid and direct plate method for enumerating *Escherichia coli* biotype I in food. J. Appl. Bacteriol. 39, 111–117.

Andersson, A., Ronner, U., Granum, P.E., 1995. What problems does the food industry have with the spore-forming pathogens *Bacillus cereus* and *Clostridium perfringens*? Int. J. Food Microbiol. 28 (2), 145–155.

American Public Health Association (APHA), 2013 Compendium of methods for the microbiological examination of foods, In: Y. Salfinger and M. Tortorello (Eds.). DOI: 10.2105/MBEF.0222.074.

Bell, C., Kyriakides, A., 2002. Salmonella: a practical approach to the organism and its control in foods. Blackwell Science Ltd, London, UK.

Bellou, M., Kokkinos, P., Vantarakis, A., 2013. Shellfish-borne viral outbreaks: a systematic review. Food Environ. Virol. 5 (1), 13–23.

Bergamini, A.M.M., Simões, M., Irino, K., Gomes, T.A.T., Guth, B.E.C., 2007. Prevalence and characteristics of shi-ga toxin producing *Escherichia coli* (STEC) strains in ground beef in Sao Paulo, Brazil. Brazilian J. Microbiol. 38, 553–556. ISSN 1517-8382.

Beuchat, L.R., 1973. Interacting effects of pH, temperature, and salt concentration on growth and survival of *Vibrio parahaemolyticus*. App. Microbiol. 25 (5), 844.

Beuchat, L.R., 1974. Combined effects of water activity, solute, and temperature on the growth of *Vibrio parahaemolyticus*. Appl. Microbiol. 27 (6), 1075–1080.

Beuchat, L.R., 1996. Pathogenic microorganisms associated with fresh produce. J. Food Prot. 59 (2), 204–216.

Bhaduri, S., Cottrell, B., 2004. Survival of cold-stressed *Campylobacter jejuni* on ground chicken and chicken skin during frozen storage. Appl. Environ. Microbiol. 70, 7103–7109.

Birollo, G.A., Reinheimer, J.A., Vinderola, C.G., 2001. Enterococci vs non-lactic acid microflora as hygiene indicators for sweetened yoghurt. Food Microbiol. 18 (6), 597–604.

Böhnel H., 2002. Household biowaste containers (bio-bins) – potential incubators for *Clostridium Botulinum* and *Botulinum*, Neurotoxins 140.

Brown, C.M., Cann, J.W., Simons, G., Fankhauser, R.L., Thomas, W., Parashar, U.D., et al., 2001. Outbreak of nor-walk virus in a Caribbean island resort: application of molecular diagnostics to ascertain the vehicle of infection. Epidemiol. Infec. 126 (3), 425–432.

Calvin, L., Flores, L., Foster, W., 2003. Safety in food security and food trade – case study: guatemalan raspberries and Cyclospora. Int. Food Pol. Res. Ins. 10.

Camu, N., De Winter, T., Verbrugghe, K., Cleenwerck, I., Vandamme, P., Takrama, J.S., et al., 2007. Dynamics and biodiversity of populations of lactic acid bacteria and acetic acid bacteria involved in spontaneous heapfer-mentation of cocoa beans in Ghana. Appl. Environ. Microbiol. 73 (6), 1809–1824.

Camu, N., De Winter, T., Addo, S.K., Takrama, J.S., Bernaert, H., De Vuyst, L., 2008. Fermentation of cocoa beans: influence of microbial activities and polyphenol concentrations on the flavour of chocolate. J. Sci. Food Agric. 88 (13), 2288–2297.

Caprioli, A., Morabito, S., Brugere, H., Oswald, E., 2005. Enterohaemorrhagic *Escherichia coli*: emerging issues on virulence and modes of transmission. Vet. Res. 36, 289–311.

Caribbean Regional Epidemiology Centre (CAREC), 2015. Unpublished data. CAREC, Port of Spain, Trinidad and Tobago, W.I.

Casey, N.H., Van Niekerk, W.A., Webb, E.C., 2003. Goats meat. In: Caballero, B., Trugo, L., Finglass, P. (Eds.), Encyclopedia of Food Sciences and Nutrition. Academic Press, London, pp. 2937–2944.

CBS News, Norovirus outbreak sickens hundreds of cruise ship passengers, crew members, 2019. < https://www.cbsnews.com/news/norovirus-outbreak-hits-royal-caribbean-cruise-ship-277-people-sick-passengers-crew/ > (accessed 24.01.2020).

CDC (Centers for Disease Control and Prevention), Surveillance of foodborne disease outbreaks – United States: 1988 – 1992. Morbidity and Mortality Weekly Reports, Surveillance Summaries, 45 (SS-5), 1996, 1-55. < https://www.cdc.gov/mmwr/preview/mmwrhtml/00044241.htm?c_cid=journal_search_promotion 2018 > (accessed 30.04.2019).

CDC (Centres for Disease Control and Prevention), Multistate outbreak of salmonella infections linked to imported Maradol papayas, 2017. < https://www.cdc.gov/salmonella/kiambu-07-17/index.html > (accessed 01.05.2019).

CDC (Centers for Disease Control and Prevention), Foods linked to food poisoning, 2018. < https://www.cdc.gov/foodsafety/foods-linked-illness.html > (accessed 08.05.2019).

CPSI (Centers for Science and Public Interest), Global and local: food safety around the world, 2005. < https://cspinet.org/sites/default/files/attachment/global.pdf > (accessed 04.02.2020).

CPSI (Centers for Science and Public Interest), Outbreak alert! 2009. < http://www.cspinet.org/new/pdf/out-breakalertreport09.pdf > (accessed 16.06.2019)

Cean, A., Stef, L., Simiz, E., Julean, C., Dumitrescu, G., Vasile, A., et al., 2015. Effect of human isolated probiotic bacteria on preventing *Campylobacter jejuni* colonization of poultry. Foodborne Pathog. Dis. 12, 122–130.

Chaibenjawong, P., Foster, S.J., 2011. Desiccation tolerance in Staphylococcus aureus. Arch. Microbiol. 193 (2), 125–135.

Colello, R., Ruiz, M.J., Padín, V.M., Rogé, A.D., Leotta, G., Padola, N.L., et al., 2018. Detection and characterization of *Salmonella* serotypes in the production chain of two pig farms in Buenos Aires Province, Argentina. Front. Microbiol. 9, 1370. Available from: https://doi.org/10.3389/fmicb.2018.01370.

Conner, D.E., Hall, G.S., 1994. Efficacy of selected media for recovery of *Escherichia coli* O157:H7 from frozen chicken meat containing sodium chloride, sodium lactate, or polyphosphate. Food Microbiol. 11, 337–344.

Cutrim, C.S., de Barros, R.F., da Costa, M.P., Franco, R.M., Conte-Junior, C.A., Cortez, M.A.S., 2016. Survival of *Escherichia coli* O157:H7 during manufacture and storage of traditional and low lactose yogurt. LWT - Food Sci. Technol. 70, 178–184.

Dadi, L., Asrat, D., 2008. Prevalence and antimicrobial susceptibility profiles of thermotolerant *Campylobacter* strains in retail raw meat products in Ethiopia. Ethiop. J. Health Dev. 22 (2), 195–200.

Dalton, C.B., Austin, C.C., Sobel, J., Hayes, P.S., Bibb, W.F., Graves, L.M., et al., 1997. An outbreak of gastroenteritis and fever due to Listeria monocytogenes in milk. N. Engl. J. Med. 336, 100–105.

Danyluk M.D., Goodrich-Schneider R.M., Schneider K.R., Harris L.J., Worobo R.W., Outbreaks of foodborne disease associated with fruit and vegetable juices, 1922–2010. Institute of Food and Agricultural Sciences (IFAS), University of Florida Extension, FSHN, 2012, 12-04. < https://ucfoodsafety.ucdavis.edu/sites/g/files/dgvnsk7366/files/inline-files/223883.pdf > (accessed 28.01.2020).

De Schrijver, K., Buvens, G., Posse, B., Van den Branden, D., Oosterlynck, O., De Zutter, L., et al., 2008. Outbreak of verocytotoxin-producing E. coli O145 and O26 infections associated with the consumption of ice cream produced at a farm, Belgium. Euro. Surveil. Bull. Eur. Sur. Mal. Transm. Eur. Commun. Dis. Bull 13, 8041.

Deurenberg, R.H., Vink, C., Kalenic, S., Friedrich, A.W., Bruggeman, C.A., Stobberingh, E.E., 2007. The molecular evolution of methicillin-resistant Staphylococcus aureus. Clin. Microbiol. Infect. 13, 222–235.

Dewey-Mattia, D., Manikonda, K., Hall, A.J., Wise, M.E., Crowe, S.J., 2018. Surveillance for foodborne disease outbreaks — United States, 2009–2015. MMWR Surveill. Summary 67 (10), 1–11. Available from: https://doi.org/10.15585/mmwr.ss6710a1.

Doménech-Sánchez, A., Juan, C., Perez, J.L., Berrocal, C.I., 2011. Unmanageable norovirus outbreak in a single resort located in the Dominican Republic. Clinic. Microbiol. Inf. 17 (6), 952–954.

Doyle, M.E., Roman, D.J., 1982. Response of *Campylobacter jejuni* to sodium chloride. Appl. Environ. Microbiol. 43 (3), 561–565.

Doyle, M.P., Beuchat, L.R., Montville, T.J. (Eds.), 1997. Food microbiology: fundamental and frontiers. ASM Press, Washington, DC.

Doyle M.E., Archer J., Kaspar C.W., Weiss R., Human illness caused by E. coli 0157:H7 from food and non-food sources, University of Wisconsin Food Research Institute Briefings, 2006. < https://fri.wisc.edu/files/Briefs_File/FRIBrief_EcoliO157H7humanillness.pdf > (accessed 30.04.2019).

Doyle, M.P., Beuchat, L.R., Brary, I. (Eds.), 2007. Food microbiology: fundamentals and frontiers. 3rd ed ASM Press, Washington, DC.

Doyle, M.P., Erickson, M.C., 2012. Opportunities for mitigating pathogen contamination during on-farm food production. Int. J. Food Microbiol. 152 (3), 54–74. Available from: https://doi.org/10.1016/j.ijfoodmicro.2011.02.037.

Duan, J., Zhao, Y., 2009. Antimicrobial efficiency of essential oil and freeze-thaw treatments against *Escherichia coli* O157:H7 and *Salmonella enterica* Ser. Enteritidis in strawberry juice. J. Food Sci. 74 (3), 131–137. Available from: https://doi.org/10.1111/j.1750-3841.2009.01094.x.

EFSA(European Food Safety Authority), 2013. Scientific opinion on VTEC seropathotype and scientific criteria regarding pathogenicity assessment, Anel Bio. Hazards (BIOHAZ). Eur. Food Saf. Auth. J. 11 (4), 3138. Available from: http://onlinelibrary.wiley.com/doi/10.2903/j.efsa.2013.3138/epdf (accessed 30.04.2019).

Elbashir, S., Parveen, S., Schwarz, J., Rippen, T., Jahncke, M., DePaola, A., 2018. Seafood pathogens and information on antimicrobial resistance: a review. Food Microbiol. 70, 85–93. Available from: https://doi.org/10.1016/j.fm.2017.09.011.

El-Malek, A.M.A., Ali, S.F.H., Hassanein, R., Mohamed, M.A., Elsayh, K.I., 2010. Occurrence of *Listeria* species in meat, chicken products and human stools in Assiut city, Egypt with PCR use for rapid identification of *Listeria monocytogenes*. Vet. World 3 (8), 353–359.

Elez-Martínez, P., Soliva-Fortuny, R.C., Martín-Belloso, O., 2006. Comparative study on shelf life of orange juice processed by high intensity pulsed electric fields or heat treatment. Eur. Food Res. Technol. 222, 321. Available from: https://doi.org/10.1007/s00217-005-0073-3.

Emmett, R., Akkersdyk, S., Yeatman, H., Meyer, B.J., 2013. Expanding awareness of docosahexaenoic acid during pregnancy. Nutrients 5, 1098–1109.

Erickson, M.C., Doyle, M.P., 2007. Food as a vehicle for transmission of Shiga toxin-producing *Escherichia coli*. J. Food Prot. 70, 2426–2449.

Eromo, T., Tassew, H., Daka, D., Kibru, G., 2016. Bacteriological quality of street foods and antimicrobial resistance of isolates in Hawassa, Ethiopia. Ethiopian J. Health Sci. 26 (6), 533–542.

Espelund, M., Klaveness, D., 2014. Botulism outbreaks in natural environments-an update. Front. Microbiol. 5, . Available from: https://doi.org/10.3389/fmicb.2014.00287287-287.

Etcheverría, A.I., Padola, N.L., Sanz, M.E., Polifroni, R., Krüger, A., Passucci, J., et al., 2010. Occurrence of Shiga toxin-producing E. coli (STEC) on carcasses and retail beef cuts in the marketing chain of beef in Argentina. Meat Sci. 86, 418–421.

Food and Agriculture Organization of the United Nations (FAO), The State of World fisheries and aquaculture. Contributing to food security and nutrition for all, Rome, 2016, 200.

FDA (Food and Drug Administration), Foodborne pathogenic microorganisms and natural toxins handbook, US FDA Center for Food Safety and Applied Nutrition 1992, 199. < http://www.cfsan.fda.gov/ ~ mow/app3a. html >

FDA (Food and Drug Administration), 2019. Fish and fishery products hazards and controls guidance, 4[th]edn US FDA Center for Food Safety and Applied Nutrition, Maryland.

Fenlon, D.R., 1985. Wild birds and silage as reservoirs of Listeria in the agricultural environment. J. Appl. Bacteriol. 59, 537–543.

Fenlon, D.R., 1986. Rapid quantitative assessment of the distribution of Listeria in silage implicated in a suspected outbreak of listeriosis in calves. Vet. Rec. 118 (9), 240–242.

Fenlon, D.R., 1999. Listeria monocytogenes in the natural environment. In: Ryser, E.T., Marth, E.H. (Eds.), Listeria, Listeriosis and Food Safety, 2nd edn Marcel Decker, Inc, New York, pp. 21–38.

Fensterbank, R., Audurier, A., Gogu, J., Guerrault, P., Malo, N., 1984. Listeria strains isolated from sick animals and consumed silage. Ann. Rech. Vet. 15 (1), 113–118.

Florman, A.L., Schifrin, N., 1950. Observations on a small outbreak of infantile diarrhea associated with Pseudomonas aeruginosa. J. Ped 36, 758–766.

Forsythe, S.J., 2002. The microbiological risk assessment of food, 212. Blackwell Publishing, ISBN 0-632-05952-2.

Friedman C.R., Neimann J., Wegener H.C., Tauxe R.V., Epidemiology of *Campylobacter jejuni* infections in the United States and other industrialized nations, 2000.

FSANZ (Food Standards Australia New Zealand), Compendium of microbiological criteria for food. food standards Australia New Zealand, 2018. (accessed Feb. 2020) < https://www.foodstandards.gov.au/publications/ Documents/Compedium%20of%20Microbiological%20Criteria/Compendium_revised-Sep%202018.pdf >

Fukushima, H., Gomyoda, M., 1986. Inhibition of Yersinia enterocolitica serotype 03 by natural microflora of pork. Appl. Environ. Microbiol. 51 (5), 990–994.

Gallimore, C.I., Barreiros, M.A.B., Brown, D.W.G., Nascimento, J.P., Leite, J.P.G., 2004. Noroviruses associated with acute gastroenteritis in a children's day care facility in Rio de Janeiro, Brazil, Brazilian. J. Med. Biol. Res. 37 (3), 321–326.

Gautam, R.K., Kakatkar, A.S., Karani, M.N., Shashidhar, R., Bandekar, J.R., 2017. *Salmonella* in Indian ready-to-cook poultry: antibiotic resistance and molecular characterization. Microbiol. Res. 8, 6882. Available from: https://doi.org/10.4081/mr.2017.6882.

Gebretsadik, S., Kassa, T., Alemayehu, H., Huruy, K., Kebede, N., 2011. Isolation and characterization of *Listeria monocytogenes* and other *Listeria* species in foods of animal origin in Addis Ababa. Ethiopia, J. Infec. Public. Health 4, 22–29.

Gibbons, I., Adesiyun, A., Seepersadsingh, N., Rahaman, S., 2006. Investigation for possible source(s) of contamination of ready-to-eat meat products with *Listeria* spp., other pathogens in a meat processing plant in Trinidad. Food Microbiol. 23, 359–366.

Girafa, G., Carminati, D., Neviani, E., 1997. Enterococci isolated from dairy products: a review of risks and potential technological use. J. Food Prot. 60, 732–738.

Gitter, M., Richardson, C., Boughton, E., 1986. Experimental infection of pregnant ewes with Listeria monocytogenes. Vet. Rec. 118 (31), 575–578.

Goh, S.G., Kuan, C.H., Loo, Y.Y., Chang, W.S., Lye, Y.L., Soopna, P., et al., 2012. Listeria monocytogenes in retailed raw chicken meat in Malaysia. Poult. Sci. 91 (10), 2686–2690.

Gordon, A., 2016. Food safety-based strategies for addressing trade and market access issues. In: Gordon, A. (Ed.), Food Safety and Quality Systems in Developing Countries: Volume Two: Case Studies of Effective Implementation. Academic Press, London, UK, pp. 21–44.

Gordon, C.L.A., Ahmad, M.H., 1990. Heat resistance of thermoduric enterococci isolated from frankfurters in Jamaica. Int. Congr. Meat, Sci. Technol. Proc. 36, 1078–1085.

Gordon, A., Ahmad, M.H., 1991. Thermal susceptibility of S.faecium strains isolated from frankfurters. Can. J. Microbiol. 37, 609–612.

Gordon, C.L.A., Ahmad, M.H., 1993. Bacteriological conditions of a chicken hatchery in Jamaica. World J. App. Microbiol. Biotechnol. 9, 282–286.

Gordon, A., Jackson, J.C., 2013. The microbiological profile of Jamaican Ackees (*Blighia sapida*). Nutr. Food Sci. 43 (2), 142–149.

Gordon Z., Gordon A., Kerr J., Microbiological map of selected Caribbean foods over the 11-year period 2004 through 2014. In IAFP 2017 Annual Meeting (July 9–12, 2017). International Association for Food Protection (Abstract), 2017.

Gravani, R., 1999. Incidence and control of Listeria in food-processing facilities. In: Ryser, E.T., Marth, E.H. (Eds.), Listeria, Listeriosis and Food Safety, second ed. Marcel Decker, Inc, New York, pp. 657–709.

Griffiths, M.W., Phillips, J.D., Muir, D.D., 1981. Thermostability of proteases and lipases from a number of species of psychrotrophic bacteria of dairy origin. J. Appl. Bacteriol. 50, 289–303.

Gunasena, D., Kodikara, C., Ganepola, K., Widanapathirana, S., 1995. Occurrence of Listeria monocytogenes in food in Sri Lanka. J. Natl Sci. Found. Sri Lanka 23, 107–114.

Haba, J.H., 1993. Incidence and control of Campylobacter in foods. Microbiologia 9, 57–65.

Hardalo, C., Edberg, S.C., 1997. Pseudomonas aeruginosa: assessment of risk fromdrinking water. Crit. Rev. Microbiol. 23, 47–75.

Headrick, M.L., Tollefson, L., 1998. Food borne disease summary by food commodity. Vet. Clin. N. Am. Food Anim. Pract. 14, 91–99.

Hailegebreal, G., 2017. A review on *Clostridium perfringens* food poisoning, Global Research. J. Public. Health Epidemiol. 4 (3), 104–109.

Henderson, A., MacLaurin, J., Scott, J.M., 1969. Pseudomonas in a Glasgow baby unit. Lancet 1, 316–317.

Henson, S., Loader, R., 2001. Barriers to agricultural exports from developing countries: the role of sanitary and phytosanitary requirements. World Dev. 29 (1), 85–102.

Hilborn, E.D., Mshar, P.A., Fiorentino, T.A., Dembek, Z.F., Barrett, T.J., Howard, R.T., et al., 2000. An outbreak of *Escherichia coli* O157:H7 infections and haemolyticuraemic syndrome associated with consumption of unpasteurized apple cider. Epidemiol. Infect. 124, 31–36.

Hitchins, A.D., Feng, P., Watkins, W.D., Rippey, S.R., Chandler, L.A., 1998. *Escherichia coli* and the coliform bacteria. In: 8th edn Merker, R.L. (Ed.), FDA bacteriological analytical manual, revision A. AOAC International, Gaithersburg, MD.

Hoadley, A., 1977. Potential health hazards associated with Pseudomonas aeruginosa in water. In: Hoadley, A., Dutka, B.J. (Eds.), Bacterial indicators/Health hazardsassociated with water. ASTM, Philadelphia, p. 80.

Hoang Minh, S., Kimura, E., Hoang Minh, D., Honjoh, K., Miyamoto, T., 2015. Virulence characteristics of Shiga toxin-producing *Escherichia coli* from raw meats and clinical samples. Microbiol. Immunol. 59, 114–122. Available from: https://doi.org/10.1111/1348-0421.12235.

Hoffmann, S., Devleesschauwer, B., Aspinall, W., Cooke, R., Corrigan, T., Havelaar, A.H., et al., 2017. Attribution of global foodborne disease to specific foods: findings from a World Health Organization structured expert elicitation. PLoSONE 12, e0183641.

Holbrook, R., Anderson, J.M., Baird-Parker, A.C., 1980. Modified direct plate method for counting *Escherichia coli* in foods. Food Tech. Aust. 32, 78–83.

Hunter, C.A., Ensign, P.R., 1947. An epidemic of diarrhea in a newborn nursery caused by Pseudomonas aeruginosa. Amer. J. Pub. Health 37, 1166–1169.

ICMSF (International Commission on Microbiological Specifications for Foods), 1986. Microorganisms in foods sampling for microbiological analysis: principles and specific applications. Aspen Publishers, Inc, Gaithersburg, Maryland, USA.

ICMSF (International Commission on Microbiological Specifications for Foods), 2000. Microorganisms in foods: meat and meat products, Microbial Ecology of Food Commodities. Aspen Publishers, Inc, Gaithersburg, Maryland, USA.

ICMSF (International Commission on Microbiological Specifications for Foods), 2002. Microorganisms in foods: microbiological testing in food safety management. Kluwer Academic/Plenum Publishers, New York, USA.

IFST (Institute of Food Science and Technology), 1997. Development and use of microbiological criteria for foods: guidance for those involved in using and interpreting microbiological criteria for foods. IFST Prof. Food Microbiol. Group. Food Sci. Technol. Today 11 (3), 137–177.

Ingham, S., Buege, D.R., Dropp, B.K., Losinski, J.A., 2004. Survival of Listeria monocytogenes duringstorage of ready to eat meat products processed by drying, fermentation and/or smoking. J. Food Prot. 67, 2698–2702.

Jackson, J.C., Gordon, A., Wizzard, G., McCook, K., Rolle, R., 2004. Changes in chemical composition of coconut (Cocos nucifera) water during maturation of the fruit. J. Sci. Food Agric. 84 (9), 1049–1052.

Jackson-Davis, A., Mendonca, A., Hale, S., Jackson, J., King, A., Jackson, J., 2018. Microbiological safety of unpasteurized fruit and vegetable juices sold in juice bars and small retail outlets. In: Ricke Steven, C., Atungulu, G. G., Rainwater, C.E., Park, S.H. (Eds.), Food and Feed Safety Systems and Analysis. Academic Press, pp. 213–225.

Jaffee, S., Henson, S., Unnevehr, L., Grace, D., Cassou, E., 2019. The safe food imperative: accelerating progress in low- and middle-income countries. Agriculture and Food Series World Bank, Washington, DC, <https://doi.org.10.1596/978-1-4648-1345-0>.

Jansen, W., Woudstra, S., Müller, A., Grabowski, N., Schoo, G., Gerulat, B., et al., 2018. The safety and quality of pork and poultry meat imports for the common European market received at border inspection post Hamburg Harbour between 2014 and 2015. PLoS ONE 13 (2), e0192550. Available from: https://doi.org/10.1371/journal.pone.0192550. accessed 30.04.2019.

Jay, J.M., Loessner, M.J., Golden, D.A., 2005. Modern food microbiology, 7th Edn Springer Science-Business Media, LLC, New York, NY.

Jeon, S.H., Kim, N.H., Shim, M.B., Jeon, Y.W., Ahn, J.H., Lee, S.H., et al., 2015. Microbiological diversity and prevalence of spoilage and pathogenic bacteria in commercial fermented alcoholic beverages (beer, fruit wine, refined rice wine, and yakju). J, Food Prot. 78 (4), 812–818.

Jiménez, M.L., Apostolou, A., Suarez, A.J., Meyer, L., Hiciano, S., Newton, A., et al., 2011. Multinational cholera outbreak after wedding in the Dominican Republic. Emerg. Infec. Dis. 17 (11), 2172–2174. Available from: https://doi.org/10.3201/eid1711.111263.

Johnston, D.W., Bruce, J., 1982. Incidence of thermoduric psychrotrophs in milk produced in the west of Scotland. J. Appl. Bacteriol. 52, 333–337.

Kabwama, S.N., Bulage, L., Nsubuga, F., Pande, G., Oguttu, D.W., Mafigiri, R., et al., 2017. A large and persistent outbreak of typhoid fever caused by consuming contaminated water and street-vended beverages: Kampala, Uganda. BMC Public Health 17 (1), 23.

Kalinowski, R.M., Tompkin, R.B., Bodnaruk, P.W., Pruett Jr, W.P., 2003. Impact of cooking, cooling, and subsequent refrigeration on the growth or survival of Clostridium perfringens in cooked meat and poultry products. J Food Prot 66 (7), 1227–1232.

KDHE (Kansas Department of Health and Environment), Outbreak of Salmonella Enteritidis infections associated with the beaches negril resort-Negril, Jamaica, 2011, KDHE, Bureau of Epidemiology and Public Health Informatics, Topeka, Kansas, USA.

Kanarat, S., Jitnupong, W., Sukhapesna, J., 2011. Prevalence of Listeria monocytogenes in chicken production chain in Thailand. Thai. J. Vet. Med. 41 (2), 155–161.

Kaneko, T., Colwell, R.R., 1973. Ecology of Vibrio parahaemolyticus in Chesapeake Bay. J Bacteriol 113 (1), 24–32.

Kang, S.J., Ryu, S.J., Chae, J.S., Eo, S.K., Woo, G.J., Lee, J.H., 2004. Occurrence and characteristics of enterohemorrhagic Escherichia coli O157 in calves associated with diarrhea. Vet. Microbiol. 98, 323–328.

Khan, A.S., Georges, K., Rahaman, S., Abdela, W., Adesiyun, A., 2018. Prevalence and serotypes of Salmonella spp. on chickens sold at retail outlets in Trinidad. PLoS ONE 13 (8), 0202108. Available from: https://doi.org/10.1371/journal.pone.0202108.

Kim, S.A., Kim, N.H., Lee, S.H., Hwang, I.G., Rhee, M.S., 2014. Survival of foodborne pathogenic bacteria (Bacillus cereus, Escherichia coli O157: H7, Salmonella enterica serovar Typhimurium, Staphylococcus aureus, and Listeria monocytogenes) and Bacillus cereus spores in fermented alcoholic beverages (beer and refined rice wine). J. Food Prot 77 (3), 419–426.

Knight O., Gordon C.L.A., Alicyclobacillus in tropical Caribbean beverages: incidentsand intervention strategies for control of this threat to the industry, In: 2004 IFT Annual Meeting, 2004, Las Vegas, NV.

Kongsaengdao, S., Samintarapanya, K., Rusmeechan, S., Wongsa, A., Pothirat, C., Permpikul, C., et al., 2006. An outbreak of botulism in Thailand: clinical manifestations and management of severe respiratory failure. Clinic. Infec. Dis 43 (10), 1247–1256.

Koodie, L., Dhople, A.M., 2001. Acid tolerance of Escherichia coli O157:H7 and its survival in apple juice. Microbios 104 (409), 167–175.

Kotloff, K.L., Nataro, J.P., Blackwelder, W.C., Nasrin, D., Farag, T.H., Panchalingam, S., et al., 2013. Burden and aetiology of diarrhoeal disease in infants and young children indeveloping countries (the Global Enteric Multicenter Study, GEMS): a prospective, case-control study. Lancet 382 (9888), 209–222. Available from: https://doi.org/10.1016/s0140-6736(13)60844-2.

Kuan, C.H., Goh, S.G., Loo, Y.Y., et al., 2013. Prevalence and quantification of Listeria monocytogenes in chicken offal at the retail level in Malaysia. Poultry Sci. 92 (6), 1664–1669.

Lakicevic, B., Stjepanovic, A., Milijasevic, M., Terzic Vidojevic, A., Golic, N., Topisirovic, L., 2010. The presence of Listeria spp., Listeria monocytogenes in a chosen food processing establishment in Serbia. Arch. Biol. Sci. 62 (4), 881–887.

Larsen L., Mexican papayas still on import alert for Salmonella contamination, Food Poisoning Bull 2019. <https://foodpoisoningbulletin.com/2019/mexican-papayas-import-alert-salmonella> (accessed 30.04.2019).

Leclerc, H., Oger, C., 1974. Les eaux uses des hopitaux et leur importance epidemiologique. Revue Epidémiologique de MédecineSociale etSanté Publique 22, 185–198.

Leclerc H., Oger C., Les eauxusees des abattoirs et leur importance epidemiologique,. 23, 1975, 429–444.

Lee, S.K., Park, H.J., Lee, J.H., Lim, J.S., Seo, K.H., Heo, E.J., et al., 2017. Distribution and molecular characterization of Campylobacter species at different processing stages in two poultry processing plants. Foodborne Pathog. Dis. 14, 141–147. Available from: https://doi.org/10.1089/fpd.2016.2218.

Lenglet, A., Khamphaphongphane, B., Thebvongsa, P., Vongprachanh, P., Sithivong, N., Chantavisouk, C., et al., 2010. A cholera epidemic in Sekong Province, Lao People's Democratic Republic. Japanese J. Infec. Dis. 63 (3), 204–207.

Liang, S.Y., Watanabe, H., Terajima, J., Li, C.C., Liao, J.C., Tung, S.K., et al., 2007. Multi-locus variable number tandem repeat analysis for molecular typing of Shigella sonnei. J. Clin. Microbiol. 45 (11), 3574. Available from: https://doi.org/10.1128/JCM.00675-07.

Lin, S., Schraft, H., Odumeru, J.A., Griffiths, M.W., 1998. Identification of contamination sources of Bacillus cereus in pasteurized milk. Int. J. Food Microbiol 43 (3), 159–171.

Long, S.C., Tauscher, T., 2006. Watershed issues associated with Clostridium botulinum: a literature review. J. Water Health 4 (3), 277–288.

Lu, J., Sun, L., Fang, L., Yang, F., Mo, Y., Lao, J., et al., 2015. Gastroenteritis outbreaks caused by norovirus GII. 17, Guangdong Province, China. Emerging Infec. Dis 21 (7), 1240.

Luo, Q., Li, S., Liu, S., Tan, H., 2017. Foodborne illness outbreaks in China. Int. J. Clin. Exp. Med. 10, 5821–5831.

Lund, B.M., Baird-Parker, T.C., Gould, G.W. (Eds.), 2000. The microbiological safety and quality of foods. Aspen Publishers, Inc, Gaithersburg, MD.

Lynch, O.A., Cagney, C., McDowell, D.A., Duffy, G., 2011. Occurrence of fastidious Campylobacter spp. in fresh meat and poultry using an adapted cultural protocol. Int. J. Food Microbiol. 150, 171–177. Available from: https://doi.org/10.1016/j.ijfoodmicro.2011.07.037.

Magnusson, M., Svensson, B., Kolstrup, C., Christiansson, A., 2007. Bacillus cereus in free-stall bedding. J. Dairy Sci. 90 (12), 5473–5482. Available from: https://doi.org/10.3168/jds.2007-0284.

Maharjan, M., Joshi, V., Joshi, D.D., 2006. Prevalence of Salmonella species in various raw meat samples of a local market in Kathmandu. Ann. N.Y. Acad. Sci 1081, 249–256. Available from: https://doi.org/10.1196/annals.1373.031.

Maijala, R., Lyytikainen, O., Autio, T., Aalto, T., Haavisto, L., Honknen-Buzalski, T., 2001. xposure of Listeria monocytogenes within an epidemiccaused by butter in Finland. Int. J. Food Microbiol 70, 97–109.

Makwana, P.P., Nayak, J.B., Brahmbhat, M.N., Chaudhary, J.H., 2015. Detection of Salmonella spp. from chevon, mutton and its environment in retail meat shops in Anand city (Gujarat), India. Vet. World 8 (3), 388–392.

Margulis, L., Jorgensen, J.Z., Dolan, S., Kolchinsky, R., Rainey, F.A., Lo, S.C., 1998. The arthromitus stage of Bacillus cereus: intestinal symbionts of animals. Proc. Nat. Acad. Sci 95 (3), 1236–1241. Available from: https://doi.org/10.1073/pnas.95.3.1236.

Masana, M.O., Leotta, G.A., Del Castillo, L.L., D'Astek, B.A., Palladino, P.M., Galli, L., et al., 2010. Prevalence, characterization, and genotypic analysis of Escherichia coli O157:H7/NM from selected beef exporting abattoirs of Argentina. J. Food Prot 73, 649–656.

Mathew, E.N., Muyyarikkandy, M.S., Kuttappan, D., Amalaradjou, M.A., 2018. Attachment of *Salmonella* enterica on mangoes and survival under conditions simulating commercial mango packing house and importer facility. Front. Microbiol. 9, 1519. Available from: https://doi.org/10.3389/fmicb.2018.01519.

McClure, P.J., Beaumont, A.L., Sutherland, J.P., Roberts, T.A., 1997. Predictive modeling of growth of Listeria monocytogenes. The effects on growth of NaCl, pH, storage and $NaNO_2$. Int. J. Food Microbiol. 34 (3), 221–232.

Mendonca A., Microbial challenge studies to evaluate the viability of four human enteric pathogens in Walkerswood Jamaican Jerk Seasoning at 25°C, Iowa State University, 2006, Submitted to Walkerswood Caribbean Foods Limited, St. Ann, Jamaica.

Mendonca A.F., Battle K.L., Viaji C.Y., Copeland M.A., Goodridge L.D., Influence of Jamaican jerk seasoning paste on growth of natural bacterial flora and *Salmonella* typhimurium on raw chicken breast meat, International Association for Food Protection Annual Meeting, 2009, Grapevine Texas.

Mendonca, A.F., 2010. Microbiology of cooked meats. In: Lynne Knipe, C., Robert Rust, E. (Eds.), Thermal Processing of Ready-to-Eat Meat Products. Blackwell Publishing, Danvers, MA, pp. 17–38.

Mendonca, A., Jackson-Davis, A., Moutiq, R., Thomas-Popo, E., 2018. Use of natural antimicrobials of plant origin to improve the microbiological safety of foods, Chapter 14. In: Ricke Steven, C., Atungulu, G.G., Rainwater, C. E., Park, S.H. (Eds.), Food and Feed Safety Systems and Analysis. Academic Press, pp. 249–272.

Menz, G., Aldred, P., Vriesekoop, F., 2011. Growth and survival of foodborne pathogens in beer. J. Food Prot 74 (10), 1670–1675.

Mishu, B., Darweigh, A., Weber, J.T., Hathewahy, C.L., El-Sharkaway, S., Corwin, A., 1991. A foodborne outbreak of type E botulism in Cairo, Egypt. Am. J. Trop. Med. Hyg. 45 (3), 109.

Molla, B., Yilma, R., Alemayehu, D., 2004. Listeria monocytogenes and other Listeria species in retail meat and milk products in Addis Ababa, Ethiopia, Ethiop. J. Health Dev 18 (3), 208–212.

Moran, L., Scates, P., Madden, R.H., 2009. Prevalence of Campylobacter spp. in raw retail poultry on sale in Northern Ireland. J. Food Prot 72 (9), 1830–1835.

Mossel, D.A.A., 1967. Ecological principles and methodological aspects of the examination of foods and feeds for indicator microorganisms. J. Assoc. Off. Agric. Chem. 50, 91–104.

Mossel, D.A.A., 1978. Index and indicator organism—a current assessment of their usefulness and significance. Food Technol. Aust. 212–219.

Mossel, D.A.A., Corry, J.E.L., Struijk, C.B., Baird, R.M., 1995. Essentials of the microbiology of foods: a textbook for advanced studies. John Wiley and Sons, Chichester, UK.

Mosqueda-Melgar, J., Raybaudi-Massilia, R.M., Martín-Belloso, O., 2008. Combination of high-intensity pulsed electric fields with natural antimicrobials to inactivate pathogenic microorganisms and extend the shelf-life of melon and watermelon juices. Food Microbiol. 25 (3), 479–491. Available from: https://doi.org/10.1016/j.fm.2008.01.002.

Musa, Z., Onyilokwu, S.A., Jauro, S., Yakubu, C., Musa, J.A., 2017. Occurrence of *Salmonella* in ruminants and camel meat in Maiduguri, Nigeria and their antibiotic resistant pattern. J. Adv. Vet. Anim. Res. 4 (3), 227–233.

Mwanyika, G.O., Buza, J., Rugumisa, B.T., Luanda, C., Murutu, R., Lyimo1, B., et al., 2016. Recovery and prevalence of antibiotic-resistant *Salmonella* from fresh goat meat in Arusha, Tanzania, African. J. Microbiol. Res. 10 (32), 1315–1321.

Nachamkin I., Blaser M.J., *Campylobacter*. 2nd ed. Washington, D.C. American Society for Microbiology Press, 121–138.

National Soft Drink Association, Quality specifications and test procedures for bottler's granulated and liquid sugar, 1975, Washington, D.C.

National Marine Fisheries Service (NMFS), Fisheries of the United States, 2010. < http://www.st.nmfs.noaa.gov/st1/fus/fus09/fus_2009.pdf > (accessed 19.06.2019).

National Marine Fisheries Service (NMFS), Fisheries of the United States, 2013. < http://www.st.nmfs.noaa.gov/Assets/commercial/fus/fus13/01_front2013.pdf > (accessed 19.06.2019)

N'cho, H.S., 2019. Notes from the field: typhoid fever outbreak—Harare, Zimbabwe, MMWR. Morbidity Mortality Weekly Rep 68, 44–45.

Neela, V., MohdZafrul, A., Mariana, N.S., Van Belkum, A., Liew, Y.K., Rad, E.G., 2009. Prevalence of ST9 methicillin-resistant Staphylococcus aureus among pigs and pig handlers in Malaysia. J. Clin. Microbiol. 47, 4138–4140.

Nidaullah, H., Mohd Omar, A.K., Rosma, A., Huda, N., Sohni, S., 2016. Analysis of *Salmonella* contamination in poultry meat at various retailing, different storage temperatures and carcass cuts-a literature survey. Int. J. Poult. Sci. 15 (3), 111–120.

Obaji, M., Oli, A.N., Enweani, I., Udigwe, I., Okoyeh, J.N., Ekejindu, I.M., 2018. Public health challenges associated with street-vended foods and medicines in a developing country: a mini-review. The J. Med. Res. 4 (6), 283−287.

O'Mahony, M., Cowden, J., Smyth, B., Lynch, D., Hall, M., Rowe, B., et al., 1990. An outbreak of *Salmonella* Saint Paul infection associated with bean sprouts. Epidemiol. Infect. 104, 229−235.

Olson, K.E., Sorrells, K.N., 2001. Thermophilic flat sour sporeformers. In: Downes, F.P., Ito, K. (Eds.), Compendium of Methods for the Microbiological Examination of Foods, 4th edn American Public Health Association, Washington, DC, pp. 245−248.

Painter, J.A., Ayers, T., Woodruff, R., Blanton, E., Perez, N., Hoekstra, R., et al., 2009. Recipes for foodborne outbreaks: a scheme for categorizing and grouping implicated foods, Foodborne Path. Dis 6, 1259−1264.

Paudyal, N., Anihouvi, V., Hounhouigan, J., Matsheka, M.I., Sekwati-Monang, B., Amoa-Awua, W., et al., 2017. Prevalence of foodborne pathogens in food from selected African countries—a meta-analysis. Int. J. Food Microbiol. 249, 35−43.

Penteado, A.L., 2017. Microbiological safety aspects of mangoes (*Mangiferaindica*) and papayas (*Carica papaya*): a mini-review. Vigil. Sanit. Debate 5 (2), 127−140. Available from: https://doi.org/10.22239/2317-269x.00779.

Pires, S.M., Vieira, A.R., Perez, E., Wong, D.L.F., Hald, T., 2012. Attributing human foodborne illness to food sources and water in Latin America and the Caribbean using data from outbreak investigations. Int. J. Food Microbiol. 152 (3), 129−138.

Premarathne, J.M.K., Anuar, A.S., Thung, T.Y., Satharasinghe, D.A., Jambari, N.A., Abdul-Mutalib, N.A., et al., 2017. Prevalence and antibiotic resistance against tetracycline in *Campylobacter jejuni* and *C. coli* in cattle and beef Meatfrom Selangor, Malaysia. Front. Microbiol. 8, 2254. Available from: https://doi.org/10.3389/fmicb.2017.02254.

Rahimi, E.A., Ameri, M., Kazemeini, H.R., 2010. Prevalence and antimicrobial resistance of *Campylobacter* species isolated from raw camel, beef, lamb, and goat meat in Iran. Foodborne Pathogens Dis 7 (4), 443−447.

Rahimi, E.A., Kazemeini, H.R., Salajegheh, M., 2012. Escherichia coli O157:H7/NM prevalence in raw beef, camel, sheep, goat, and water buffalo meat in Fars and Khuzestan provinces, Iran. Vet. Res. Forum 3 (1), 13−17.

Ray, B., Bhunia, A., 2013. Fundamental Food Microbiology, Fifth Edn. CRC Press, Boca Raton, FL, USA.

Ribeiro, C.M., Stefani, L.M., Lucheis, S.B., Okano, W., Cruz, J.C.M., Souza, G.V., et al., 2018. Methicillin-resistant *Staphylococcus aureus* in poultry and poultry meat: a meta-analysis. J. Food Prot 81 (7), 1055−1062.

Richter, L., Du Plessis, E.M., Duvenage, S., Korsten, L., 2019. Occurrence, identification, and antimicrobial resistance profiles of extended-spectrum and AmpC β-lactamase-producing Enterobacteriaceae from fresh vegetables retailed in Gauteng Province, South Africa. Foodborne Pathogens Dis 16 (6), 421−427.

Roberts, T.A., 1980. Contamination of meat: the effects of slaughter practices on the bacteriology of the red meat carcass. Royal Society of Health 100, 3−9. < https://doi.org/10.1177/146642408010000205 >.

Robertson, K., Green, A., Allen, L., Ihry, T., White, P., Chen, W.S., et al., 2016. Foodborne outbreaks reported to the U.S. food safety and inspection service, fiscal years. J. Food Prot 79, 442−447. < https://doi.org/10.4315/0362-028X.JFP-15-376 >.

Rodrigo, S., Adesiyun, A., Asgarali, Z., Swanston, W., 2006. Occurrence of selected foodborne pathogens on poultry and poultry giblets from small retail processing operations in Trinidad. J. Food Prot 69 (5), 1096−1105.

Rosow, L.K., Strober, J.B., 2015. Infant botulism: review and clinical update, Ped. Neurol 52, 487−492.

Ryser, E.T., 1999. Foodborne listeriosis. In: Ryser, E.T., Marth, E.H. (Eds.), Listeria listeriosis and food safety, 2nd edn Marcel Decker, Inc, New York, N.Y.

Ryser, E.T., Arimi, S.M., Donnelly, C.W., 1997. Effects of pH on distribution of Listeriaribotypes in corn, hay and grass silage. Appl. Environ. Microbiol 63, 3695−3697.

Ryu, J.H., Beuchat, L.R., 2005. Biofilm formation and sporulation by *Bacillus cereus* on a stainless steel surface and subsequent resistance of vegetative cells and spores to chlorine, chlorine dioxide, and a peroxyacetic acid-based sanitizer. J. Food Prot 68 (12), 2614−2622.

Sakazaki, R.I.I.C.H.I., 1983. Vibrio parahaemolyticus as a food-spoilage organism. Academic Press, New York.

Salim, A.P., Canto, A.C., Costa-Lima, B.R., Simoes, J.S., Panzenhagen, P.H., Franco, R.M., et al., 2017. Effect of lactic acid on *Escherichia coli* O157:H7 and on color stability of vacuum-packaged beef steaks under high storage temperature. J. Microbiol. Biotechnol. Food Sci. 6, 1054−1058.

Scallan, E., Hoekstra, R.M., Angulo, F.L., Tauxe, R.V., Widdowson, M.A., Roy, S.L., et al., 2011. Foodborne illness acquired in the United States-major pathogens. Emerg. Infect. Dis 17, 7−15. < https://doi.org/10.3201/eid1701.P11101 >.

Sebsibe A., Sheep and goat meat quality characteristics, In: A. Yami and R.C. Merkel (Eds.), Sheep and Goat Production Handbook for Ethiopia, 2008, 325-340. < http://dawog.net/Goats/Sheep%20&%20Goat%20Hdbk% >

Shafi, R., Cotterill, O.J., Nichols, M.L., 1970. Microbial flora of commercially pasteurized egg products. Poult. Sci 49 (2), 578−585. < https://doi.org/10.3382/ps.0490578 >.

Shane, S.M., 1992. The significance of *Campylobacter jejuni* infection in poultry: a review. Avian. Pathol. 21 (2), 189−213.

Sheridan, J.J., 1998. Sources of contamination during slaughter and measures for control. J.Food Saf. 18, 321−339.

Shonhiwa, A.M., Ntshoe, G., Essel, V., Thomas, J., McCarthy, K., 2019. A review of foodborne diseases outbreaks reported to the outbreak response unit, national institute for communicable diseases, South Africa. Int. J. Infec. Dis. 79, 73.

Silliker, J.H., 1963. Total counts as indices of food quality. Microbiological Quality of Foods. Academic Press, New York.

Silliker, J.H., Gabis, D.A., 1976. ICMSF methods studies. VII. Indicator tests as substitutes for direct testing of dried foods and feeds for *Salmonella*. Can. J. Microbiol 22, 971−974.

Silva, J., Leite, D., Fernandes, M., Mena, C., Gibbs, P.A., Teixeira, P., 2011. *Campylobacter* spp. as a foodborne pathogen: a review. Front. Microbiol. 2, 200. Available from: https://doi.org/10.3389/fmicb.2011.00200.

Skarp, C.P.A., Hänninen, M.L., Rautelin, H.I.K., 2016. , Campylobacteriosis: the role of poultry meat. Clin. Microbiol. Infect 22, 103−109. Available from: https://doi.org/10.1016/j.cmi.2015.11.019.

Skovgaard N., Microorganisms in foods, use of data for assessing process control and product acceptance, 2012, International Commission on Microbiological Specifications for Foods (ICMSF). ISBN 978-1-4419-9374-8.

Spector, M.P., Kenyon, W.J., 2012. Resistance and survival strategies of *Salmonella*enterica to environmental stresses. Food Res. Int. 45, 455−481.

Splittstoesser, D.F., Queale, D.T., Bowers, J.L., Wilkison, M., 1980. Coliform content of frozen blanched vegetables packed in the United States. J. Food Saf. 2, 1−11.

Srinivasa, H., Baijayanti, M., Raksha, Y., 2009. Magnitude of drug resistant Shigellosis: a report from Bangalore. Indian J. Med. Microbiol. 27 (4), 358.

Statistica, Broiler meat production worldwide in 2019 by country, 2019. < https://www.statista.com/statistics/237597/leading-10-countries-worldwide-in-poultry-meat-production-in-2007/ > (accessed 15.06.2019).

StenforsArnesen, L.P., Fagerlund, A., Granum, P.E., 2008. From soil to gut: *Bacillus cereus* and its food poisoning toxins. FEMS Microbiol. Rev. 32 (4), 579−606. Available from: https://doi.org/10.1111/j.1574-6976.2008.00112.x.

Stevens, M.P., Humphrey, T.J., Maskell, D.J., 2009. Molecular insights into farm animal and zoonotic *Salmonella* infections. Philosophical Transactions of the Royal Society B 364, 2709−2723.

Stiles, M.E., 1989. Less recognized or presumptive foodborne pathogenic bacteria. Foodborne Bacterial Pathogens. Marcel Dekker Inc, New York, pp. 673−733.

Strawn, L.K., Danyluk, M.D., 2010. Fate of *Escherichia coli* O157:H7 and *Salmonella* on fresh and frozen cut pineapples. J. Food Prot 73, 18−24.

Swaminathan, B., 2001. Listeria monocytogenes. In: Doyle, M.P., Beuchat, L.R., Montville, T.J. (Eds.), Food Microbiology: Fundamentals and Frontiers, 2nd edn ASM Press, Washington, DC, pp. 383−409.

Syne, S.M., Ramsubhag, A., Adesiyun, A.A., 2013. Microbiological hazard analysis of ready-to-eat meats processed at a food plant in Trinidad, West Indies. Infec. Ecol. Epidemiol 3 (1), 20450. Available from: https://doi.org/10.3402/iee.v3i0.20450.

Tadesse, G., Tessema, T.S., 2014. A meta-analysis of the prevalence of *Salmonella* in food animals in Ethiopia. BMC Microbiol. 14, 270. < http://www.biomedcentral.com/1471-2180/14/270 >.

Thomas, A., Lallo, C.H.O., Badhre, N., 2006. Microbiological evaluation of broiler carcasses, wash and rinse water from pluck shops (cottage poultry processors) in the county Nariva/Mayaro, Trinidad, Trinidad and Tobago, West Indies. Tropicultura 24 (3), 135−142.

Thomas-Popo, E.R., Mendonca, A., Dickson, J., Shaw, A., Coleman, S., Daraba, A., et al., 2019. Isoeugenol significantly inactivates *Escherichia coli* O157:H7, *Salmonella enterica*, and *Listeria monocytogenes* in refrigerated tyndallized pineapple juice with added *Yucca schidigera* extract. Food Control 106, 106727. < https://doi.org/10.1016/j.foodcont.2019.106727 >.

Todd, E.C., 1997. Epidemiology of foodborne diseases: A worldwide review. World Health Statistics Quarterly 50, 30−50.

Toro, M., Rivera, D., Jimenez, M.F., Díaz, L., Navarrete, P., Reyes-Jara, A., 2018. Isolation and characterization of non-O157 Shiga toxin-producing *Escherichia coli* (STEC) isolated from retail ground beef in Santiago, Chile. Food Microbiol. 75, 55−60.

Ugarte-Ruiz, M., Dominguez, L., Corcionivoschi, N., Wren, B.W., Dorrell, N., Gundogdu, O., 2018. Exploring the oxidative, antimicrobial and genomic properties of *Campylobacter jejuni* strains isolated from poultry. Res. Vet. Sci. 119, 170−175. Available from: https://doi.org/10.1016/j.rvsc.2018.06.016.

USDA Food safety and Inspection Service, Shiga toxin-producing *Escherichia coli* in certain raw beef products, 2012. < www.fsis.usda.gov/OPPDE/rdad/FRPubs/2010-0023FRN.pdf. >

United States Meat Export Federation. 2017. New records for U.S. beef export value, pork export volume in 2017 < https://www.usmef.org/news-statistics/press-releases/new-recordsfor-u-s-beef-export-value-pork-export-volume-in-2017-2/ >. (accessed 20.04.2019)

USDA Economic Research Service. 2019. Livestock and Meat International Trade Data. < https://www.ers.usda.gov/data-products/livestock-and-meat-international-trade-data/livestock-and-meat-international-trade-data/#Annual%20and%20Cumulative%20Year-to-Date%20U.S.%20Livestock%20and%20Meat%20Trade%20by%20Country > (accessed 24.04.2019).

USDHEW (U.S. Department of Health, Education and Welfare), 1965. National Shellfish Sanitation Program, Manual of Operations. Part 1. Sanitation of shellfish growing areas. U.S. Government Printing Office, Washington, D.C.

Van, T.T.H., Moutafis, G., Taghrid, I., Tran, L.T., Coloe, P.J., 2007. Detection of *Salmonella* spp. in retail raw food samples from Vietnam and characterization of their antibiotic resistance. Appl. Environ. Microbiol. 73 (21), 6885−6890.

Vaskas, T., Dahesht, E., Seifi, S., Rahmani, M., Motaghifar, A., Safanavaee, R., 2012. Study and comparison of the bacterial contamination outbreak of chicken meat consumed in some cities of Mazandaran province (Iran), Iran. Afr. J. Microbiol. Res. 6, 6286−6290.

Vokaty, S., Torres, J.G.R., 1997. Meat from small ruminants and public health in the Caribbean. Rev. Sci. Tech. Off. Int. Epiz. 16 (2), 426−432.

Waldner, L.L., MacKenzie, K.D., Köster, W., White From, A.P., 2012. Exit to Entry: Long-term survival and transmission of *Salmonella*. Pathogens 1, 128−155. Available from: https://doi.org/10.3390/pathogens1020128.

Wang, F., Jiang, L., Yang, Q., Han, F., Chen, S., Pu, S., et al., 2011. Prevalence and antimicrobial susceptibility of major foodborne pathogens in imported seafood. J. Food Prot 74 (9), 1451−1461.

Washam, C.J., Olson, H.C., Vedamathu, E.R., 1977. Heat resistant psychrotrophic bacteria isolated from pasteurized milk. J. Food Prot 40, 101−108.

WHO (World Health Organization). WHO Estimates of the Global Burden of Foodborne Diseases. World Health Organization, Switzerland, p. 268. < https://apps.who.int/iris/bitstream/handle/10665/199350/9789241565165_eng.pdf?sequence = 1 > (accessed 26.10.2019).

Williams, E.P., Cameron, B.A., 1897. Upon general infection by Bacillus pyocyaneusin children. J. Path. Bacteriol. 3, 344−351.

Woldemariam, T., Asrat, D., Zewde, G., 2009. Prevalence of thermophilic *Campylobacter* species in carcasses from sheep and goats in an abattoir in Debre Zeit area, Ethiopia, Ethiop. J. Health Dev 23 (3), 229−233.

Wong, H.C., Chen, M.C., Liu, S.H., Liu, D.P., 1999. Incidence of highly genetically diversified *Vibrio parahaemolyticus* in seafood imported from Asian countries. Int. J. Food Microbiol. 52, 181−188.

Worku, A., Dereje, M., 2017. A review on current status of small ruminant meat production comparison yield and carcass characteristics in Ethiopia. Adv. Life Sci. Technol. 58, 25−33.

Wurfel, S.F.R., da Silva, W.P., de Oliveira, M.G., Kleinubing, N.R., Lopes, G.V., Gandra, E.A., et al., 2019. Genetic diversity of *Campylobacter jejuni* and *Campylobacter coli* isolated from poultry meat products sold on the retail market in Southern Brazil. Poultry Sci. 98, 932−939. Available from: https://doi.org/10.3382/ps/pey365.

Yücel, N., Balci, S., 2010. Prevalence of *Listeria, Aeromonas*, and *Vibrio* species in fish used for human consumption in Turkey. J. Food Prot 73 (2), 380−384.

Zarrazquin, I., Torres-Unda, J., Ruiz, F., Irazusta, J., Kortajarena, M., Hoyos Cillero, I.H., et al., 2014. Longitudinal study: lifestyle and cardiovascular health in health science students. Nutr. Hosp. 30, 1144−1151.

Zdolec, N., Dobranić, V., Filipović, I., 2015. Prevalence of *Salmonella* spp., Yersinia enterocolitica in/on tonsils and mandibular lymph nodes of slaughtered pigs. Folia Microbiol. 60, 131−135. Available from: https://doi.org/10.1007/s12223-014-0356-9.

CHAPTER

6

The role and importance of packaging and labeling in assuring food safety, quality & compliance with regulations I: Packaging basics

Melvin A. Pascall

Department of Food Science and Technology, The Ohio State University, Columbus, OH, United States

OUTLINE

Packaging basics

The use of packaging to contain, transport, protect, regulate, preserve, store, market, and provide information about food, has become commonplace in the modern world in which we live. Although the size, shape, design and materials used to make food packaging have evolved over the last few decades, the purpose of packaging has remained the same. What has changed significantly over time is the technology that has been injected into food packaging. This has increased the versatility of packaging for applications relating to foods from the production of primary commodities such as fresh fruits and vegetables, roots, herbs and spices, tubers and other produce, to highly processed products, including acidified and low acid canned foods, snacks, dairy products, seafood, meats, and beverages. Many of these foods are marketed and consumed by individuals in locations sometimes far removed the processing/packaging location. The five main materials used to fabricate packaging for foods are glass, metal, plastic, paperboard and wood. For many applications, selected combinations of plastics and/or paperboard are used and they are called composite materials.

Types of packaging

Glass packaging

Glass has been used for centuries to fabricate bottles and jars for food packaging applications. Just before the turn of the last century, the use of glass as a material for food packaging was significantly reduced and replaced by plastic. Many factors account for this occurrence. These include but are not limited to lower cost, lighter weight, ease of fabrication, the unbreakable nature of plastics, ready availability, and plastic being inert in many food product applications. Notwithstanding these advantages of plastic packaging, glass is still considered to be the gold standard by consumers for food packaging because of its inert nature, impermeability to gases, and zero leaching of chemicals and/or scalping of flavors to and from packaged food products. As a result, glass does not affect the taste of packaged food, and it maintains its integrity while in contact with almost all foods for a very long time. In addition to this, glass is tolerant to the heat of food processing and the cold of refrigeration and freezing without a loss of integrity. It is also environmentally friendly since it is easily recycled. It is very versatile since it comes in different shapes, formats and sizes (Fig. 6.1).

Notwithstanding these advantages, glass is breakable when subjected to excessive mechanical or thermal shock. It is relatively heavy and increases the cost of transportation when compared with other types of packaging. It is also expensive to produce and requires a significantly higher amount of energy to manufacture. To reduce the cost of producing glass packaging for food, manufacturers have significantly reduced the thickness of the containers and have almost eliminated the use of returnable bottles and jars. As a result, there is more recycling of single use containers and this has served to reduce the cost of the raw materials and the energy required during the glass smelting process.

FIGURE 6.1 Glass containers used in food packaging. *Source: Pascall, M.A. 2019. Personal photographs taken my this author.*

Metal packaging

The use of metal to fabricate packages goes back to the era of Napoleon during the 19th century. Today, the manufacture and use of metal cans for food packaging is a global multibillion-dollar business. Containers made from steel, tin and aluminum are considered to be rigid packaging. This is also the case with glass containers because they are not considered to be flexible without irreparable damage. However, the rigidity of metal packaging is one of its advantages (Fig. 6.2). It offers excellent protection to the packaged food since it is not breakable like glass, and it is less susceptible to damage caused by mechanical and thermal shock. As a result, metal packaging is extensively used for heat processed foods and others that are packaged under positive and negative pressures. Consequently, properly processed foods can be stored in metal containers for an extended period of time without refrigeration, as long as there is no interaction between the food and the metal surface. However, the disadvantages experienced with metal packaging is its susceptibility to corrosion that could be caused by certain foods, if it is not adequately protected. In addition to this, metal is more expensive than plastic, for example, and is relatively heavier in weight. Like glass, metal is expensive to manufacture, and the smelting process used to make it is energy intensive. Notwithstanding this, the use and disposal of metal packaging is friendly to the environment since almost all metal containers can be recycled. If not recycled, they are capable of being broken down in the environment (e.g. in a landfill) in a relatively shorter time when compared with plastic and glass packaging materials.

Plastic packaging

Of all the material types, plastic is the most varied, and even though it has been extensively researched, new and modified material are still coming onto the market, even at

FIGURE 6.2 Metal containers used in food packaging. *Source: Pascall, M.A. 2019. Personal photographs taken my this author.*

this time of writing. As a result, plastics have significantly replaced glass and metal packaging in several applications. This accelerated during the 1990s and into the 21st century. Plastics used for food packaging generally fall into two main categories. These are synthetic petroleum-based and bio-based plastics. Almost all synthetic plastics used in food packaging are thermoplastic in nature, although in a small number of cases, thermosets are used. Thermoplastic materials are capable of being formed into a desired shape but can be melted again and reformed into new shapes. Thermosets on the other hand, cannot be melted and be reformed after initially being shaped. Any attempt to do so would destroy the chemical morphology of the material.

Packages fabricated from plastics include but are not limited to bottles, jars, trays, cups, pouches, envelopes, bags, drums, and pails in all shapes and sizes (Fig. 6.3). No other material type has the versatility of plastics for the fabrication of different packaging types and end use applications. Very few food items in modern grocery stores are incapable of being packaged in plastic containers. Plastic packages have now been designed to withstand the heat of retorting, the bombardment of irradiation and microwave energy, the chill of frigid conditions, the polarity of acidic, alcoholic and aqueous liquid foods, the non-polarity of fats, oils and other lipids, and the pressures of gases in the headspace or the bulk phase of various foods. Plastic is now commonplace for the packaging of agricultural products such as fruits, vegetables and root crops, dairy products of all kinds, meats, poultry, seafood, cereals, grains and spices. It is also used for processed foods such as bakery products, beverages, and snacks of all types, as examples.

In recent times, the use of bio-based plastics has gained popularity. Although higher in cost when compared with petroleum-based plastics, this trend is influenced by consumers' increasing concerns about the negative impact of plastics on the environment and recent legislative actions by governments in several countries around the world. Consumers are also becoming increasing concerned about the migration of chemicals from synthetic

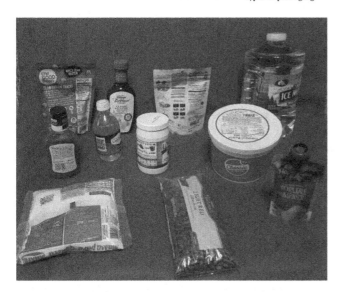

FIGURE 6.3 Plastic containers used in food packaging. *Source: Pascall, M.A. 2019. Personal photographs taken my this author.*

plastics to packaged foods and the potential negative health impact of these compounds. On the contrary, since bio-based plastics are mainly made from polysaccharides and proteins, they are considered to be sustainable, and easily degrade naturally when disposed in landfills or when discarded in the environment. Additionally, they are made with less chemical additives when compared with synthetic (petroleum-based) plastics. Biodegradable plastics can be made from the metabolic by-products of bacteria and from extracts and by-products from insects, shellfish, plants and animals. Examples of polysaccharides derived from these sources and used to make biodegradable plastics include pectin, cellulose, tapioca, chitosan, carrageenan and alginate. Examples of proteins used to make bio-based plastics include whey, corn zein, collagen and gelatin. Because these materials are food-grade, this has also given rise to their use to make *edible plastics*. These offer the additional benefit of being suitable for making coatings which can be eaten as part of the food, in addition to facilitating the extension of shelf life for many products, as well as a variety of new marketing and display options.

Paper packaging

Paper and paper-based materials are extensively used in food packaging. Paper is made from wood derived from trees, stems of plants, straw, cotton and linen. Recycled paper is another source, but this chapter will discuss virgin paper and paper products. Trees and stems are the main sources for paper of all types. For food packaging applications, paper is used to make sheets, bags, envelops, pouches, folding cartons, corrugated shipping containers, tubes, drums, protective dividers, cushioning materials and labels (Fig. 6.4). Paper is also used as a food contact material as plates, cups, bowls, etc. Paper that is made from trees tend to have longer and stronger fibers. These are better suited for paper that is used in applications where mechanical strength is required. Examples of hardwood trees used

FIGURE 6.4   Paper-based containers used for food packaging. *Source: Pascall, M.A. 2019. Personal photographs taken my this author.*

to make paper include Poplar, Aspen and Maple, while softwood trees include Spruce, Hemlock and Pine. Paper that is made from softwood tends to have longer and stronger fibers and is used in making linerboard. Paper made from hardwood has fibers that are shorter but stronger in tensile strength and tend to make a better paper for printing purposes and corrugated flutes. Paper that is made from plant stems tends to have shorter fibers and are better suited for applications such as tissue paper. Paper made from shorter fibers tend to be smoother and have more consistency in texture. Examples of plant stems used to make paper include jute, flax, kenaf, banana stocks and sugar cane bagasse.

Irrespective of the source of the paper, the main ingredient in its manufacture is cellulose. Cellulose is the main compound in the walls of plant cells and vegetable fiber. It is an insoluble complex polysaccharide and is held together in wood by lignin. In the manufacture of paper, it is essential that the cellulose be separated from the lignin and other carbohydrates. This is usually done by either mechanical hammering or by chemical treatment, or a combination of both. Once the fiber or pulp is separated, it is initially mixed with water and other chemicals, the resultant mixture being called the "furnish." To remove the water and other liquids from the furnish, and add mechanical strength to the paper, three main types of papermaking machines are used. These are the Twin-wire, Cylindrical and Fourdrinier Machines. These machines are designed to remove most of the water from the furnish before it is fed to a series of rollers that heat and squeeze out the remainder of the water. This causes an increasing amount of mechanical strength to develop in the paper. Some of the rollers perform a process known as calendaring of the now relatively dried paper. During this process clay, kaolinite, bentonite, calcium chloride or titanium dioxide or other compounds are used to aid in polishing the paper, depending on its end use applications. Paper that is used for printing purposes such as labels, tend to be more highly polished when compared with lower quality paper.

Paper that is used to make folding cartons and corrugated shipping containers are made with long and strong fibers. This facilitates mechanical strength and the ability to crease or score the material without the development of cracks. Corrugated cartons are made with linerboard containing flutes that are sandwiched between sheets of Kraft paper. Depending on the weight, density, shape and size of the product to be held in shipping cartons, the flutes in the linerboard may be single, double or even triple lined. Natural Kraft paper is one of the strongest sheets of paper that are in commercial use. In addition to its use in making linerboard, it is widely used to make packages such as grocery bags for example, where mechanical strength is essential.

For applications where paper products will encounter moisture, water resistant compounds such as waxes, resins, and polymers such as polyethylene are used. In some applications, the paper is laminated with polyolefins or Mylar and used to make moisture resistant materials for packages such as brick-type and gable-top containers that are used for containing juices and other beverages. In some applications, the laminated materials may also contain aluminum foil that provides a barrier against gases such as oxygen, water vapor and carbon dioxide.

Wood packaging

Wood is also used in food packaging but is mainly used for wholesale trade to make boxes and crates and for making pallets for shipping and handling purposes. Due to its relatively higher cost, when compared with other packaging material types, it is seldom used as a food contact material. Exceptions to this are in high quality, unique packaging that are emerging in some markets (discussed in Chapter 7). An example of this is seen with high priced products like aged alcoholic beverages such as rums, wines and whiskeys, where the product may be shipped in casks (barrels or kegs) made of wood (Fig. 6.5). In some applications, wood is used as sawdust or wood chips for cushioning. Wood used to make crates and boxes can be in the form of solid sheets of boards, particle board or plywood. Pallets are almost always made of solid sheets of wood because of the rough handling that they encounter during the movement, warehousing and

FIGURE 6.5 Wooden barrels used for holding, aging and shipping alcoholic beverages. *Source: Gordon, A. 2018. Personal photographs taken by book editor Gordon, A.*

transportation of food products. All wooden packaging requires special treatment in order to comply with international regulations, if they are to be used for products exported from one country to another (see Chapter 7 for further details).

Composites

Composite packaging materials refer to those made from more than one type of packaging material. In this type of packaging, the sublayers are usually bonded together to form laminates. Thus, paper, metal and plastic materials can be bonded together to make a composite material with properties superior to the individual sublayers. This is facilitated by combining the strengths of the sublayers. For example, paper-based materials have poor barrier properties to gases and moisture but are light weight, have excellent printing properties and hold their shapes well, when compared with other types of flexible materials. Aluminum foil laminated to plastic makes excellent pouches that are lightweight, low cost and provide good barrier to moisture, flavors and other gases and vapors. In some cases, the aluminum is bonded to the polymer by a method called sputtering. Other methods to bond the metal to plastic may include using surface treatments such as flame, corona, plasma or other ionization methods. A wax coated folding carton is another example of a composite material that is widely used for frozen packaging of dairy products and other liquids. Some materials are by nature easily recyclable while others are not so easily recycled. Composite materials are usually difficult to recycle because of the challenge in separating the individual component layers.

Different packaging options

Packaging containers are classified depending on their use in the food industry as primary, secondary and tertiary packaging. A primary package is one that is in direct contact with the food, irrespective of the food being liquid, semi-solid, powdered, solid or gaseous at room, refrigerated or frozen temperatures. Primary packaging is used to protect, preserve, contain and transport food, including the provision of information about the product. These types of packaging are intended to offer protection from accidental or intentional biological, chemical and physical contamination. These contaminants include items such as dirt, microorganisms, various industrial and household chemicals, gases like oxygen, carbon dioxide and water vapor, plus other concerns such as light, infestation and pilferage. In the area of preservation, primary packaging prevents a loss of content, including weight, flavor, nutrients, texture and aromas. Primary packaging offers the convenience of taking food items to desired locations and making them available on demand by the consumer. Materials used to make primary packaging include glass, metals such as steel, tin and aluminum, plastics, paper-based products and composites. The main packaging types that are fabricated from these materials include bottles, jars, cans, pouches, bags, trays, cups, bowls, tubes and tubs (Fig. 6.6). In many instances, primary packages also serve the purpose of labels, being printed with relevant user or consumer information. Without the printing of information about the product on the label of primary packaging,

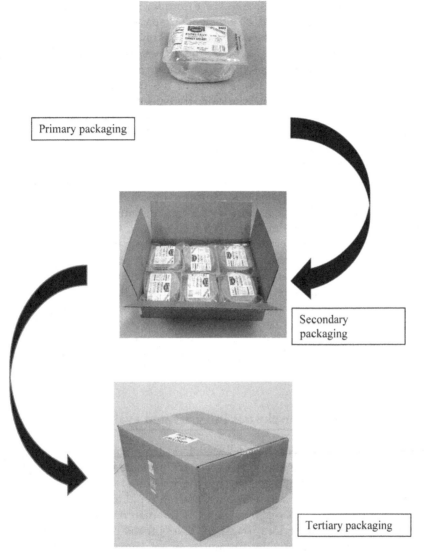

FIGURE 6.6 Primary, secondary and tertiary packaging. *Source: Pascall, M.A. 2019. Personal photographs taken my this author.*

in many cases, consumers would be clueless about the identity, manufacturer, nutritional content, ingredients, handling instructions and storage of many foods that are ingested each day. The labeling aspects of packaging are discussed further in Chapter 7.

A secondary package is designed to contain primary packaging (Fig. 6.6). These could be single or multiple primary packages. Secondary packaging plays a significant role in the marketing of the product. In some cases, it is difficult to display certain types of primary packages because of their shape. For example, in grocery stores, it is difficult to display some products in tubes and be efficient in space utilization without putting the tubes in secondary package made of paperboard. Secondary packaging helps to increase the sales of products. For example, a six-pack of beer, or twelve soft-drink containers within a secondary package increases the sale of the product because it limits the option of the consumer from purchasing single units of the commodity. Secondary packaging increases the convenience of purchasing and transporting small quantities of primary packages at the retail level. It also facilitates the bulk packing of many processed products onto pallets for warehousing, shipping and handling. Materials used to make secondary packaging are mainly paper-based and plastics. These can be in the form of folding cartons and shrink wraps as examples.

Tertiary packaging is mainly used for shipping, warehousing and handling of products packaged in primary and secondary packs during wholesale trade. In many cases, tertiary packaging is designed to offer a certain level of cushioning to the primary product, in addition to acting as an implement for bulk transportation. Corrugated containers are the main packaging type used as tertiary packages (Fig. 6.6). In most developed countries, manufacturers of corrugated containers must provide buyers with a box certificate which indicates the maximum load that the package is designed to contain. The certificate also provides information on the strength of the container walls and it puncture resistance properties. This is designed to minimize accidents by serving as a guide to food manufacturers and shipping agents, and to prevent overloading and subsequent breakage of the container. Countries not having similar regulations should consider this when contemplating exporting products to developed countries that require box certificates on shipping containers.

Container closure integrity and evaluation

In order to adequately protect the packaged food from contamination and loss of contents, it is essential that the integrity of the primary package be maintained from the time of filling and sealing until it is opened by the consumer. Container integrity focuses mainly on the quality of the closure (capping/sealing type) and the method used to close the package. There is also focus on the quality of the material used to fabricate the package. At the food manufacturing location, there is less focus on the material itself because most food manufacturers are not the producers of packaging materials. In some operations, the same company that processes the food, fills and seals the containers may also manufacture the primary packaging. For example, some processors of beverages may also have extruders that produce empty plastic bottles.

Methods used to close packages include caps on bottles and jars, double seams on cans, the use of adhesives and the heat sealing of plastic pouches, trays, cups, bowls, etc. Closures used on bottles and jars include continuous and multiple-start threaded caps, lug-types, roll-on, press-on twist-off, crowns, and pressure plug caps (Fig. 6.7). These may be fitted with a liner or they could be liner-less caps. The main methods used to close plastic containers include conduction, impulse, ultrasonic and induction sealing equipment. Although used in limited applications, other types of plastic sealing methods in

FIGURE 6.7 Bottle and jar closures. *Source: Pascall, M.A. 2019. Personal photographs taken my this author.*

commercial use include spin welding, chemical solvent methods, etc. Container integrity has a direct influence on the safety and quality of the packaged product and thus on its shelf life. As a result, it is essential that food manufacturers invest in container integrity quality control as a normal part of food manufacturing and processing.

Quality control testing can be done using on-line or off-line methods. Off-line integrity testing involves statistical sampling of the production output at predetermined times during the food processing and packaging operations. In some cases, it involves using destructive test methods to determine the quality of the package closure. On-line testing on the other hand, is designed to test each package produced using non-destructive testing methods. In programming the equipment for on-line non-destructive testing, it is essential to establish minimum and maximum limits for defect identification and rejection criteria by the equipment. This is crucial if false positives and/or false negatives are not to be produced by the equipment. On-line testing only works well if there is consistency in the shape and dimensions of the package. This is so because the characteristics of an acceptable container must be imaged and stored in the memory of the computer in the equipment so that it could compare each package and decide which one to accept and which to reject. Irrespective of the type of package and the detection method used, defects fall into three main categories. These are critical, major and minor defects (FDA, 2014), where containers with critical defects must be discarded, those with major defects may be discarded depending on the nature of the defect, and those with minor defects are considered not significant to the shelf life, quality and safety of the packaged product.

Integrity testing of glass containers

Integrity testing of empty glass containers can be divided into different categories depending on the location of the defect and its nature. These include defects on the finish, neck, shoulder, and sidewall or on the bottom of the container. Fig. 6.8 shows these different parts of a glass container. Most defects in glass containers display themselves in the form of cracks of various types. Depending on the nature and location of the cracks, they are given specific names in the glass manufacturing industry (Bucher Emart Glass, 2017).

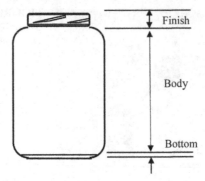

FIGURE 6.8 The parts of a glass container showing finish, body and bottom. *Source: Pascall, M.A. 2019. Personal photographs taken my this author.*

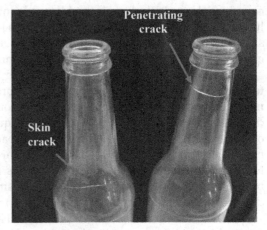

FIGURE 6.9 A glass container with a stone and bottles with two different types of cracks. *Source: Pascall, M.A. 2019. Personal photographs taken my this author.*

Examples of these include cracks called checks. These are usually very small cracks that are located around the finish of the container. Various types of cracks can occur on the shoulder and body of the container, some are called skin cracks when they do not go all the way through the wall of the container, while others penetrate the wall (Fig. 6.9). Some are also called hairline cracks when they appear on the body, shoulder or base of the container and are very narrow in dimension.

Other types of defects in glass containers show themselves as blemishes in the glass. Examples of these include small circular inclusions called seeds. Larger sized air bubbles are called blisters and small pebbles that are unintentionally left in the raw materials and are not melted during the smelting of the glass are called stones (Fig. 6.9). Inclusions in the glass could also come from lubrication obtained from the bottle forming machine or conveyor system before the glass solidifies. Some defects in glass bottles are dimensional in nature. Examples of these include bulged ring, choke bore and incorrect measurements

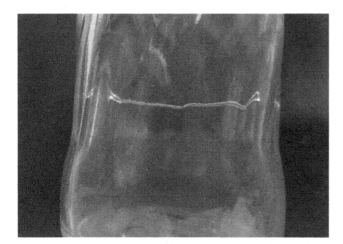

FIGURE 6.10 A bird-swing defect in a glass container. *Source: Pascall, M.A. 2019. Personal photographs taken my this author.*

of the threaded finish on the bottle. Defects that could fall into this category also include incorrect capacity of the container due to uneven wall distribution or the height of the bottle. Another category of defect is related to the molecular structure of the glass that went into making the container. If not proper smelted, glass could develop internal strains, and these can show as inhomogeneity when viewed under a polarized microscope. Other types of strains within the glass could come from uneven cooling of the glass after it comes from the bottle forming (molding) machine. To solve this problem, a process known as annealing is performed. This process heats the glass up to a predetermined degree above its glass transition temperature, then slowly cools it back to room temperature. Some defects can be caused by glass inclusions within the container. An example of this is a bird-swing. This appears as a cord-like piece of glass that bridges one side of the inner wall of the container to that of the opposite wall, hence the name "bird-swing" (Fig. 6.10).

Various types of quality control tests are routinely performed on glassware to ensure that good and defect-free containers are manufactured and delivered to customers. These include a series of physical, optical and chemical tests. Liu et al. (2008) reported on online non-destructive machine vision testing for glass containers. Physical tests include dimensional analyses, internal pressure, thermal shock and impact resistance, capping worthiness, and abrasion resistance as examples. Optical testing includes color measurements, annealing and Polaroid microscopy analysis. Examples of chemical testing include the quantification of compounds from the raw materials — sodium, calcium, aluminum, iron, silica, etc. These analyses are done by dissolving the glass in hydrofluoric acid before performing the quantification using instruments such as inductively coupled plasma or atomic absorption spectroscopy. The density of the glass is usually measured on a daily basis since it gives an indication of the ionic composition of the glass.

After the glass containers are delivered to the food processor and they are filled and sealed with the food product, various tests are usually done to ensure that the containers are hermetically sealed. This is to ensure that there are no leaks. If the containers are retorted or used for aseptic packaging of foods, governmental regulations require certain

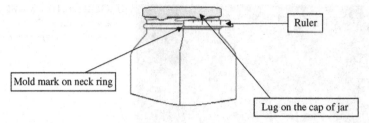

FIGURE 6.11   The pull-up test on a glass jar. *Source: Pascall, M.A. 2019. Personal photographs taken my this author.*

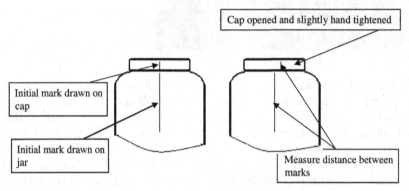

FIGURE 6.12   The security tests on a glass container. *Source: Pascall, M.A. 2019. Personal photographs taken my this author.*

types of capping tests. Examples of these include the vacuum, pull-up and security tests (Figs. 6.11 and 6.12). During the pull-up test, the containers are taken from the filler and the distance for the leading edge of the lug on the cap to the mold mark on the finish is measured without removing the cap. For the security test, a pen is used to place a vertical mark on the cap and another vertical mark on the body of the container as shown in Fig. 6.12. The cap is then removed from the container and replaced, but only finger tightened. The distance from the two marks is then measured. The tolerances for distances measured for the pull-up and security tests are usually provided by the container manufacturer and should be used to determine if the capper is applying an appropriate torque to the containers (Lin et al., 2001).

Integrity testing of metal containers

Two types of metal containers are mainly used for food packaging. These are two and three-piece cans. These can be mainly made from tin-coated steel, tin-free steel or from steel coated with a corrosion resistant compound. Aluminum cans are almost all made as two-piece cans. Both types of cans are used for beverage packaging, but both two and

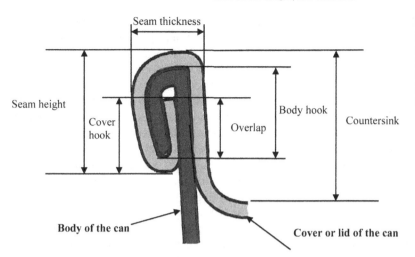

FIGURE 6.13 Cross-section of double seam showing sealing of the lid to the body of metal can. *Source: Pascall, M.A. 2019. Personal photographs taken my this author.*

three-piece steel cans are also used to package a variety of foods that are retorted[1] or asepti-cally processed[2]. Both two and three-piece cans are sealed with a cover or lid after they are filled with product at the food processor's location. Two-piece cans are made from two pieces of metal sheets. One piece is hammered and stretched to form the body of the can and the other piece stamped to form the lid. Three-piece cans are made from three pieces of metal sheets. One piece is cut and welded to form a cylinder. This is then shaped to form the body of the can while a second piece of the metal sheet is stamped into the lid and attached to the cylindrical body of the can. This occurs at the can manufacturer's facilities, so this end is called the "manufacturer's end". The third piece of metal sheet is also stamped into a lid, but this is put on to the can after it is filled with food by the food processor (the "canner's end"). Irrespective of the type of can, it is essential that the seal on the cylinder that forms the body, and the seal on the lids of the can must be defect free in order to form a hermetical seal[3] after the container is closed. Most can manufacturers perform various types of machine vision and pressure differential testing to ensure that the empty cans are free from leaks and metal frag-ments, and that the welding of the body seal is defect free.

The joining of the lid to the body of the can, irrespective of it being two or three-piece cans, is done through the formation of a "double seam" (Fig. 6.13). If not done properly, vari-ous types of defects can result (Weddig et al., 2007). These can be caused by incorrect dimen-sions of the lid or the body where they join together to form the double seam. Thus, a short or long body hook or cover hook could lead to a short overlapping of these two parts of the can and this could result in a weak seal that could fail under the pressure of retorting. Other

---

[1] Retorting is the thermal processing of food in a hermetically sealed container using steam under pressure.

[2] Aseptic processing is a process in which a product and its package are sterilized separately and then combined and sealed under sterile (aseptic) conditions. This can be done by form, fill and seal technology such as Tetra Pak or using other systems.

[3] A hermetic seal is an airtight seal that precludes the passage of air, oxygen or other gases and also excludes microorganisms and prevents their entry into the container.

types of defects could result from incorrect joining of the lid to the body by the rollers of the can seaming machine. This could give rise to defects such as loose, tight or false seams, dead-head, knock-down flange or droop as examples. Other defects could result from too much pressure of the chuck on the can seaming machine and this could lead to a short overlapping of the cover to the body of the can. To ensure that cans are free of can seam defects, they are usually tested destructively by a process called can seam "teardown". In this process, the cans are disassembled and the cover hook, body hook, overlap, seam thickness and seam width (Fig. 6.13) are optically (using an instrument) or manually measured. The containers are also examined for other types of defects including dimensions, cuts, shear stresses, abrasions, and damage to the containers caused by incorrect pressures within the retort, if so heat treated. Fig. 6.14 shows the use of a can seam micrometer to measure the seam thickness and the seam height, while a caliper is used to measure the depth of the countersink (Fig. 6.14). These measurements can also be taken by digital instruments that could directly feed the data to a computer software designed by several can equipment testing manufacturers. Companies in North America and Europe that produce these digital instruments include CMC-Kuhnke, Quality By Vision, Waco Wilkens-Anderson Company, One Vision, etc. Other companies can also be found in Asia.

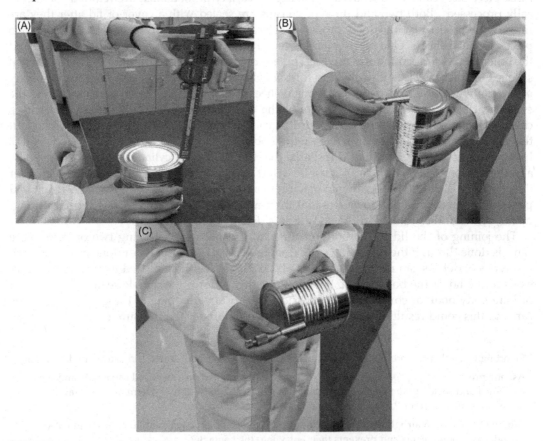

FIGURE 6.14 Measuring countersink (A), seam thickness (B), seam height (C) using a micrometer. *Source: Pascall, M.A. 2019. Personal photographs taken my this author.*

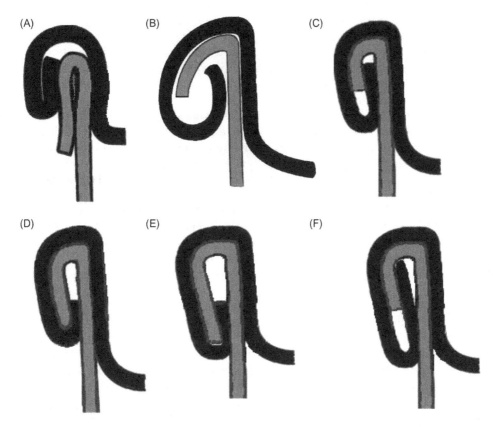

FIGURE 6.15　False seam (A), loose seam (B), short body hook (C), short cover hook (D), long body hook (E), and long cover hook (F) defects. *Source: Pascall, M.A. 2019. Personal photographs taken my this author.*

Fig. 6.15 shows false seam, loose seam and short body hook, short cover hook, long body hook and long cover hook defects that could occur in the double seam of cans used for food packaging. If these and other types of defect occur in cans filled with food product that is treated by retorting, the hermetic seal could be lost, and the product will spoil. These types of defects could cause serious cases of outbreaks if pathogenic bacteria enter and contaminated the product.

Because of the widespread use of canning for food preservation and processing in developing countries, further details on canning and quality assurance are discussed in Chapter 7.

Integrity testing of plastic containers

The integrity testing of plastic containers depends on the type of package that was fabricated or manufactured from the polymeric material. Plastic pouches and bags are almost all closed by some form of heat-sealing method. The type of sealing method is determined by the nature of the product, available equipment, cost, the sealing surfaces themselves and the processing conditions that the packages are expected to experience. Irrespective of the type of sealing method used, four or five factors are responsible for good closure of the container. These are sufficient heat, pressure, dwell time, the compatibility of the sealing materials, and/or the lack of contamination between the sealing surfaces. The sealing types used to close plastic containers include but are not limited to conduction, impulse, induction and ultrasonic sealing methods.

To test the quality of plastic seals, various leak and seal strength testing methods are used. These include destructive tests such as dye and electrolytic testing. Other types of non-destructive testing that can be used for the detection of leaks include pressure differential, acoustic and infrared, for example. Some of these techniques are also capable of detecting inclusions and voids in the seal and cracks in the material. The method of sealing pouches and bags is sometimes different from that of sealing plastic packages such as cups, trays and bowls with peelable seals (Fig. 6.16). The welded seal on pouches and bags are not designed to be pulled apart by consumers when they try to access the food within the package. In the case of peelable seals, they are designed to be pulled apart and the materials must be synthesized to accommodate this feature.

Because plastic by its nature is not as mechanically strong as metal or glass, seal quality testing of plastic containers also focuses of the strength of the seal. This is so because the package could be free of leaks but may fail during transportation and handling if the strength of the seal is not above a minimum threshold. Non-destructive seal testing such as infrared, ultrasonic, electrical impedance tomography, MRI and other imaging techniques can be modified to correlate the strength of the seal with tensile and compression testing. In these techniques, the imaging method could be used to test "good" packages and the images generated used as pass/fail criteria for subsequently tested packages. These test methods could also be used to test glass containers for defects. Techniques such as electrical impedance and X-rays can be used to test for defects in metal cans. The test methods for plastic bottles and jars are similar to those for glass bottles but since plastic is subjected to creep, caps made from thermoplastic materials are routinely tested for application and removal torque to ensure adequate sealing.

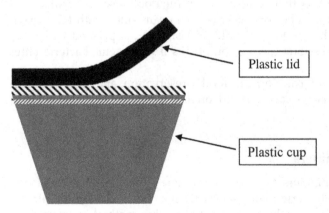

FIGURE 6.16 Plastic cup with a peelable seal. *Source: Pascall, M.A. 2019. Personal photographs taken my this author.*

Plastic lid

Plastic cup

Integrity testing of paper-based containers

Paper-based materials used as packaging for shelf stable food applications are used in composite materials. This is so because of the inability of paper (by itself) to provide barrier against moisture, gases (such as oxygen), and vapors (as flavors, for example). To provide barrier in composites, the paper-based component must be coated with hydrophobic compounds, laminated with high barrier materials such as aluminum foil or silicon dioxide, or compounded with various polymers. Recent research activities have reported on the use of nanoparticles to increase the gas barrier properties of paper-based materials (Song et al., 2014; Iyer et al., 2016). To close paper-based packaging with hermetic seals, it is either necessary to use an adhesive or a heat sealable thermoplastic material laminated to the inner food contact layer. When used to make composite can and tubes, the paper-based components are joined to metal structures that provide the necessary sealing surfaces. Paper-based packaging used for aseptic and retorted products are usually made of laminated structures with polyethylene as the heat-sealing layer for aseptic applications and polypropylene when the packages are retorted (Fig. 6.17).

For rectangular packages, several creases and folds are necessary to form the required shape during fabrication. This sometimes creates opportunities for cracks and subsequent leaks in these types of packages. The non-rigid nature of paper-based materials make them susceptible to punctures from impact with sharp objects. If this occurs after the product is processed, packaged and warehoused on pallets, leaks in packages in the upper part of the pallet could cause wetting of other packages lower down on the pallet, and this could cause the entire pallet to collapse, if the paper absorbs the liquid and softens in the process. On-line quality control pressure differential testing of paper-based packages is reported by Sivaramakrishna et al. (2007). Dye and electrolytic testing could be the statistical destructive testing of choice done to assess the quality of seals on these packages. These tests are capable of indicating and identifying the location of leaks in the packages, should they exist. For composite cans and tubes, where the paper-board material is joined to the metal, care must be

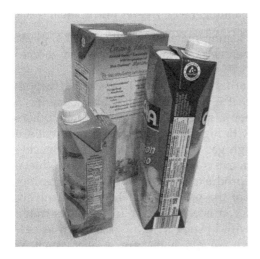

FIGURE 6.17 Brick-type multi-layered laminated packages designed for aseptic processing. *Source: Pascall, M. A. 2019. Personal photographs taken my this author.*

taken to limit the fastening pressure on the metal component. If the pressure is too high, the metal component could cut the paperboard and cause the package to leak (Weddig et al., 2007).

Packaging considerations in new product development

In developing new food products that are intended to be containerized, it is necessary to consider the package/product interaction and its impact on the safety, shelf life, quality, and marketing of the product. This begins by considering the factors that are responsible for a loss of shelf life and how the package could be used to reduce this loss. For this to be successful, there must be compatibility between the product, the package, the processing method used and the storage environment. The end of a product's shelf life occurs when it no longer meets the desired expectation that it had when initially processed. The shelf life of the product can be defined through issues such as the contaminating microbial types and loads that can cause spoilage, sensory characteristics, and chemical, mechanical, physical and nutrient changes that take place in the product during storage. These factors promote spoilage in combination with environmental factors such as temperature, light, time, water activity and partial oxygen pressure. Thus, in the selection of the most appropriate packaging material for a particular product, factors that must be considered include barrier properties to:

- Oxygen
- Moisture
- Light
- Flavor loss
- Migration of packaging additives to the food
- Carbon dioxide
- Microorganisms
- Filth
- Mechanical damage (cushioning, loss of product homogeneity, freezer burns, etc.).

These factors are directly influenced by the packaging material selection (metal, glass, paper or plastic), the method of closing the package, the material dimensions and geometry, and the integrity of the package itself.

By its nature, plastic is porous and the degree of its permeability to gases and vapors varies depending on the type of polymeric units that it is made from. A plastic having a high degree of crystals tends to have a higher barrier to gases when compared with others having a higher amorphous fraction. Thus, foods with high lipid components and that are susceptible to rancidity, should be packaged in high gas barrier plastics or impermeable materials such as glass or metal. The same situation exists for moisture sensitive products. If the package has a low barrier to moisture, these products will absorb water that permeates through the material and it will cause a loss of the product's quality in a relatively short time. Carbonated beverages will keep their consistency longer in packages made with high gas barrier materials than in packages with lower barrier to gases such as carbon dioxide. Composite paper-based materials that contain high barrier layers such as aluminum, will provide gas and flavor barrier comparable to that of glass or metal. Because of its organic chemical composition, plastic packaging materials are prone to withdraw flavors from products such as fruit juices. This is known as flavor scalping and it is capable of significantly reducing the shelf life of such products. A flavor such as limonene that is the active chemical in citrus fruits, is notorious for being scalped by plastic materials, especially by polyolefins such as polyethylene and polypropylene (Licciardello et al., 2009; Mauricio-Iglesias et al., 2011; Číhal et al., 2015).

Plastics are also known to have several organic compounds that are necessary for its synthesis and properties. These include but are not limited to plasticizers, antistatic moieties, flame retardants, defoaming agents, surface treatment compounds, coupling and cross-linking chemicals, etc. One problem with these compounds is that some are compatible with and have affinities for some of the chemicals in food products. This causes some of these compounds to migrate from plastic containers to packaged foods. If the level of this migration exceeds certain legal and toxic limits, the food could then be considered unsafe for human and animal consumption. This could also cause an end to the shelf life of the packaged food. Irrespective of the container being made of 100% plastic or if it has a plastic food contact layer in a composite structure, or it serves as the protective plastic coating in a metal can, the issue of chemical migration is still the same. Glass packaging, on the other hand, due to its inert nature, does not create this problem. In the case of metal containers, this problem can be created if uncoated materials begin to corrode in the interior of the can, resulting from chemical reactions between the base metal and water and acidic components in the food (Waldman, 2015).

Foods which contain lipids are also susceptible to photooxidation. These foods require protection from light for an extended shelf life. Light protection is also essential to extend the shelf life of other foods such as milk-based and meat products, and beer. To solve this problem, these foods are usually packaged in colored or light impermeable materials, in clear plastics with ultraviolet absorbing additives, or in containers wrapped with light blocking labels. These are among the important issues that product development teams must consider when choosing processing methods and appropriate packaging for their new products.

Packaging considerations in risk and hazard analysis and HACCP plans

The purpose of a hazard analysis and critical control points (HACCP) plan is to systematically design strategies that would outline a preventative approach to ensuring the safety of processed foods. The plan requires the articulation of potential biological, chemical, and physical hazards that could be associated with the production activities linked to the given food from the input of raw materials until the delivery of the finish product. The plan also considers the critical points in the production process and sets minimum and maximum limits with appropriate monitoring, to ensure that the process is maintained within these boundaries. Corrective actions, if the process exceeds the limits, regular audits and the keeping of adequate records are other essential elements of the HACCP plan.

A HACCP plan that focuses on the packaging of a particular food should consider the risk of using the chosen packaging material and type. All packaging types with critical and even major defects should be discarded in order to minimize the hazards associated with a loss of hermetical seals. If glass containers are the chosen packaging type, the physical hazard of broken glass falling into the products should be considered. This could occur during the handling of the empty containers, filling or capping of the package. Both glass and plastic bottles and jars could be susceptible to leaks and a loss of the hermetical seal if defects are found in the sealing surface and or threaded region of the container and/or the cap itself. Improper sealing could also result from problems with the capping machine, the inadequate formation of a proper vacuum, or the incorrect application of the required torque for cap sealing.

In the case of metal cans, it is essential that the retorting process is designed to avoid paneling or buckling[4] of the containers during the cool of the retort prior to its opening. If this is not done properly, it could cause temporary leaks to develop in the double seam of the cans, particularly in the case of buckling, and this could cause the cooling water to enter and contaminate the container, especially if chlorinated cooling water is not used in the process. As was discussed above, certain foods packaged in plastic containers can be susceptible to chemical migration under certain conditions. This can create an unsafe condition if the migration exceeds the minimum toxic limit of the permeating chemical compounds.

References

Bucher Emart Glass, 2017. Glass container defects causes & remedies. Bird swing defect. <https://www.emhart-glass.com/system/files/download_center/BEG_BR0060Defect_Guide_0.pdf> (accessed 22.04.19.).

Číhal, P., Vopička, O., Pilnáček, K., Poustka, J., Friess, K., Hajšlová, J., et al., 2015. Aroma scalping characteristics of polybutylene succinate-based films. Polym. Test. 46, 108–115. 2015.

FDA, 2014. Guide to Inspections of Low Acid Canned Food Manufacturers: Part 3. <https://www.fda.gov/ICECI/Inspections/InspectionGuides/ucm109763.htm> (accessed 02.04.19.).

Gordon, A., 2018. Personal photographs taken by book editor Gordon, A.

Iyer, K.A., Lechanski, J., Torkelson, J.M., 2016. Green polypropylene/waste paper composites with superior modulus and crystallization behavior: optimizing specific energy in solid-state shear pulverization for filler size reduction and dispersion. Compos. Part. A: Appl. Sci. Manuf. 83, 47–55.

Licciardello, F., Del Nobile, M.A., Spagna, G., Muratore, G., 2009. Scalping of ethyl octanoate and linalool from a model wine into plastic films. LWT - Food Sci. Technol. 42 (6), 1065–1069.

Lin, R.C., King, P.H., Johnston, M.R., 2001. Examination of Containers for Integrity. In: Bacteriological Analytical Manual, 8th Edition, Revision A, 1998. Chapter 22B. <https://www.fda.gov/Food/FoodScienceResearch/LaboratoryMethods/ucm072694.htm>. (accessed 02.04.19.).

Liu, H., Wang, Y., Duan, F., 2008. Glass bottle inspector based on machine vision. Int. J. Electr. Comput. Eng. 3, 462–467.

Mauricio-Iglesias, M., Peyron, S., Chalier, P., Gontard, N., 2011. Scalping of four aroma compounds by one common (LDPE) and one biosourced (PLA) packaging materials during high pressure treatments. J. Food Eng. 102 (1), 9–15.

Pascall, M.A., 2019. Personal photographs taken my this author.

Sivaramakrishna, V., Raspante, F., Palaniappan, S., Pascall, M.A., 2007. Development of a timesaving leak detection method for brick-type packages. J. Food Eng. 82, 324–332.

Song, Z., Xiao, H., Zhao, Y., 2014. Hydrophobic-modified nano-cellulose fiber/PLA biodegradable composites for lowering water vapor transmission rate (WVTR) of paper. Carbohydr. Polym. 111, 442–448.

[4] Paneling and buckling are can defects that occur as a result of inadequate pressure management during thermal processing. These are discussed along with other can defects in Chapter 7.

Waldman, J., 2015. Coating the can. Rust the Longest War. Simon and Schuster, New York, pp. 72–118.
Weddig, L.M., Balestrini, C.G., Shafer, B.D., 2007. Closures for double seamed metal and plastic containers, Canned Foods: Principles of Thermal Process Control, Acidification and Container Closure Evaluation, seventh ed. GMA Science and Education Foundation, Washington, DC, p. 143163.

Further reading

Gaonkar, S., 2014. Marketing - Project packaging CBSE. Marketing, Business, Entertainment & Humor. <https://www.slideshare.net/Saish97/marketing-packaging-project-cbse> (accessed 15.04.19.).

The role and importance of packaging and labeling in assuring food safety, quality and regulatory compliance of export products II: Packaging & labeling considerations

André Gordon[1] and Rochelle Williams[2]

[1]Chairman & CEO, Technological Solutions Limited, Kingston, Jamaica, West Indies
[2]Technical & Regulatory Compliance, Technological Solutions Limited, Kingston, Jamaica

The importance of packaging for exporting products from developing countries to developed country markets

Background

Packaging and labeling are important to exporters as they both play a critical role in the process of getting products into the importing country and to the consumer. While some thought is often given to both, the label and its importance in meeting market requirements often gets the main focus and the choice and details of the packaging is often left up to the supplier. With the rapid and ongoing changes in regulatory requirements in the major export markets, a focus on labeling and related regulations and compliance with these is absolutely critical as the label is the first point of contact with the product for the regulator in the importing territory, the buyer/distributor and, of course, the consumer. This will be covered in detail for the European Union (EU), United States of America (USA) and Canadian markets later in the chapter. However, even if all labeling and related requirements are met and the product is not properly packaged to be able to attract and meet the needs of the buyer and then the consumer, the product is not likely to be successful in the chosen market. If the packaging of the product does not fit with the system of handling and distribution of the importing entity or, if it does not meet the storage, display, handling and shelf life requirements of the target retail outlets, it will not succeed in the market. Finally, if the packaging is in breach of stipulated regulatory handling and disposal requirements or does not fulfill or meet the unmet needs of the final consumer (information, convenience, reusability, robustness, environmentally responsible, portion size and shelf life, among others), the product will not survive in the export market. Consequently, these technical and market-based considerations of the role and importance of packaging need further examination, particularly in light of the dramatic changes that are, and will continue to occur with food packaging in keeping with changes in populations, demographics, habits, trends and innovation.

The lack of attention and focus on the nature and role of packaging has often resulted in issues that limit the success of the exported product because of a variety of important considerations. In the previous chapter, a comprehensive introduction to the role and importance of packaging was presented. It discussed several issues, including the primary functions of packaging vis-à-vis the food being sold, including protection from the external environment and from damage, as a barrier from contamination and gas (oxygen, moisture) transfer, providing information, and convenience in handling and use. This chapter will discuss the secondary and equally important functions of packaging, food/packaging interactions, novel packaging and innovations, trends driving the sector and the impact of packaging on shelf life. Important considerations in the production and use of flexible packaging for export products and a more detailed examination of container closure and defects examination will also be discussed.

Special considerations for wood and flexible packaging

The discussions in Chapter 6, Packaging Basics, provided and introduction to plastic packaging, composites and wood-based packaging, as well as glass, metal and paper-based packaging. Traditional packaging focused heavily on glass, metal and paper packaging, with plastic-based packaging and wood often getting less attention. While glass, metal cans and paper packaging (including shipping cartons) are important for the food industry in developing countries, being among the major packaging options available, wood and, increasingly, flexible packaging are becoming more important as the industry seeks to move further up the value chain and enter into more sophisticated, mainstream areas of the markets in destination countries. Both of these packaging formats require further attention which is provided in this chapter with special considerations for wood-based packaging materials and flexible packaging, which comprises plastics, plastic-based laminates and films, also getting additional discussion. Also discussed below in detail because of its increasing importance in Europe, North America and globally, is the issue of biodegradability of packing, mainly of plastics and composites. Issues such as ensuring food safety in the manufacture of packaging, in general, using flexible packaging as an example is explored. For ease of reference and the purposes of the rest of this chapter, non-rigid plastic packaging and films will be referred to as flexible packaging.

Secondary functions of packaging

The secondary functions of packaging include playing a critical role in the marketing of the product, providing appropriate portion control for buyers, traceability of the product through the coding on the packaging and providing security for the food contained within it. In this regard, the presence of tamper evident components of the package has attained a level of importance on par with the physical ability of the package to protect and secure the food. Another very important secondary function of packaging is in shelf life extension. This is particularly so as the packaging itself and any accompanying technology associated with it play a central role in shelf life determination, a topic to be discussed in more detail below.

Among the secondary functions of packaging, the ability of the package to secure the product from intentional tampering is important as there have been several prominent cases where intentional contamination of packaging occurred. Consequently, the tertiary and secondary packaging should secure the product from the factory, farm or production site, during transportation or shipping through to arrival at the point of sale where these are likely to be opened so that the product could be displayed in its primary packaging. As such, the nature, design, strength and integrity of any seals included with the secondary or tertiary packaging, including shrink-wrapping, are critical considerations for the functionality and security of the product/packaging combination that is taken to comprise the product as a whole. Of equal, if not more commercial importance is the packaging's role in the marketing of the product. Very attractive packaging such as is shown in Fig. 7.1

FIGURE 7.1   Attractive packaging play an important role in helping to market products. *Source: André Gordon, 2019.*

serve to attract customers to more closely examine the product, if they are not familiar with it and, oftentimes to buy and try the product to see whether it meets their needs. A range of options in convenience, shapes and portion control as is evident in the range of aseptic packaged product (Fig. 7.1B) also helps to encourage the consumer to purchase the product. With a significantly extended shelf life vs. traditionally packaged product, aseptic packaged product such as those from the Middle East shown here (Fig. 7.1B), the major proponents of which are Tetra Pak and Combibloc, have shown significant growth, particularly in Asia/Pacific, Africa and the Middle East and also in South and Central America, including the Caribbean region.

Another increasingly important role that packaging is playing is delivering the kind of portion control that consumers are looking for. This is being achieved through more flexibility and versatility being provided with multiple options for consumers in the sizes in which products that they like are available. By offering different portion sizes, the package allows consumers to purchase in accordance with their needs and circumstance: single use packaging and multiple sub-packages, as appropriate provide size control, facilitate inventory control and convenience. This helps to build demand and loyalty among consumers who want products such as those being offered in the packaging that is being sold. When combined with the informational component of the role of packaging (conveying information about the contents), as well as facilitating proper inventory management, traceability and, if required, effective recall, the secondary roles of packaging add significantly to the merchantability of the product being offered for sale.

A critical role that packaging plays in the marketing and commercial success of many products is in shelf life extension. In addition to the composition and critical factors for the specific product, which are themselves major determinants of the stability of the product, its shelf life and its ability to meet the consumers' unmet needs, packaging greatly influences the overall shelf life of the product. A low acid food such as milk has a completely different shelf life when packaged aseptically after Ultra High Temperature (UHT) processing versus being packaged in a gable top paperboard laminate box which has to be kept refrigerated. The UHT product is shelf stable at room temperature while the milk packed in the box will not last for more than a few days if left at room temperature. Similarly, meat (e.g. frankfurters) will keep for well over a year if packed under a hermetic seal in a can, as against being presented in a vacuum sealed laminate package, even when refrigerated. In developing countries where producers typically market their foods first on the domestic market before getting into exports, the choice of packaging and packaging technology will significantly influence the shelf life, and therefore the commercial opportunities for the product in the export market.

Developing country exporters have begun to utilize a much wider range of packaging options as they gain an increasingly greater share of the markets in developed countries where expectations for the presentation, utility, performance, safety and informational content of packaging may vary significantly from their home market. While it is not always the case, legislation in many developing countries as regards labeling, consumer protection and food safety, particularly the latter (Stier et al., 2002), tends to lag significantly behind their developed country counterparts, thereby oftentimes requiring a completely different approach to the packaging and labeling of products destined for export markets. This, along with the established and developing trends in metropolitan markets, often leads to changes in the packaging technology employed for developing country products being exported to these markets and also benefit the domestic markets as firms produce for all markets using the same technology.

Trends in food packaging for exporters

Consumers in the major metropolitan markets into which developing country exports are going in greater numbers have growing expectations that the food that they consume and the packaging in which it is conveyed will reflect their changing tastes, needs and values. This has continued to drive the changes in consumer preferences, expectations and purchasing patterns, all of which favor those producers who understand and cater for the needs of these consumers. As has been emerging over the last 5 years or so, consumers in Japan, the United Kingdom, Mainland Europe, Australia, New Zealand, Canada and, more latterly, the USA have continued on a path where choices have been driven by the following factors:

- The environment
- Logistics
- Product protection
- Health
- Utility/Usability
- Economy

among others. This has resulted in their expectations of their food and its packaging being colored by specific considerations. In terms of *the environment*, more and more developed country consumers are expecting that their packaging will be made with a minimal use of resources from the environment, that they will be easy to empty and use and that they will be easy to dispose of in a manner that is respectful of the environment when they are finished with them. They expect their packaging to be *utilitarian* and therefore easy to open, hold and close with easy to understand and complete information about the product contained in it. *Logistically*, they expect that the packaging will be easy to handle, transport and store, it will be *economical* in terms of cost, will provide adequate protection from contamination and will preserve the healthy nature of the food it contains. Consequently, this means that food and beverage manufacturers and producers trading in the global market, will have to ensure that the food products that they are offering for sale and the packaging in which it is contained are:

- effective in delivering the taste, freshness and nutrition expected by the consumer
- aligned with their expectations for "clean" ingredients delivering more healthful food
- packaged responsibly, with minimal waste and
- delivered in a manner that shares their values.

Among the key drivers for the growing trend in food packaging options is the "Evolving Consumer". Today's consumers are liberated, confident and assertive in seeking the value propositions they want and are also self-expressive. With the change in attitude

TABLE 7.1 Changes in packaging preferences by consumers over the last 30 yrs[a].

Product	Conventional	Current
Milk	Glass, metal	Film, Pouches
Beverage	Glass	PET
Pharmaceuticals	Paper & glass	PVC, HDPE, PP, Glass
Toothpaste Tubes	Metal	HDPE, LLDPE
Soaps	Paper	Polyester Film, PVC
Fertilizer	Jute	PP woven sacs
Retail Carrier Bas	Paper, jute bags	LDPE, HM HDPE

[a]Definitions: PET – polyethylene terephthalate; PVC – polyvinylidene chloride; LLDPE – Linear low-density polyethylene; LDPE – low density polyethylene; HDPE high density polyethylene; HM HDPE – high molecular weight high density polyethylene; PP – polypropylene.

and increase in the number of working women, there has been a shift in the trends for the packaging of food in the last three decades. This is summarized below.

It is clear from Table 7.1 that the trend for packaging materials has changed from rigid to flexible and from heavy to light and elegant. From the manufacturers' and producers' perspective, satisfying these needs requires greater flexibility of equipment and the ability to scale rapidly if the growth in number of stock keeping units (SKUs) and in each SKU explodes as a result of increased demand. They will also need to have a greater understanding of their products, their processes, the regulatory requirements of the market and the trends driving them. These are among the major changes over the last few years as reflected in the findings of industry specific and other research which has shown both a growing and coalescing of trends across several markets. This can drive significantly greater value to food industry practitioners if they understand them and satisfy the needs of consumers. These and other emerging trends are discussed below.

Making a statement

Consumers in Europe have long used their purchasing power to strongly support causes to which they were partial such as fair trading practices, sustainable fishing and farming practices and waste reduction, among others. North American consumers have now begun more and more to buy products that say what they believe in and stand for. More than 70% of Americans are now more loyal to purpose-driven brands and are willing to defend them (Scroggins, 2019). For an increasing number of consumers, it is no longer acceptable for companies to simply make money with just under 80% expecting firms to positively impact society as well. Further, more than 60% of American consumers today would change brands better aligned to their issues and more than 55% are willing to pay more for products made by these companies (Scroggins, 2019). Globally, therefore for the foreseeable future, the opportunity exists for products from developing countries, properly marketed and packaged, to reach consumers willing to pay better prices for them if the story of their production and the value chain involved aligns with the beliefs and values of the consumers.

Sustainability

Consumers are pushing hard for their governments, retailers and the food industry in general to take urgent and sustained steps to employ sustainable production practices that minimize or eliminate negative impacts on the environment. The North American industry is beginning to follow Europe's lead with the retail trade taking greater responsibility for the management of recycling of packaging and products as well as insisting on smaller package sizes (Evergreen Packaging, 2018; Scroggins, 2019). Governments will also move more and more to impose punitive taxation on packaging that is not easily re-usable or recyclable.

Telling a story/sharing values

Consumers are now making choices based on the corporate social responsibility of producers or the perceived environmental impact of the production, handling and distribution process for the products being bought. European consumers led the trend of modifying purchasing decisions to favor firms and products that could demonstrate environmental responsibility through a limited carbon footprint and the employment of sustainable production practices. Wiley (2018) reported that 67% of grocery shoppers surveyed agreed that retailers should choose products to be stocked based on the environmental friendliness of both the product and its packaging. Further, more than 50% of consumers avoid purchasing products of producers that were not socially responsible or who did not have environmentally responsible practices (Evergreen Packaging, 2018; Wiley, 2018). By these means, consumers in the foreseeable future in North America and Europe will live their stories through their purchases and seek to align with firms and products that share their values.

Convenience

A major driver now and in the foreseeable future for consumers in more developed markets is convenience. This includes portability, being able to have food on the go, have smaller and more convenient portion sizes, and be able to transport, store and handle food in a manner compatible with their lifestyle. These properties of the food are mediated mainly by the packaging and delivering these features should therefore be a major focus of the industry value chain.

Healthier packaging & cleaner labels

As consumers focus more on eating healthier, more natural foods, they are demanding packaging that is better aligned with the ingredients of the product which it contains (Evergreen Packaging, 2018; Wiley, 2018). Consumers who want healthier ingredients also want healthier packaging. A survey found that 71% of grocery shoppers agree that healthier foods and beverages should use healthier packaging materials (Evergreen Packaging, 2018), including 61% of who want to see greater use of recyclable packaging. Further, 65% expect organic products and healthy beverage brands to offer more and better alternatives to non-biodegradable plastic packaging (Wiley, 2018).

The rise of private labels

One of the trends that has emerged is a rise in opportunities for the manufacturing of a range of food products for principals under their own private labels (LaCroix, 2017; Scroggins, 2019). Distributors such as Carrefour (France), Albert Heijn (The Netherlands), ASDA, Sainsbury, Tesco and Marks and Spencer (the United Kingdom), Loblaws (Canada) and Walmart, Publix, Whole Foods and Trader Joe's (the United States of America) all significantly expanded the shelf facing taken by their own house and other private brands. Contract packing opportunities currently abound if the producer can meet the Food Safety and Quality Systems (FSQS) requirements stipulated by the principals for whom the contract packaging is to be done. This is a major opportunity for the agribusiness sector in developing countries.

Vintage packaging

Investigators have seen an increasing trend towards consumer preference of packaging that has a vintage look and feel, inclusive of graphic elements reminiscent of the past but with a modern twist. This trend is particularly important for artisanal products or products that have a homemade quality and appeal. Products packaged such as this convey a sense of authenticity (another trend for food) and exclusivity, all of which consumers are willing to pay higher prices for. All of this augurs well for authentic, traditional flavorful products from developing countries that are packaged in a manner that meets these articulated needs of these discerning consumers.

Responsible packaging

The nature and amount of packaging that are part of products being offered for sale has become one of the driving factors in consumer purchasing decisions in North America, this is finally beginning to catch up with a long existing trend in the EU. Consumers indicate that they are now both choosing what they buy and changing buying decisions based on the type and amount of packaging involved with the product (Wiley 2018). Consumers are paying attention to the quantity and recyclability of the packaging used, whether it uses renewable material, inclusive of plant-based material such as bio-based, biodegradable packaging (Evergreen Packaging, 2018), the demand for which is an accelerating trend.

Incorporation of technology into packaging

In an effort to meet the requirements for traceability, better management of product handling, documentation of product history, easier provision of information to the

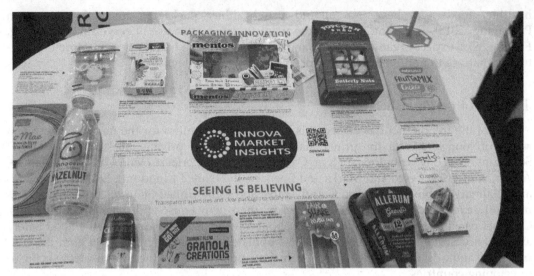

FIGURE 7.2   Transparent and see-through packaging on display at IFT 2019. *Source: Innova Market Insights.*

consumer and increasing regulatory requirements for labeling, producers have been making increased use of technology incorporated into packaging (LaCroix, 2017). This includes radio frequency identification (RFID), printed quick response (QR) codes that take the consumer to a video/website providing information about the product, company, country, its cause, etc. and time/temperature (TT) sensors, among others. Some of these are discussed further under the section on innovations in packaging (later on in this chapter).

Transparent packaging

Consumers want to see what they are buying or at least get a sneak peak of the product. This is driving a trend towards transparent packaging. This is evident in an increasing range of products, building on a trend which counts Sonoco's TruVue among its pioneers. This trend continues to grow and the range of packaging options satisfying it also continues to expand (Fig. 7.2).

Stricter labeling regulations

As consumers demand to know more about the foods they are eating and governments increase the requirements for declarations and disclosures for foods products globally, stricter laws governing labeling are being enacted. All major markets have had upgrades to their labeling regulations within the last three (3) years, with the attendant implications for the food industry. One consequence of this is that exporters now need to develop separate labels for Canada, the UK/EU and the US as a single label can no longer suffice for more than one market. This will be discussed further in the section on labeling below.

Hollowing of the middle class

A major emerging trend globally, and one which is accelerating more rapidly in the United States of America is the decline of the middle class (LaCroix, 2017). This transition in the socio-economic structure of consumers is creating a scenario in which packaging will need to be differentiated such that one variant of the product being offered is affordable (for the working class), while the other is packaged to be sold to the super affluent.

Container closure evaluation

As indicated in Chapter 6, container closure evaluation is an area that is critical, particularly for producers in developing countries seeking to export products packaged in glass, metal or composite containers and selected flexible packaging who must comply with the same requirements as domestic producers. There are several developing countries in which local regulations do not require any assessment of packaging containers and in which there is no training available or skilled personnel to undertake the assessments. It is therefore important that producers from these countries are able to understand what the regulations are, where to find the requisite information and how to undertake the evaluations required by the importing country food laws and regulations. The United States FDA regulations (Food and Drug Administration, 1998) and the Canadian Food Inspection Agency (CFIA) under the Safe Food for Canadians Regulations (SFCR) (Canadian Food Inspection Agency CFIA, 2019a,b,c) require the development of a scheduled process for thermally processed food as well as the accompanying container closure evaluations required to verify that commercial sterility is continuously being attained by said process. Consequently, provided below are more details in a summarized form of how selected container closure evaluation and examination of defects which comply with the requirements can be done. For more details, the Grocery Manufacturers Association's Science Education Foundations' (GMA-SEF) Better Process Control School (BPCS)[1] manual (Black and Barach, 2015) is an excellent reference. Likewise, the FDAs website (Food and Drug Administration, 1998) and CFIAs SFCR webpages (Canadian Food Inspection Agency CFIA, 2019c) also provide excellent guidance, further information and clarifications on this.

Can seam evaluation for metal and semi-rigid containers

The approach to the overall evaluation is described in a point-wise manner in the sections below.

Preliminary examination

- Remove labels and before removing any product sample, perform complete external can examination, observing such defects as evidence of leakage, pinholes or rusting, dents, buckling, and general exterior conditions.

---

[1] Dr. André Gordon has been a lecturer on this program for over 22 years.

Hard swell   *Soft swell*

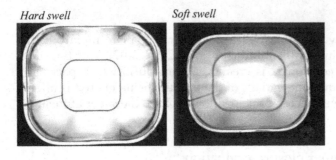

FIGURE 7.3 Hard and soft swells in metal cans. *Source: GMA Science and Education Foundation.*

- Classify each can as (a) flat, (b) flipper, (c) springer, (d) soft swell, or (e) hard swell (Fig. 7.3) according to established criteria.
- Use your hand as well as eye to evaluate the cans that present with either a hard or soft swell or which appear to be normal from a lot in which hard or soft swells have been recovered. A magnifying glass with proper illumination is helpful.
- Run your thumb and forefinger around seam on inside and outside of seam to locate any roughness, unevenness, or sharpness. Examine by sight and touch for the defects listed below that may result in leakage of the can through a faulty seam.

Visual examination

As a part of the container evaluation process, visual assessments are to be made during which the absence or presence of the features listed below should be noted and recorded. The definitions of these defects are also provided below.

- sharp seam
- cutovers or cut-throughs
- false seam (although some false seams may not be detected by external examination)
- dents
- deadheads (incomplete seam)
- excessive droop
- jump over
- excessive scuffing in chuck wall area
- knocked-down flange
- cable cuts on double seam
- excess solder

The can is then opened as per the procedure (see Canadian Food Inspection Agency CFIA, 2019c), the contents of the cans should then be emptied, following which the cans should be washed and dried, and tested can for leakage. The end is then removed, beginning at cutout strip closest to side seam, moving clockwise until entire cover hook is removed. If rocker-like motion does not work, pull out and away. Take care to prevent any injury or cut. Partially remove coded end with bacteriological can opener. The tools required to do a tear down evaluation were shown in Chapter 6 and are available on the CFIA or FDA websites (Canadian Food Inspection Agency CFIA, 2019c; Food and Drug Administration, 1998).

TABLE 7.2   Measurement required on metal cans during can seam evaluation.

Required	Optional
Cover Hook (CH)	Overlap (by calculation)
Body Hook (BH)	Countersink
Width (length, height)	
Tightness (observation for wrinkle)	
Thickness	

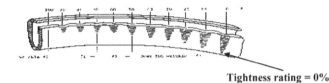

Tightness rating = 0%

FIGURE 7.4 Tightness rating assessment for can double seams. *Source: GMA Science and Education Foundation.*

The measurements to be taken during can seam teardown and evaluation are outlined in Table 7.2.

The tightness rating is a measure of the extent of pressure delivered during the double seam formation and is an index of the safety of the seam. It is measured by estimating the extent of wrinkling on the cover hook as indicated by the length of the wrinkles present. A tightness rating close to 0 indicates a well-formed double seam (Fig. 7.4). For a definition of tightness rating see below. For a fuller treatment and description of this and other measurements see Black and Barach (2015) or Food and Drug Administration (1998).

Can seam defects (definitions)

Definitions of selected defects include:

- Cutover — A cutover is a sharp seam that has fractured and is considered a serious seam defect. A sharp seam is considered a minor seam defect. Cutovers are best detected by running a finger around the inside of the seam.
- Droop — a smooth projection of the end hook of the double seam below the bottom of the normal seam.
- False Seam — A false seam is considered a serious seam defect due to the absence of overlap. A defect where a portion of the body flange is bent back against the body, without being engaged with the end hook, but does not protrude below the bottom of the end hook radius.
- Insufficient overlap — Any portion of the double seam having an optical overlap of less than 25% of the internal seam length is considered to contain a serious double seam defect.
- A knocked-down curl is considered a serious double seam defect due to the absence of overlap.

Can seam defects (visual examination and testing)

The suitability of the particular food to be preserved also affects the performance of the container. Some of the instances in which can defects are observed include:

- Hydrogen swells and sulfide stains caused by chemical corrosion sometimes occur.
- In addition, prolonged storage of cans at elevated temperatures promotes corrosion and may result in perforation.
- Improper retorting operations, such as rapid pressure changes, may cause can deformation and damage seam integrity.
- Post-process contamination by non-chlorinated cooling water or excessive buildup of bacteria in can-handling equipment may also cause spoilage, and abusive handling of containers may result in leaker spoilage.
- Container examinations associated with food spoilage are usually accompanied by pH determination of the product, gas analysis of can headspace, and microbiological testing of the product.
- Spoilage within the can may be caused by leakage, under-processing, or elevated storage temperatures.
- Leaker spoilage occurs mainly from seam defects and mechanical damage.
- Improper pressure control during retorting and cooling operations may stress the seam, resulting in poor seam integrity and, possibly, subsequent leaker spoilage.

Figs. 7.5–7.7 show some of the more common can seam defects that may occur occasionally.

Packaging considerations for food safety and quality

Food/packaging interactions

The quality of a packaged food is directly related to the characteristics of the food and packaging material. When manufacturers are selecting the types of packaging suitable for the food product, it is important to consider the characteristics of the product and the possible

FIGURE 7.5 False seam or knocked down flange. *Source: GMA Science and Education Foundation.*

FIGURE 7.6 Buckling of a metal can. *Source: GMA Science and Education Foundation.*

FIGURE 7.7 Paneling of a metal can. *Source: GMA Science and Education Foundation.*

interactions that may occur in the packaging. The quality of a packaged food is directly related to the characteristics of the food and packaging material and any interactions that may take place. Some of the possible food-package interactions include *adsorption, absorption, migration* and *permeation of volatile compounds*. Each of these can play a role in the transloca-tion of material associated with the packaging into the food which it contains.

In terms of transfer of components from one area of a flexible, printed package to another each of the above processes can play a role. *Adsorption* is the process by which components of one phase are extracted and concentrated at the surface of another phase of

TABLE 7.3 Types of migration across packaging films.

Types of migration

Diffusion migration	Set off migration	Migration due to heating
Migration of substances through the walls of the packaging into the food	Migration from the printed to the unprinted side of a flexible package laminate due to both sides coming into contact with each other.	Migration of gaseous substances or vapors into the food during heating.
E.g. Soapy flavors into a drink in a PET bottle through the walls of the bottle.	*E.g. where packaging film is stacked one on top of the other during storage.*	*E.g. vapors entering a flexibly packaged product during microwaving when the pores of the package open up.*

the packaging material. This may happen where a package is immersed in a liquid or a gaseous environment and a component of the liquid or gas adsorbs and concentrates on the exterior of the packaging film. *Absorption,* on the other hand is where the soluble component goes across the surface and enters into the other phase. An example of this would be the passage of a gas from the external environment into a package through the film surrounding the package or the loss of carbon dioxide (CO_2) from a carbonated beverage as it escapes from the PET bottle to the atmosphere over time. While absorption directly results in phase transfer and entry of the component into the package, entry of the adsorbed component requires the additional processes of either migration or permeation for entry of the component into the product. Table 7.3 shows the three mechanisms by which migration can occur.

As indicated in Table 7.3, because many packaging materials are permeable to varying degrees, small molecules like gases, water vapor, organic vapors and other compounds may pass through the packaging and mix with and into the contents. Alternatively, this may only happen momentarily during a particular phase of the process (e.g. heat sealing, microwaving, pressure processing), but may be enough to ensure that a component become a part of the food being consumed that was not expected to be present. For this reason, it is critically important that producers understand the components that comprise the packaging being used, as well as the label applied and any inks or other material used to complete the design, comply with labeling requirements or simply color the packaging. While this is a consideration for all packaging, for glass and metal containers which are more inert and have low permeability, it is less of a concern. For paper and paperboard packaging, wood and composites, it is more so, while for flexible (plastic) packaging, including semi-rigid containers, it is often a major concern. It is in this context that there is now a focus on food safety of primary, secondary and tertiary packaging, including the handling, storage and transportation practices.

Important considerations for flexible packaging for exports

As has been seen, the demand for flexible packaging is increasing significantly as it gives the producers and marketers of foods a wide range of options in the marketplace. Flexible packaging, however, also has an impact on the safety of the foods being displayed and sold and this has come under greater scrutiny as food safety systems have been extended to include packaging suppliers who must now meet the same requirements as suppliers of other inputs. For flexible packaging, issues such as the composition of the films, resins and processing aids used, food/packaging interaction and food safety of package components are considered. Other important considerations are the handling of primary food contact film during production and post-production handling. This should be done so as to protect the primary food contact surface from contamination with potential pathogen and/or harmful chemicals. Critical considerations include the environment in which the packaging is made (Fig. 7.8) and how it is handled and stored, post production (Fig. 7.9).

The composition of the ink or dyes used to print the labels or information directly on the non-food contact surface of the packaging film is also important. While there are many printing options available to the manufacturers of the various different kinds of flexible packaging, in all instances they will need to have a function in which the design to be printed on the package is typeset and the specific colors to be used are prepared, stored and dispatched to the production line as required. The nature of the inks used, their management, storage, handling and traceability should there be a problem are an important, if subsidiary aspect of the production of film or laminate to be used as an important part of a final package (Fig. 7.10). Important characteristics of other components of the laminated films that comprise many flexible packaging solutions include the nature of the resin/plastic used, the overall barrier properties of the film, including oxygen transmission rates (OTR) and permeability to moisture (moisture transition rates (MTR)). For the ink, the questions that arise include: What is its composition? Is it food safe? Does it transit the film and enter the product being packaged (i.e. film/product interactions) through the film by any of the means discussed previously?

FIGURE 7.8   Production area for packaging film in a packaging plant. *Source: A. Gordon (2019).*

FIGURE 7.9   Storage of packaging film post production in a packaging plant. *Source: A. Gordon (2019).*

FIGURE 7.10   Handling of inks, dyes and colors used for printing in a packaging plant. *Source: A. Gordon (2019).*

These are among the considerations that producers in developing countries need to be cognizant of as they continue to upgrade their technology in keeping with the trends required by their export market. Whether the flexible packaging being used is basic, relatively simple single or multilayer films or more complex composite materials to be used for more robust processing (including thermal processing), the same issues arise as regards food safety, traceability and ensuring high quality packaging inputs to a high quality, safe food product that meets all of the requirements in its target markets.

Wood packaging: considerations

Wood is used as packaging material for foods in a variety of ways. As was noted in Chapter 6, wood can be used for primary packaging, but this is typically limited to a select group of products. When used as a primary packaging material, the wood is typically heated or otherwise treated to eliminate any risk of direct contamination with pests, chemicals or undesirable microorganisms. As such, the items that wood is used to package are often those that benefit directly from the properties of the wood itself, such as the tannins or other organics that impart flavor and/or color to beverages stored or aged in them such as rums (Fig. 7.11), wines, whiskey and selected other alcoholic, natural and specialty beverages. In these instances, the treatment used for the wood would typically be heat treatment (International Plant Protection Convention IPPC, 2016), as the use of chemical treatment would likely compromise the safety and organoleptic profile of the products being held, conveyed, stored or aged in the wooden containers. It may also be interesting to note that among other products that are held and aged in wooden barrels are pepper sauces, such as A. Smith Bowman Distillery's hot sauce and Jamaica's Pickapeppa Sauce, as well as teas, vinegar, mustard, maple syrup, vanilla and pickles (Covington, 2014). These are, however, not sold in the wooden barrels, as are some bulk wines, whiskeys and rum, among other spirits.

FIGURE 7.11 Wooden drums used for aging and displaying Guyana's award-winning El Dorado Rum. *Source: A. Gordon (2019).*

Regulation of the export of wooden packaging materials

Developing countries exporting a range of food products into Europe, North America, Japan, the Middle East, China and the more developed countries of the Pacific Rim, need to be aware that the use of wooden packaging material in the form of boxes and related packaging, as well as pallets is prohibited unless these products meet new regulatory requirements formally adopted in 2013. These products are governed by the International Plant Protection Convention (IPPC) of the United Nations Food and Agriculture Organization (FAO) which adopted new regulations for the use of wood in international trade, International Standard for Phytosanitary Measures (ISPM) 15. This was adopted in 2013, having been implemented in most countries by the mid-2000s. The standard requires that all wood-based products must be demonstrably free of targeted pests to be approved for use in international trade.

The most common use of wood in packaging is for the making and repair of wooden pallets or crates that may be used to convey foods items. Pallets are widely used throughout the developing world because of their cost (relatively inexpensive), the ability to produce them locally and, where necessary, repair and reuse them, thereby also saving on costs. In some instances, wood is used as a container for products such as where some produce, fruits and vegetables may be packed and shipped in wooden crates which are more sturdy and protect them against damage. These crates, however, are typically made of manufactured wood such as plywood, particle board, oriented strand board or thin wood that would have been created by a process involving heat, or pressure or a combination of both and therefore are typically fully compliant. The pallets, on the other hand would not be compliant with ISPM 15 unless they are treated in accordance with its requirements. The phytosanitary treatment methods approved and described in the standards are heat treatment or the application of methyl bromide (MB) as per the guidelines, manufacturing for selected wood products and marking as per the standard. Of the treatments allowed, MB treatment is the least preferred as its use is being phased out. The heat treatments allowed involve the use of a conventional steam or dry kiln heat chamber or dielectric heating. All three treatments have their own designation (mark) that must be imprinted on each piece of wood that is treated. These are HT for conventional or dry kiln heating, DH for dielectric heating and MB for methyl bromide treatment (Health Canada, 2016).

Each type of heat treatment must meet specific requirements with conventional heat chamber technology (HT) requiring the attainment of 56 °C in the core of the wood for at least 30 minutes while dielectric heating requires 60 °C for one (1) continuous minute, the temperature to be reached within 30 minutes of commencement. Wood that is not treated in either of these ways or treated with MB, cannot be marked as fit for use in international trade and could face rejection at port of entry in major export markets. In countries where wooden packaging may be re-used, repaired and/or re-manufactured, re-use of previously treated and marked wood is allowed as long as it has not been repaired or remanufactured. In these cases, the wood may/will need to be retreated and re-marked, the mark as always only being applied after treatment is complete to prevent accidental by-pass of treatment and thus reduce/limit the risk of untreated wood getting into international commerce.

Other wooden packaging materials are exempt because they have been shown to be at low risk for contamination with the pests that are being targeted. These include wood packaging material made entirely from thin wood (6 mm or less in thickness), wood packaging made wholly of processed wood material (e.g. plywood, particle board or oriented strand board) that has been made using glue, heat, pressure, or a combination of heat and pressure. Other exempt wood products are barrels for wine and spirit that have been heated during manufacture and gift boxes for wine or other food products that are made from wood that has been processed and/or manufactured in such a way that it is rendered free of pests. In meeting the demands of their customers for innovative or natural packaging, developing country food producers and exporters must be cognizant of ISPM 15.

Impact of packaging on shelf life

Extending shelf life

Packaging plays an important role in determining the shelf life of the products they contain. Traditional packaging such as glass bottles and cans were, and remain, widely used because they are able to not only maintain a hermetic seal, thereby excluding microorganisms and extending the shelf life of products, they also exclude gases and, in the case of cans, exclude light. Light and the presence or absence of moisture and selected gases impact the quality and therefore the shelf life of a wide range of products, including fruits and vegetables, beverages, dairy products, meats, snacks, among others. Other types of packaging, such as paper in the form of paperboard packaging (Fig. 7.12) and corrugated cardboard, wood and some flexible packaging including metalized pouches and

FIGURE 7.12 Examples of paperboard packaging presented by Innova market insights. *Source: Innova Market Insights.*

other composite packaging which has a component with specific barrier properties, such as laminates with an opacity agent (paper, titanium dioxide, others), also help to extend shelf life. Consequently, all aspects of the production and delivery of foods from manufacturing, through storing, handling, transporting and displaying for retail need to be considered in the selection of packaging for optimal shelf life.

The need for greater convenience has fueled a growing demand for flexible packaging, particularly plastics with high barrier properties for various applications in the food industry. These plastics are used to protect the contents of the packaging against deterioration caused by contamination by microorganisms or occasioned by the entry of air (more specifically oxygen) or water vapor into the product, or loss of the gases used to create an inert environment (i.e. by gas flushing) through the packaging. Flexible packages are also required to protect the product against the loss of specific characteristics of its ingredients (e.g. flavors, color), while protecting its integrity. In addition to the demand for high-barrier films, there is also an increasing demand for biodegradable packaging, aspects of the biodegradability of packaging being discussed later on in this chapter. This is driving demand for alternative materials, including "green" polymers. These varied needs have seen the food industry experiencing an increase in the availability of flexible film-based packaging that offers a rapidly expanding array of properties, providing greater choice for the consumer and enhancing the options available to the producer in meeting their needs. Producers in developing countries need to be cognizant of these options when developing products for these markets or seeking to enhance the market opportunities for existing products.

All foods lose quality and other attributes over their shelf life during distribution by virtue of microbiological growth, enzymatic activity, physical damage and biochemical reactions. Some of this loss is due to product interaction with oxygen in the air and also oxygen dissolved and/or entrained in the product. Microbiological deterioration is best prevented by product formulation and processing but is often heavily influenced by storage, handling and, critically, the type and expected role of the packaging material used. Effective vacuum packaging, for example, reduces oxygen and eliminates or significantly reduces the growth of microorganisms that need oxygen to grow. Physical damage can only be averted or reduced by proper handling and the use of effective primary, secondary and tertiary packaging options. Enzymatic activity and biochemical reactions can be minimized by product design, formulation and processing. It can also be controlled by the exclusion of oxygen, reducing its availability to facilitate unwanted changes such as the enzymatic browning of products, among others.

Controlling moisture

Moisture (water vapor) migration into or out of the packaged product can also have a deleterious impact on shelf life. This is because of the following effects of the migration:

a. Migration in: increased oxidation, other biochemical reactions, browning, wetness, agglomeration of product (hardening, "staleness"), etc.
b. Migration out: hardening, staleness, loss of texture, flavor modification, etc.

Ensuring that the packaging chosen has the required barrier properties to moisture (expressed as its moisture transmission rate − MTR) is therefore important in assuring optimal shelf life for susceptible products. A thorough discussion of the role of moisture in the deterioration of foods can be found in Labuza (1982).

Controlling oxygen

Oxygen in the package headspace, oxygen entrained or dissolved in the food, and that which migrates from package exterior to the inside through the package as described above (Table 7.3) are the principal sources of oxygen in packaged foods. As with unwanted water vapor, oxygen can have negative consequences on the shelf life of products. Because of this, several methods have been developed to reduce oxygen in foods, including the use of oxygen scavengers, vacuum packaging, the use of high oxygen-barrier packaging and flushing with inert gasses (such as nitrogen), among others. Current processing and packaging technologies are capable of decreasing oxygen concentration in products to as low as 0.01%, significantly enhancing shelf life in susceptible products. Packaging that offer an effective oxygen barrier reduces:

- fat oxidation − this is accomplished by controlling or reducing environmental oxygen which causes deterioration (including rancidity) of lipids;
- nutrient loss − oxidative reactions cause nutrient loss in some foods (e.g. loss of vitamin E);
- color loss/changes − oxidation causes the loss of color or *browning* reactions in fruits, some vegetables, some juices and seafood.

Effective control of the presence and availability of oxygen in packaged foods will enhance the shelf life of products and should be among the options considered when seeking to market products in more competitive markets.

In summary, producers seeking to attain optimal shelf life performance from their products must consider the various modes by which product deterioration can occur and, in addition to the application of effective processing technology, seek to use available packaging technology to extend shelf life. These technologies include those which exclude or prevent the entry into products of water vapor and oxygen, allow for the creation of an environment in the package or container post-processing which is supportive of extended shelf life or which facilitate the application of processing technology that produce extended shelf life products. Whatever the means by which these desirable objectives are achieved, food industry participants should be cognizant and make full use of packaging technology to help drive value for their products through extended shelf life.

Packaging innovations driving changes in food safety, marketing and distribution

As consumers become more demanding, producers and retailers have had to respond to the demands and developing trends by developing or employing packaging innovations that can help them to meet the needs as articulated. This has resulted in a number of innovations as food industry participants seek to satisfy, and also to capitalize on, consumer demands and the changing tastes and needs of different segments of the market. In addition to the trends already mentioned, the packaging industry has responded to other existing and developing consumer needs. These have included greater *environmental awareness* among consumers (as previously mentioned) which has driven the trend towards the use of *alternative materials*, including biodegradable packaging, and other materials with much lower carbon footprints, including those that *result in a reduction in packaging volume/weight*. Recycleable and renewable (reusable) packaging such as those being used by Elopak®, a liquid food package producer, and Sabic®, a manufacturer, who collaborated on the development of a "virtually 100% renewable" polyethylene carton are also in vogue (Elopak and Sabic, 2019), particularly in European countries where reduction in packaging waste is a major trend and also a part of EU regulations (The European Parliament and the Council of the European Union, 2018).

Packaging manufacturers are also responding to the ongoing demand for greater *food safety* as well as increasing demand for Ready-to-Cook (RTC) and Ready-to-Eat (RTE) meals. Meals-on-the-go, home meal delivery, smaller portion sizes, products that are easy to open and re-seal when used are among the industry's response to the need to deliver more convenient food choices, all while ensuing greater safety of the foods with various tamper-evident closures. These include innovations such as Ziploc® bags for storing foods, a range of flexible packaging based on re-sealable slide or press-to-close zipper closure technology, and resealable spouts on flexible packaging pouches (Fig. 7.13). Other more convenient food options are made possible by innovations in packaging technologies that have produced packaging with easy peel and easy open closures, resealable packaging, and a range of other

FIGURE 7.13   Resealable spouts on flexible Pouches. *Source: A. Gordon (2018).*

TABLE 7.4 Types of active and intelligent packaging.

Active	Intelligent
Oxygen scavenging	Time-temperature history
Ethylene scavenging	Light protection (photochromic)
Odor and flavor absorbing/releasing	Leakage, microbial spoilage indicating
Moisture absorbing	Physical shock indicators
Anti-microbial	Microbial growth indicators
Heating/cooling	Hot/cold indicators

technological responses to meet consumer needs. Other major technological advances in the RTE category are the development of *alufoil* for use in microwaveable, foil packed meals and technology-intensive active and intelligent (or smart) packaging.

Active and intelligent packaging

Active packaging is packaging that plays other roles and functions beyond the basic passive containment and protection of the product. Intelligent (smart) packaging is that which has been designed to sense or measure one or more attributes of the product, the internal environment of the package or the external environment. Active and intelligent packaging are innovations that help to extend the shelf life of the packaged product. Active packaging actively eliminates one or more of the conditions that cause spoilage or loss of quality. Basic examples of these would be packaging that includes a desiccant to absorb excess moisture and keep the food dry or those using oxygen scavengers. In both cases, the shelf life of the product would be extended. On the other hand, intelligent packaging monitors and indicates specific internal conditions which affect the product. An example of this would be the use of thermochromic ink as part of Time Temperature Indicator (TTI) technology that can be used to reveal the condition of the food since preparation. The latter depends on the reversible or irreversible (depending on the nature of product and the product handling system along the value chain) change in color of special ink depending on its exposure to predetermined temperatures or specific time/temperature combinations. In the case of irreversible thermochromic ink, once the set temperature or time/temperature combination has been reached, the color of the package (or a part thereof) will change to indicate the temperature history of the product. Examples of active and intelligent packaging are shown in Table 7.4, with typical characteristics outlined below.

Active packaging

This consists of a matrix polymer, such as polyethylene terephthalate (PET), an oxygen scavenging/absorbing component and a catalyst. There are also active food packaging systems that use oxygen scavenging and anti-microbial technologies. These provide significant extensions to shelf life, improved quality throughout the distribution chain and greater convenience to the consumer.

Intelligent packaging

Intelligent packaging has functionality which switches on and off in response to changing external/internal conditions and can include a communication to the customer or end user as to the status of the product. A simple definition of intelligent packaging is 'packaging that senses and informs' (Day, 2008).

Edible packaging

With the potential to reduce or replace conventional packaging, much interesting research is being done on the development and use of edible packaging (Wang and Kerry, 2018). Edible packaging is intended to be eaten as a part of the product and as such is inherently biodegradable (Janjarasskul and Krochta, 2010). Edible packaging generally takes the form of films, sheets, coatings or pouches, with the major forms being coatings and films, the latter having an average thickness of less than 254 μm (Cerqueira et al., 2016a,b) and the sheets having thickness greater than 254 μm. The films and sheets are usually applied to the food product or components while coatings on the other hand are thin layers of edible materials formed directly on the surface of the food products (Janjarasskul and Krochta, 2010). Among the factors that determine the applicability of edible packaging are gas migration through the packaging, relative humidity, pH and the mechanical property of the packaging as well as other factors such their relative strength and durability during handling when compared to conventional packaging. While edible packaging has many potential applications and is likely to experience continued growth in its use, the opportunities which exist will be tempered by the challenges which arise, particularly with handling, shelf life, gas permeability and other considerations. More details on edible packaging is available in Cerqueira et al. (2016a).

Sustainable (biodegradable) packaging

A major driver of innovation in the packaging industry is the imperative for the food industry to significantly reduce the use of one-way, non-recyclable, non-reusable packaging. Efforts to develop biodegradable packaging were prompted by several factors, including the accumulation of plastic in the environment, the regulatory changes mentioned, evidence of the release of hazardous gases during incineration and greater consumer activism. More than ten (10) countries in Africa, France, India and Mongolia are among the countries banning the use of single-use plastic bags (McCarthy, 2018). Columbia, Jamaica and St. Vincent and the Grenadines in the Americas, and Papua New Guinea in Asia-Pacific have also instituted a ban, with others set to follow. Several states in the United States, EU member states and others have legislated the reduction, reuse or recycling of all packaging or have imposed taxes on packaging. These trends have made the environmental conservation and hence the biodegradability of packaging a major driver of changes in the packaging industry. It is therefore evident that as "great importance is given to products from renewable sources, for their positive impact on nature" (Ivanković et al, 2017), this drive for packaging biodegradability will continue for the foreseeable future.

There are different forms of bio-packaging adapted to the varying requirements for the packaging and storage of different types of products. They can exist as biodegradable gels, bags, films, fruit/vegetable trays and boxes with lids (Ivanković et al, 2017), among others. Many are composed of biopolymers which may be made of 100% biodegradable materials (biomaterials) or combinations of biomaterials and synthetic materials. The increasing importance of biodegradable packaging, the range, nature and speed with which research and development in this area is proceeding, as well as the differing and sometimes involved nature of the regulatory requirements demand further attention of producers, exporters and food industry practitioners. Further discussion on biodegradable packaging is therefore presented below, with extensive treatment on the subject covered in Ramos et al. (2018).

Personalization and digital printing

Another new and interesting innovation is the development of personalized and interactive packaging driven by advances in digital printing and compatible technologies, including mobile phones. In 2015, a Belgian brewer printed interactive characters on beer bottles (News Desk, 2015). The characters on the beer bottles are brought to life using a smart phone app through which the characters deliver special performances. If two bottles are brought close to each other, the app creates a dialog between the two characters, providing entertainment for consumers of the product. This kind of innovation where food becomes a part of actively providing entertainment may be of interest to exporters seeking a unique selling proposition for their product in highly competitive developed markets.

Biodegradable packaging

The trend in recent years towards the use of polymeric materials for food packaging has seen them increasingly being used to replace glass, metal and paperboard-based products. This is because of the lower cost of polymers, their light weight and ease of use in fabricating packaging units, resilience, ease of coloring, inertness, and consumer demand, among other considerations. However, polymeric materials are generally slow to biodegrade, except those made from plant, animal or microbial sources, or those that have been deliberately altered chemically. Recent legislative actions by several countries around the world are now requiring food processors who use polymeric packaging to select those that are biodegradable, compostable, reusable or which can be recycled. This has helped to continue driving the trend that has seen the global biodegradable packaging industry being projected to reach US$21.6 billion by 2026, with Europe being the largest market, followed by North America and the Asia Pacific region (Kenneth Research, 2019). The industry is divided in polymeric plastic and paper packaging, with plastics holding the largest market share. Plastics include bacterial cellulose and cellulose esters, poly(hydroxyalkonoates) (PHA), poly(hydroxybutyrates) (PHB), polylactic acid (PLA), protein-based plastics and starch-based plastics (Ramos et al., 2018) which have different applications depending on their specific properties and can also incorporate a range of bioactive compounds, extending the shelf life and improving the quality of food products (Robertson, 2012). The availability of these forms of packaging has facilitated the accelerating move towards biodegradable packaging.

The European Union has been leading the trend towards biodegradable packaging and in Europe, companies using polymeric materials are often required to prove that the material meets established standards and are compliant with EU directives. One such standard is Directive 94/62/EC of 20 December 1994 on packaging and packaging waste (The European Parliament and the Council of the European Union, 1994). This standard seeks to encourage recycling, reuse and a general reduction in waste materials from reaching city and municipal landfills. The EU standard EN 13432 (Deutsches Intitut fur Normung DIN, 2000) outlines the test methods and the assessment criteria to be used to establish if a package meets the minimum compostable threshold. The key requirements of the EN 13432 Standard are:

1. Minimum concentration limits for volatile compounds, fluorine, and heavy metals such as copper, zinc, nickel, cadmium, lead, mercury, chromium, selenium, arsenic and molybdenum in recycled packaging.
2. At least 90% of the packaging materials must be broken down into carbon dioxide and water by biological action within 6 months.
3. After 12 weeks of composting, at least 90% of the packaging material should disintegrate into pieces that are small enough to pass through a 2×2 mm mesh.
4. The quality of the compost should not decline because of the added packaging material. An ecotoxicity test should be used to determine if the germination and biomass production of plants are not adversely affected by the influence of the composted packaging.

The EN 13432 Standard specifies that packaging may be deemed to be compostable only if all the constituents and components of the packaging are compostable. During the certification procedure, an assessment is made not only of the basic materials, but also of the various additives and other product properties incorporated into the package. Biodegradability of packaging is also addressed by ISO 14855-1:2012, ISO 15270:2008 and ASTM D6400-04. These criteria can be used to determine the most appropriate materials when an assessment as to which packaging is biodegradable is to be done.

Selection of biodegradable packaging

Product development, manufacturing and marketing are among the departments within a company that have to consider various aspects of packaging material when new or existing product packaging is being reviewed. While each function will have its own specific considerations, the package selected must be compliant with the regulations in the destination market. Because the area of biodegradable packaging and the regulations governing them are so dynamic, many jurisdictions do not have settled criteria or regulations in place. This makes the selection and use of appropriate packaging that meets all of the various functional, esthetic, handling and regulatory requirements quite involved.

Biodegradable packaging can be made from 100% natural compounds, including cellulose, natural materials derived from animal or plant sources such as pulp (made of sugar cane bagasse, bamboo and other natural fibers) and cotton and wood, or from additives incorporated into a range of other films, such as low-density polyethylene (LDPE) or linear low density polyethylene (LLDPE). In the case where the firm makes its own LDPE or LLDPE packaging or has this packaging made for them, then the additive that the supplier proposes to be used should be evaluated against the same criteria for packaging being bought directly from a supplier. The data should show that the additive(s) imparted acceptable biodegradability to packaging materials made with it, that it met the required strength and other performance criteria, and also that these packages met the requirements for being non-toxic. There are approaches and specific considerations that can make selecting the right packaging options more manageable for non-packaging specialists and these are discussed below, as well as considerations that should also advise the choices made.

In selecting between packaging options for biodegradability and functionality, and compliance with regulations, the following approach may be useful:

- Undertake research of the area to get an understanding of the requirements of the specific market and the technical specifications and performance criteria of various packaging options
- Review the technical material presented by the potential suppliers of various packaging options for the specific products to be packaged
- Review the relevant standards which should be referenced in the technical documents presented by the supplier
- Review reports on the performance of the packaging and, where the offer is for an additive to be used, of packaging made with the additives
- Where required, source additional materials to verify the information being provided by the potential suppliers
- On the basis of the review undertaken, select the most suitable packaging or packaging ingredient/additive supplier.

The documentation to be provided by the potential supplier should include test reports and studies that show that their packaging meets the required technical standards mandated by the markets in which the product is to be sold. Studies presented should include the details of test methods and results for tests done on the packaging/films, as well as the controls, to determine if the films met the standard for biodegradability as required by the EN13432 standard. They should also include a series of reports on biodegradability and related tests carried out on packaging material made specifically for the firm.

Assessment and testing of biodegradable packaging

Among the tests directly required by the EN 13432 Standard for biodegradable packaging made from polyolefin polymers (including LDPE and LLDPE) are the following:

1. *Plant seedling emergence and seedling growth test using the OECD[2] 207 and 208 Guidelines.* This test is used to determine if the addition of an additive to the polyolefin samples that would degrade in landfills will have an adverse effect on the growth of vegetation. The seedlings used in studies such as these include wheatgrass and onions.

---

[2] Composite packaging is made from combining the full surface of at least two different materials. It has also been defined as an outer packaging and inner receptacle combined so that they form an integral unit. Essentially it is multi-layer packaging with at least one inner and one outer layer such that each layer provides different functionalities that imbue the package with properties not obtainable from one type of flexible packaging layer alone.

2. *Biodegradation testing of the polyolefins.* This is accomplished by one of three methods or combinations of them. These are: 1) abiotic oxidative breakdown of the material; 2) microbial digestion of the materials themselves; or 3) microbial digestion of fragments of the materials. For #1 above, the test method used is typically ASTM 6954-04 Tier 1. In the case of #3, initial breakdown of the materials could be from oxidative breakdown, as an example. The degradation testing can be done according to the ASTM D-6954 and D-5510 standard test methods or to the ASTM D-6954 and D-5208 standard test methods. During the testing, the materials are expected to degrade into carbon dioxide, water and biomass. Organizations such as the Swedish National Testing and Research Institute, the Swiss Federal Laboratories for Material Science and Technology, and the National Centre for the Evaluation of Photoprotection, Clermont-Ferrand, France, among others, have expertise in these kinds of tests.

3. *Analysis of the materials for the presence of heavy metals.* This can be done according to the ASTM D1976-01 method using inductive coupled argon plasma spectroscopy.

The packaging film should also be assessed for shelf life stability, thermal degradation, degradation in the presence of ultraviolet light, thickness determination, as well as being examined by X-ray florescence (XRF) spectroscopy, and Fourier-transform Infra Red (FTIR) spectroscopy. Additional mechanical tests should be performed on the materials to investigate the influence on the film's properties from the incorporation of recycled virgin polyolefins blended with and without the additive being assessed. These mechanical tests include tensile, elongation and tear resistance of the materials before and after the addition of the additive. These evaluations should be done by testing the mechanical properties of the samples according to ASTM D882-09 for tensile, ASTM D1922-09 for elongation, and ASTM D1922-09 for tear strength (puncture and tear resistance), respectively. The intent of this second set of testing is to determine if the materials would be susceptible to a loss of mechanical properties after blending virgin, with recycled polyolefins containing the additive. An outline of these tests are below.

Stability testing

This test is designed to determine the ability of the material to maintain its inherent properties after exposure to a given quantum of sunlight and heat, not exceeding 30 °C, for a maximum of 700 hours. Degradation occurring within the material was estimated by measuring the concentration of carbonyl compounds such as aldehydes, ketones and carboxylic acids that developed in the material during the time of exposure at the testing conditions. These compounds are known to develop during the oxidative breakdown of polymers such as LDPE. Thus, the absence of carbonyl compounds in the sample is viewed as a sign of material stability. This test is done under accelerated conditions. This test method is designed to shorten the testing time by speeding up the rate of degradation. It is a well-accepted way of reducing the duration of this test while still being able to produce accurate results.

Thermal degradation testing

This test measures the rate at which the polymer degrades in conditions of high temperature that simulate exposure to the sunlight. This test is done in the presence and absence of UV light. Thus, it simulates exposure to heat in the presence and absence of direct sunlight. The control samples, without the additive, should show no increase in carbonyl concentration during the test.

Testing for the presence of metal ions

This test is done using energy-dispersive x-ray fluorescence spectroscopy. This is a nondestructive method that will assay for the presence of, and quantify, metal ions in polymeric samples such as the LDPE films. The intent of this test is to determine the presence and concentration of the additive compound in the films and its absence in the control.

Thickness testing

The thicknesses of all sample materials should be measured before the ageing tests are done.

Conclusion

As the importance of providing packaging options demanded by the market continues to increase, professionals along the food value chain, including producers and exporters from developing countries, as well as the professionals that support them, will need to avail themselves of packaging that meets market requirements. This will mean the introduction of increasing amounts of biodegradable packaging as an important component of the final packaged product delivered to the market. This brief summary of some of the key considerations should allow food industry professionals to have a good grasp of the issues that will need to be addressed, as well as well as the criteria that the packaging will have to meet to comply with regulatory and market requirements.

Additional sources of information on biodegradable plastics include the Oxo-biodegradable Plastics Association, American Society for Testing and Materials (ASTM) standards, the EU Directives, major packaging companies such as Kruger Inc., the Mondi Group, Amcor, BASF SE and Reynolds Group, among others. Smaller, specialist manufacturing entities which provide additives to first generation packaging companies making traditional LDPE packaging, including Symphony Environmental Technologies (with their d_2w packaging additive) and Eco Poly Solutions Inc. (with their *Oxo Elite* packaging additive) also provide detailed information on their products. These, along with the information provided in this section, should allow decision-makers to have the tools needed to effectively engage with the opportunities presented by employing the right packaging technologies to get food products from developing countries in the hands of the consumers in all selected markets.

The importance of labeling for exporting products from developing countries to selected developed country markets

Introduction

Many of the products produced in developing are destined for the United States of America (USA), Canada and European Union (EU), the major export markets for these products. Each of these markets have their own regulatory infrastructure and their own requirements for labeling for products being sold in their countries. As is the case for packaging, the food industry in each jurisdiction, including exporters wishing to have their products sold there, must meet all of the legal requirements to be able to retail their products in the market. In addition, where there are recommendations, traditional practices or new trends occasioned by changing consumer demands or the major firms within the country, it is the norm that products being offered for sale also seek to comply with these as well.

In the recent past it used to be possible to do a label that had compliant information on it for more than one major destination market. That has changed with the updating of their labeling requirements by the USA, Canada and the EU such that now each market requires specific declarations that are not in keeping with the regulations in other jurisdictions. For example, it was possible to prepare a label for a product being exported to Canada and also have it comply with the requirements in the USA. However, changes due to the Safe Food for Canadians Act as well as changes to the US labeling requirements regarding consumption sizes and the attendant declarations, when combined with the requirement for multi-lingual labels in Canada make it virtually impossible for one label to comply in both markets. The same applies to Canada and the EU and the USA and the EU. Consequently, and very importantly, companies offering products for sale in any of the major markets need to have an understanding of the current status of the regulations regarding labeling in each and also to keep abreast of the changes and the attendant implementation dates, where relevant. If not, they will be unable to avoid having their products be found to be non-compliant in the respective markets and therefore at risk of being barred at port of entry or withdrawn from the shelves, if they managed to enter the territory and get displayed in the market.

Labeling basics

The regulatory bodies that monitor label compliance are FDA and Canadian Food Inspection Agency (CFIA) in the US and Canada respectively and the Food Standards Agency (FSA) and European Food Safety Authority (EFSA) in the United Kingdom (UK) and the EU, respectively. While the EUs regulations would (and still do) normally apply in the UK, impending changes in the relationship may mean changes in regulatory compliance as regards foods being sold. It is germane, therefore, to consider EU and UK requirement differences, if any, if a firm intends to trade in both, in the same way as they are now impelled to if they trade in the US and Canada. Each territory has specific information that is required to be on the label of packaged food entering and produced there. Generally, the information presented on package labels should include:

- The name of the product and, if relevant, a statement of identity
- Net contents
- The name and address of the manufacturer/distributor/importer
- Directions for safe use of the product (including storage and preparation)
- The list of ingredients
- Any warning statement(s) that may be required

These are among the basic components of the information that must be present on the label or packaging for each product. In addition to this, each market also has specific requirements for advising consumers about the contents of the package and composition of the product, including its nutritional content. As such, "Nutrition Facts" labeling is a very important tool for providing information to the consumers on the packaged item. With the prevalence of lifestyle diseases, including diabetes and hypertension on the rise, both linked to obesity, regulatory bodies across the world, including the US FDA, CFIA, FSA and EFSA require that food manufacturers present detailed information on the composition of their food products by way of a Nutrition Facts Label. Additionally, the significant increase in allergen-related foodborne illnesses has resulted in regulators also requiring that food products are specifically labeled as to the potential allergens it may contain. These requirements apply to packaged foods produced domestically with the US, Canadian, UK or EU markets, as well as those imported from other countries, the compliance of which is routinely checked on randomly select food samples in the marketplace. These sampled products are checked for compliance to several compulsory regulatory requirements, including nutrition analyzes. This chapter will seek to bridge the divide between producers and exporters from developing countries and the requirements of importing countries that has seen a significant increase in rejections at port of entry over the last several years (Gordon, 2016), a trend which has continued.

Labeling requirements for export to the UK and EU

The labeling requirements in the EU, as with other markets, have been undergoing change over the last several years. The EUs "Food Information to Consumers (FIC)" Regulation 1169/2011 became applicable on December 13, 2014 and presented new obligations and changes to the previously existing rules (United States Department of Agriculture, 2019). Nutrition labeling, for example, became mandatory in the EU on December 13, 2016. Despite this single European-wide approach to labeling, it is important for exporters to be aware that there may be some variation among the different Member States in implementing the harmonized EU legislation and they should ensure that there are not additional requirements in the specific market in which they seek to trade. In such a case where harmonization is not yet finalized, it is the importer's responsibility to ensure that the existing requirements in that Member State are met.

In the EU, the terminology used have differences in meaning and understanding these is important in navigating the requirements and ensuring compliance. Specific EU terms are used in the EU legislative framework and will be mentioned frequently in this section of the chapter. The terms and their meanings are outlined below:

"An EU *Directive* is a form of legislation that is **"directed" at the Member States**. It will set out the objective or policy which needs to be attained. The Member States must then pass the relevant domestic legislation to give effect to the terms of the Directive within a time frame set in the directive, usually two years" (Europa EU, 2019). As this direct quote indicates, Directives define the results that must be achieved by the application of a particular legislative requirement in each Member State but leaves each Member State with choices as to how to translate the directives into their national laws.

"Regulations", on the other hand, are binding and once they come into force on a specific date, all member states are required to implement them in their entirety (Europa EU, 2019). There is no leeway for interpretation for inclusion in the domestic law of each Member State as the Regulation has to be implemented as documented, without exception or modification.

A *"Decision"* is only applicable to those to whom it is specifically addressed (e.g. individuals or a country) but is binding on them.

A *"Recommendation"* is not binding and has no legal consequence. It allows institutions to make their views known and suggest an approach towards implementation but does not impose a legal obligation on them to implement it.

All of the above relate to all categories of food produced or sold in the EU. There are some types of food, however, that have specific additional labeling requirements which are handled and published separately. These include beef, cocoa and chocolate products, coffee and chicory extracts, fortified foods, fruit jams, jellies and marmalades, fruit juice, GMO products, honey, organic products, sugars, wine, wholly dehydrated preserved milk, dietetic or special use foods and foods labeled with nutrition and health claims. The discussion in this section will deal with all relevant categories of food and expand on the general requirements.

Food information to consumers

Packaged food sold in the EU must provide consumers with enough information for them to be able to make informed choices. This is captured in Regulation (EU) No. 1169/2011 (the FIC Regulation). In addition to the basic information indicated above, the label for food and beverages must also provide:

- Allergen highlighted in the list of ingredients (not in a "contains" box on the label)
- Quantitative Ingredient Declaration (QUID) for certain ingredients or category of ingredients
- Nutrition declaration
- Appropriate durability indication (shelf life/best before date)
- Indication of alcoholic strength per volume, if appropriate
- Country of Origin Labeling (COOL), if required
- Contact information for the manufacturer, packer or seller

The FIC regulation established new horizontal labeling requirements and repealed labeling Directive 2000/13/EC, as well as nutrition labeling Directive 90/496/EEC and warning labels Directive 2008/5/EC. Article 13 of the FIC regulation requires the mandatory information to be easily visible, indelible and clearly legible without being obscured by written or pictorial material. The size of the font should be greater than or equal to 1.2 mm as determined by the height of the letter 'x'. When the packages are smaller than 80 square centimetres (cm^2), the minimum font size should be 0.9 mm. Nutrition labeling is not required for packages with a surface area less than 25 cm^2 and packages smaller than 10 cm^2 are not required to bear nutrition labeling nor a list of ingredients.

It is important to note that the mandatory information should be presented in a language easily understood by the consumers of the Member State where the food is to be marketed, i.e. for France the required language is French while for the Netherlands it is Dutch, etc. Member States may specify which information is required to be present in one or more official languages. To avoid non-compliance, translations of the mandatory information must be accurate. The ingredients should be listed in descending order of weight. However, in the EU, substances or products causing allergies must be indicated in the list of ingredients with the source of the allergen as listed in Annex II of Article 21 of the FIC Regulation. The name of the allergen should be highlighted by using a typeset that clearly distinguished it from the remainder of ingredients (Fig. 7.14), for example "tofu (soya)" and "whey (milk)". Bold type or a background color can be used. Below is the list of allergens that should be declared when present in food and beverage, including alcoholic beverages:

- Cereals containing gluten
- Crustaceans
- Eggs
- Fish
- Peanuts
- Soybeans
- Milk
- Nuts
- Celery
- Mustard
- Sesame seeds
- Sulfur dioxide and sulfites at concentrations of more than 10 mg/kg
- Lupine
- Molluscs

The allergenic ingredients should be listed in the ingredients list as required. The voluntary use of warning boxes or statements such as "contains X" to repeat the presence of allergenic ingredients is not permitted. Only in the absence of an ingredients list (as in the case of small packages mentioned above) can the presence of allergens be indicated using the word "contains" followed by the name of the substance or product. Allergen advice may be included to direct the attention of the consumer to the highlighted allergens in the ingredients list as seen in Fig. 7.14.

The EU has a requirement for information about the quantity of the major or characteristic ingredient in the product, the Quantitative Ingredient Declaration (QUID), to be indicated as required for the following cases:

- It is highlighted on the label, emphasized in words (e.g. *made with butter*) or by the use of pictures and graphics
- Mentioned in the name of the product e.g. *"15% Strawberries" on strawberry ice cream*
- The ingredient is essential to characterize the product and to distinguish it from other products

The QUID declaration should be expressed as a percentage (%) and must appear in or immediately next to the name of the food or in the list of ingredients (EUR-Lex, 2017).

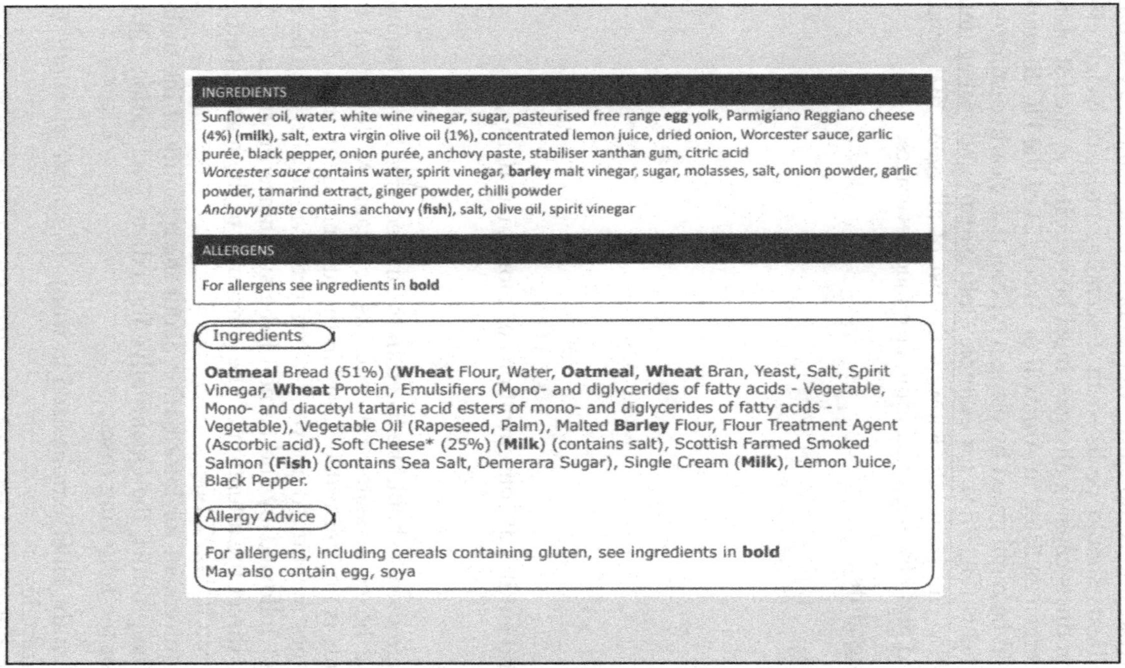

INGREDIENTS

Sunflower oil, water, white wine vinegar, sugar, pasteurised free range **egg** yolk, Parmigiano Reggiano cheese (4%) **(milk)**, salt, extra virgin olive oil (1%), concentrated lemon juice, dried onion, Worcester sauce, garlic purée, black pepper, onion purée, anchovy paste, stabiliser xanthan gum, citric acid
Worcester sauce contains water, spirit vinegar, **barley** *malt vinegar, sugar, molasses, salt, onion powder, garlic powder, tamarind extract, ginger powder, chilli powder*
Anchovy paste contains anchovy **(fish)**, *salt, olive oil, spirit vinegar*

ALLERGENS

For allergens see ingredients in **bold**

Ingredients

Oatmeal Bread (51%) **(Wheat** Flour, Water, **Oatmeal, Wheat** Bran, Yeast, Salt, Spirit Vinegar, **Wheat** Protein, Emulsifiers (Mono- and diglycerides of fatty acids - Vegetable, Mono- and diacetyl tartaric acid esters of mono- and diglycerides of fatty acids - Vegetable), Vegetable Oil (Rapeseed, Palm), Malted **Barley** Flour, Flour Treatment Agent (Ascorbic acid), Soft Cheese* (25%) **(Milk)** (contains salt), Scottish Farmed Smoked Salmon **(Fish)** (contains Sea Salt, Demerara Sugar), Single Cream **(Milk)**, Lemon Juice, Black Pepper.

Allergy Advice

For allergens, including cereals containing gluten, see ingredients in **bold**
May also contain egg, soya

FIGURE 7.14 Examples of ingredient list with allergens highlighted and allergen advice. *Source: British Retail Consortium.*

If ingredients are natural constituents of the food (as caffeine is in coffee), the QUID requirement does not apply. The QUID is not required in certain cases such as where an ingredient is used in small quantities for flavoring purposes.

Nutrition labeling

All the elements mandatory for nutrition declaration should be presented in the same field of vision on the food label or packaging. Nutrition declarations are to be done on a "per 100 g or per 100 milliliters" basis, however per portion or per consumption units can be presented in addition to the declaration per 100 g or milliliters provided that the number of portions or consumption units is clearly specified on the package. The energy value should be expressed in kilojoules (kJ) and kilocalories (kcal). The declaration must list the constituents in this particular order:

- Amount of energy
- amounts of fat
- saturates (*i.e. saturated fats*)
- carbohydrate
- sugars
- protein
- salt

Constituents should be expressed in grams (g), milligrams (mg) or micrograms (μg) per 100 g or per 100 milliliters, as appropriate. Voluntary declarations can be made as well to supplement the mandatory information by including values for monounsaturates, polyunsaturates, polyols, starch, fiber, the vitamins and minerals listed in Part A of Annex XIII of the FIC regulation (European Commission, 2019).

For voluntary declarations of vitamins and minerals, the nutrient should be present in significant amounts. Declarations should be presented as percentage *nutrient reference values* (NRV) or *reference intakes* (RI) based on an average sized adult doing average amounts of physical activity[3]. Significant amounts are defined as follows:

- for products other than beverages, 15% of the nutrient reference values per 100 g or 100 ml
- for beverages, 7.5% of the nutrient reference values per 100 ml
- for single portion packages, 15% per portion.

Once information on % RIs per 100 g or per 100 ml is provided, it is a requirement that the additional statement "*Reference intake of an average adult (8400 kJ/2000 kcal)*" be presented in close proximity to the information on reference intakes (Fig. 7.15). On the other hand, the additional statement is not required if the information provided is on % RIs *per portion* and/or *per consumption unit* only basis.

---

[3] *Nutrient reference values* are defined as a set of recommended daily nutrient targets based on current available scientific knowledge; *reference intake* is defined as a means of communicating maximum recommended nutrient intake to the public.

Nutrition Information

	Per 100 g	
Energy	485 kJ / 117 kcal	
Fat	8 g	
Of which Saturates	3,7 g	
Carbohydrate	9 g	
Of which Sugars	8 g	
Protein	1,4 g	
Salt	0,02 g	
Vitamin C	14,81 mg	19% RI*

Salt content is exclusively due to the presence of naturally occurring sodium.

*Reference intake of an average adult (8 400 kJ / 2 000 kcal)

INGREDIENTS:Mandarin Oranges (37.9%), Light Whipping Cream (Milk), Pears (12.4%), Peaches (7.7%), Thompson Seedless Grapes (7.6%), Apple (7.5%), Banana (5.9%), English Walnuts (Tree Nuts)

Nutrition Information

	Per 100 g %Reference Intake RI	
Energy	485 kJ / 117 kcal	6% RI
Fat	8 g	11% RI
Of which Saturates	3,7 g	19% RI
Carbohydrate	9 g	3% RI
Of which Sugars	8 g	9% RI
Protein	1,4 g	3% RI
Salt	0,02 g	0% RI
Vitamin C	14,81 mg	19% RI

Salt content is exclusively due to the presence of naturally occurring sodium.

Reference intake of an average adult (8 400 kJ / 2 000 kcal)

INGREDIENTS:Mandarin Oranges (37.9%), Light Whipping Cream (Milk), Pears (12.4%), Peaches (7.7%), Thompson Seedless Grapes (7.6%), Apple (7.5%), Banana (5.9%), English Walnuts (Tree Nuts)

Nutrition Information

	Per 100 g %Reference Intake RI		Per portion of 249 g %Reference Intake RI	
Energy	485 kJ / 117 kcal	6% RI	1 181 kJ / 284 kcal	14% RI
Fat	8 g	11% RI	19 g	27% RI
Of which Saturates	3,7 g	19% RI	9,2 g	46% RI
Carbohydrate	9 g	3% RI	23 g	9% RI
Of which Sugars	8 g	9% RI	21 g	23% RI
Protein	1,4 g	3% RI	3,4 g	7% RI
Salt	0,02 g	0% RI	0,06 g	1% RI
Vitamin C	14,81 mg	19% RI	36,91 mg	46% RI

Salt content is exclusively due to the presence of naturally occurring sodium.

Reference intake of an average adult (8 400 kJ / 2 000 kcal)

INGREDIENTS:Mandarin Oranges (37.9%), Light Whipping Cream (Milk), Pears (12.4%), Peaches (7.7%), Thompson Seedless Grapes (7.6%), Apple (7.5%), Banana (5.9%), English Walnuts (Tree Nuts)

FIGURE 7.15 Examples of nutrition declaration in different formats. *Source: https://www.esha.com/products/genesis-rd-food-labeling-software/labels-and-labeling/european-union-nutrition-facts-label/.*

TABLE 7.5 Limits and suggested declarations for negligible nutrients.

Nutrient	Negligible amount	Nutrition declaration
Fat		
Carbohydrate	No detectible amount present	"0 g"
Sugars	≤ 0.5 g per 100 ml or 100 g	" ≤ 0.5 g"
Protein		
Saturates	No detectible amount present	"0 g"
	≤ 0.1 g per 100 ml or 100 g	" ≤ 0.1 g"
Salt	No detectible amount present	"0 g"
	≤ 0.0125 g per 100 ml or 100 g	" ≤ 0.5 g"

Previously, most nutrition declarations were provided on the back of packages. However, this is not required and manufacturers can place the declaration on any surface of the package provided that the font size requirements are adhered to. Once space permits, the declaration should be done in a tabular form. Otherwise, the declaration must appear in the linear format.

For the EU, instead of sodium the declaration of *salt* content in used so that this is clear to consumers at the point of purchase. Salt content is based on the total amount of sodium in the packaged product (both added and natural) and is calculated by multiplying the sodium content by 2.5. When salt is not added to the product, this can be indicated as seen in Table 7.5. If the presence of salt in the product is due exclusively to the natural occurring sodium, statements like "This product contains no added salt" or "Salt content is due to naturally occurring sodium" may be used.

Front of pack labeling

Front of pack (FoP) labeling allows for the voluntary repeat of mandatory information for pre-packaged food (Department of Health, UK, 2016a). Elements that are allowed on FoP are:

- Energy value (kJ/kcal)
- Energy value plus the amount in grams of *fat, saturates, sugars* and *salt*. These are commonly referred to as *"energy + 4"*.

Foods offered for sale to the final consumer or mass caterers without pre-packaging, foods packed on sales premises at the consumers' request and foods packaged for direct sales are considered non-pre-packaged food. There are no requirements for nutrition information to be presented on these items but the information can be provided on a voluntary basis (Department of Health, UK, 2016b). When this is done, the full mandatory nutrition declaration, the energy value only or "energy + 4" should be provided. The declaration cannot omit any of the 4 nutrients in energy + 4 if that format is used. Strict adherence to any of the three formats must be followed.

TABLE 7.6 Representation of minimum durability.

Food type	Acceptable format
Foods that will keep for longer than 3 months	Best Before followed by Day and Month
Foods that will keep for longer than 3 mo but less than 18 months	Best Before End followed by Month and Year
Foods that will keep longer than 18 months	Best Before End followed by Year only or Month and Year

Appropriate durability indication

The labels of pre-packaged food also require an indication of the minimum durability (commonly known as the *shelf life*) of the product. There are different formats required depending on the nature of the product. Minimum durability is referred to as the date until which the product will retain its specific qualities when properly stored. Two ways to indicate this date is by using *"best before"* when date includes an indication of the day and *"best before end"* in all other cases. Examples of appropriate declarations are given in Table 7.6.

The term *"Use by"* is another means of indicating the minimum durability, however this is specifically for foods considered perishable and the food is deemed unsafe for consumption after the date indicated. Whenever there are individual pre-packaged portions each should bear the use by date on the package. Food that is sold frozen should bear information on the date it was frozen indicating the day, month and year preceded by the term "frozen on".

Warning on labels

The FIC regulation *Food Additives Regulation 1333/2008* presents a list of products that are required to bear a warning label. These include the following:

- foods when the durability has been extended by use of packaging gases
- foods which contain sweeteners
- foods which contain added sugar and sweeteners
- foods which contain aspartame
- foods which contain more than 10% added polyols
- confectionery and beverages which contain liquorice (glycyrrhizinic acid or its ammonium salt)
- beverages which contain more than 150 mg/l of caffeine and foods with added caffeine
- foods or food ingredients with added phytosterols, phytosterol esters, phytostanols or phytostanol esters
- foods that have been irradiated
- foods that contain genetically modified ingredients, unless their presence is accidental and $\leq 0.9\%$
- foods that contain sulfur dioxide in levels above 10 mg/l.

Additionally, it is important for manufacturers to investigate the additives that are permitted for use in the destination market to avoid the risk of noncompliance and possible destruction or return of non-compliant products. There are six color additives, according to Annex V to *Food Additives Regulation 1333/2008* (EUR-Lex, 2019a), that when used in the manufacturing of a product would be accompanied by the statement "may have an adverse effect on activity and attention in children". They are sunset yellow (E110), quinoline yellow (E104), carnosine (E122), allura red (E129), tartrazine (E102) and ponceau 4 R (E124). The numbers in brackets are "E numbers" which are specific codes for substances allowable for use as food additives in the EU. All additives used in food products must be declared and indicated by its assigned E number on the package.

Alcoholic beverages must have a warning on their label indicating the alcohol content when the content is above 1.2%. The alcoholic strength is required to be indicated by a figure with maximum one decimal place followed by the symbol "% *vol.*." This information should be presented in the same field of vision as the product name and net quantity. Allergen labeling is also mandatory on all alcoholic beverages, EU FIC exempts alcoholic beverages with >1.2% alcohol content from mandatory nutrition labeling (Department of Health, UK, 2017). There are established standards for wine and guidelines available for the labeling of these products as EU and non-EU wines may be labeled differently.

Country of origin labeling (COOL)

Country of Origin Labeling (COOL) is primarily required in the EU for specific products and in specific instances. Mandatory COOL applies:

- whenever failure to indicate the country of origin might mislead the consumer
- to honey, fruit and vegetables, olive oil, fishery and aquaculture products and beef

Mandatory COOL has also been extended to *fresh, chilled and frozen pork, sheep and goat meat* and *poultry* in addition to the products for which COOL was already mandatory. Further, it is expected that COOL will also be required when the country of origin is given voluntarily but the origin of the primary ingredient in the product is different from that of the food product.

Organic food labels

The use of the EU organic logo became mandatory on all pre-packaged organic products produced in the EU since July 1, 2012. Organic products imported from outside EU may carry the EU organic logo if they comply with the EU production rules and the indication of the place of farming becomes mandatory. "Organic" and all its derivate terms or diminutives such as "bio" and "eco" may be used only to label products that comply with EU organic production rules and if at least 95% of the ingredients of agricultural origin are organic. Products containing less than 95% organic ingredients should indicate and make reference to the individual organic ingredients in the list of ingredients. The total percentage of organic ingredients must be indicated when reference is made to the organic production method in the ingredients list.

Using the e mark

Packages meeting the requirements of the Packaged Goods Regulations and are between 5 g and 10 kg for food and 5 ml − 10 L for beverages can bear the *e mark*. When used, manufacturers are declaring their compliance with the requirements of the "average system" under European Union Directive 76/211/EEC (EUR-Lex, 2019b). The mark is a metrological passport to trade allowing free access within European Economic Area (EEA) and respective markets including France, Netherlands and the UK. The products bearing the mark are not subjected to further weight and measurements regulation. Its use is optional, however packages that do not display the **e mark** must meet the regulations of the destination country.

Signposting

Traffic light food labels, more formerly called "signposting", was first introduced in the United Kingdom with the intent to provide consumers with a clearer indication of the amount of salt, sugar or fat the products have. Like traffic light, the colors used in signposting are red, amber and green and are based on the quantity of the nutrients in the product. It was found that consumers were able to more easily decide on the products to be bought based on their content when signposting was used (Michalopoulos, 2017). In 2011, however, the traffic light system for food labeling was rejected at EU level

TABLE 7.7   Traffic light signposting criteria for 100 g of food.

Text	LOW[δ]	MEDIUM	HIGH	
			Red	
Colour code	Green	Amber	>25% of RIs	>30% of RIs
Fat	≤ 3.0g/100g	> 3.0g to ≤ 17.5g/100g	> 17.5g/100g	> 21g/portion
Saturates	≤ 1.5g/100g	> 1.5g to ≤ 5.0g/100g	> 5.0g/100g	> 6.0g/portion
(Total) Sugars	≤ 5.0g/100g	> 5.0g to ≤ 22.5g /100g	> 22.5g/100g	> 27g/portion
Salt	≤ 0.3g/100g	> 0.3g to ≤ 1.5g/100g	>1.5g/100g	>1.8g/portion

Note: Portion size criteria apply to portions/serving sizes greater than 100 g.
Source: https://www.food.gov.uk/sites/default/files/media/document/fop-guidance_0.pdf.

TABLE 7.8 Traffic light signposting criteria for 100 ml of drink.

Text	LOW[§]	MEDIUM	HIGH	
Colour code	Green	Amber	Red	
			>12.5% of RIs	>15% of RIs
Fat	≤ 1.5g/100ml	> 1.5g to ≤ 8.75g/100ml	> 8.75g/100ml	>10.5g/portion
Saturates	≤ 0.75g/100ml	> 0.75g to ≤ 2.5g/100ml	> 2.5g/100ml	> 3g/portion
(Total) Sugars	≤ 2.5g/100ml	> 2.5g to ≤ 11.25g/100ml	> 11.25g/100ml	> 13.5g/portion
Salt	≤ 0.3g/100ml	>0.3g to ≤0.75g/100ml	> 0.75g/100ml	> 0.9g/portion

Note: portion size criteria apply to portions/serving sizes greater than 150 ml.
Source: https://www.food.gov.uk/sites/default/files/media/document/fop-guidance_0.pdf.

as part of negotiations on the food information to consumer regulation. It remains a voluntary front of pack labeling element based on realistic size portions of the food or beverage.

The traffic light signposting requires a definition for the colors (Department of Health, UK, 2016a,b) and the criteria are set out in Table 7.7 for food and Table 7.8 for beverages. In both tables, low cut off is based on the "low" nutrition claim for fat, saturates, total sugars and salt in the EU Nutrition & Health Claims Regulation (EC) 1924/2006.

Labeling requirements for export to the United States of America

In the United States of America (USA), the regulations indicate that packages have two main types of panels on which specific information should be presented. The front area of the package that will most likely be seen by the consumer is called the Principal Display Panel (PDP). To the right of the PDP is the Information Panel – IP (Fig. 7.16). The PDP should contain the statement of identity or the name of the product as well as the net contents of the package, while the IP should contain information that should not be interrupted by intervening material but instead be placed together such as the:

1. name and address of the manufacturer, packer or distributor
2. ingredients list
3. nutrition labeling
4. allergen labeling

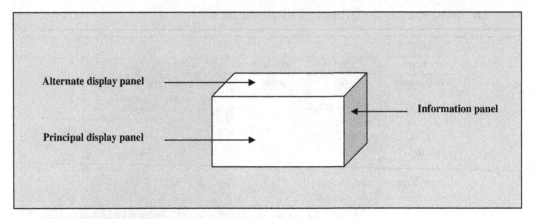

FIGURE 7.16   Principal display and information panels for US labels.

The print size and type should make the information be easy to read, conspicuous and adequately contrasted with the background of the packaging, especially on the information panel which houses the information that many consumers use to determine whether to consume the food, particularly those who may have allergies or be on restrictive diets. There are also specific requirements for the font sizes for the statement of identity and net contents of the product on the PDP.

The statement of identity should be prominent and in **bold type** and should be presented as the common/usual name of the food. The net quantity should be located within the bottom third of the PDP, have sufficient contrast and be expressed as weight for solid foods, volume measure for liquids and numeric counts for small individual units, for example cookies or cherries.

The ingredients should be listed in descending order of predominance, before or after the nutrition label and contact information for the manufacturer, packer or distributor, with a type size no less than 1/16th inch in height as measured by the lower case letter "o". It should be prominent and easy to read, with the common names of the ingredients being used unless a different term is provided by the regulations. Major food allergens should be declared as well as any additives or colors (both certified and non-certified) used during the processing or production of the packaged items.

Food allergens

In order to make it easier for consumers with food allergies and their caregivers to identify and avoid foods that contain major food allergens, the Food Allergen Labeling and Consumer Protection Act (FALCPA) was passed by the US Congress in 2004. It requires that the label of a food that contains an ingredient belonging to the identified group of food allergens or which contains protein from a "major food allergen" declare the presence of the allergen in the manner described by the law. It became effective on January 1, 2006 and all labels for foods to be consumed in the US can be assessed for their compliance with the regulations. Even though there are more than 160 foods components that have been identified as causes

Option 1

Contains Wheat, Milk, Egg, and Soy

Option 2

Ingredients: Enriched flour (**wheat** flour, malted barley, niacin, reduced iron, thiamin mononitrate, riboflavin, folic acid), sugar, partially hydrogenated soybean oil, and/or cottonseed oil, high fructose corn syrup, whey (**milk**), **eggs**, vanilla, natural and artificial flavoring) salt, leavening (sodium acid pyrophosphate, monocalcium phosphate), lecithin (**soy**), mono-and diglycerides (emulsifier)

FIGURE 7.17   Options for declaring the major food allergens on packaged food labels.

of food allergies in sensitive individuals, only 8 have been defined as major food allergens. They are so characterized because they account for greater than 90% of all documented food allergies in the US which have been shown to have severe or life-threatening consequences.

FALCPA describes the major food allergens as ingredients that contain one of the following or proteins from of the following:

- Milk
- Wheat
- Fish (e.g. salmon, flounder)
- Peanuts
- Tree nuts (e.g. walnuts, pecans, almonds)
- Soybeans
- Egg
- Crustacean Shellfish (e.g. lobster, crabs, crayfish)

There are two options for declaring the allergens in a product. The first is to place the word "contains" followed by the name of the food source from which the major allergen is derived. This should be placed in the vicinity of (i.e. immediately after or adjacent to) the ingredients list *in a font size no less than the ingredients list*. Alternatively, the allergens may be included in the ingredients list as the name of the food ingredient which is the source of the allergen followed by the common or usual name of the allergen (in brackets). Examples are shown in Fig. 7.17 below which was adopted from the FDAs website.

Upon entry to the United States of America, imported food packages may be assessed at port of entry for their compliance with Nutrition Labeling and Education Act (NLEA) of 1990, which provides FDA with specific authority to require nutrition labeling of foods regulated by the agency. It also requires that all nutrient contents claims and health claims on the labels should be consistent with the regulations. There may be flexibility for "small packages" and/or exemptions for small businesses[4].

---

[4] Small businesses are defined by the regulations as businesses that had gross food sales or $50,000 or less and those who have total annual gross sales (food and non-food) of $500,000 or less.

TABLE 7.9 List of information required in the nutrition label in the United States of America (USA) prior to the release of updated regulations in May 2016.

Mandatory	Voluntary
Total Calories	Calories from saturated fat
Calories from fat	Polyunsaturated fat
Total fat	Monounsaturated fat
Saturated fat	Potassium
Trans fat	Percent of vitamin A present as beta-carotene
Cholesterol	Soluble fiber
Sodium	Insoluble fiber
Total carbohydrate	Sugar alcohol
Dietary fiber	Other carbohydrates
Sugars	Other essential vitamins and minerals
Protein	
Vitamin A	
Vitamin C	
Calcium	
Iron	

The regulations require that the nutritional information must be set off in a box with the heading "Nutrition Facts" stretched across the width of the box. The label may be oriented perpendicularly or parallel to fit the design of the packaging, however, FDA urges manufacturers to strive for consistency of presentation of nutrition information in the market and to place the Nutrition Facts label so that it is readily observable and legible to the consumer at the point of purchase.

There are some nutrients (vitamins and minerals) and other information that must be included in the nutrition labeling (Table 7.9). This list was updated in 2016, when the FDA released the new labeling requirements on May 20. The appearance of the label required has also been redefined in order to make it easier for consumers to find the information they need when purchasing/consuming a product.

Once a nutrient claim that characterizes the level of a nutrient directly or by implication is made, the nutrient should be included in the list whether it falls in the list of mandatory or voluntary nutrients. Any voluntary information included should be placed immediately following the mandatory nutrient, maintaining the grouping of carbohydrates, minerals and vitamins i.e. voluntary minerals should be placed following mandatory minerals. There are 5 nutrients that should appear in bold type on the label: total fat, cholesterol, sodium, total carbohydrate and protein. The information presented in bold in Table 7.9 should also be presented in bold font in the nutrition facts panel (NFP).

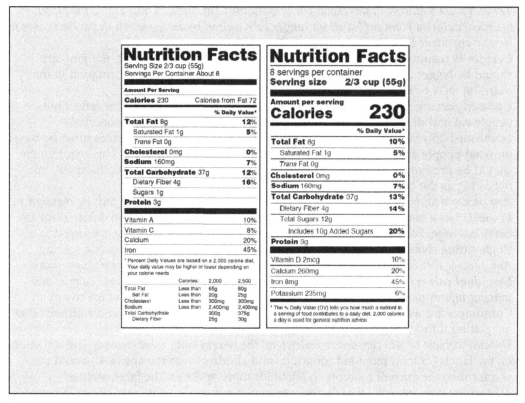

FIGURE 7.18 Current and new US label layout for packaged food respectively. *Source: The FDA website — www.fda.gov*

There are specific rounding rules for listing the calories defined within the regulations. Disclosure statements are required for nutrients that may increase the risk of disease or health related conditions i.e. where the nutrient exceeded the prescribed level as defined by the Daily Values (DV) (determined by the Reference Daily Intakes for vitamins and minerals and the Daily Reference Values for the other nutrients).

As previously indicated, changes will be made to what is required in the nutrition labels starting in January 2020. The changes required (shown in Fig. 7.18) will include the following:

- *The inclusion of "added sugars".* Many experts recommend consuming fewer calories from added sugar because this may decrease the intake of nutrient-rich foods while increasing overall caloric intake. It is recommended that the daily intake of calories from added sugars not exceed 10% of total calories.
- *Updated daily values for nutrients like sodium, dietary fiber and Vitamin D which are used to calculate the percentage daily values (DV) on the labels.* This helps the consumer to understand the nutrition information in the context of a total daily diet.
- *The declaration of the amount of **potassium and Vitamin D** on the label, because they are new "nutrients of public health significance".* Calcium and iron are still required, but Vitamins A and C will not be but could be included on a voluntary basis.

- *The continued requirement for "Total Fat," "Saturated Fat," and "Trans Fat" on the label, however "**calories from fat" will no longer be required** because research shows the type of fat is more important than the amount.*
- Calories is required to be in bold as per current regulations however, the font size should be bigger and more prominent than the remainder of the information in the nutrition facts box.
- Updated Serving Size Requirements Package Sizes which would better reflect how people eat and drink today, which has changed since serving sizes were first established 20 years ago. By law, **the label information on serving sizes must be based on what people actually eat, not on what they "should" be eating**. This requirement should be present in bold and font size bigger than for the current requirement, but not as big as the calories declaration.
- *Any packaged foods, including drinks, that are typically eaten in one sitting **will be required to be labeled as a single serving** and that calorie and nutrient information declared for the entire package. For example, a 20-ounce bottle of soda that is typically consumed in a single sitting should be labeled as one serving rather than as more than one serving.*
- *Any package that is larger and could be consumed in one sitting or multiple sittings, should bear "**dual column**" labels to indicate both "**per serving**" and "**per package**" calories and nutrient information. Examples are 24-ounce bottles of soda or a pint of ice cream. Consumers are will be better able to understand how many calories and nutrients they are getting if they eat or drink the entire package at one time.*
- *Updated footnote to help consumers understand the percent daily value concept.* The statement on the labels (of food intended for adults and children over the age of 4) would be shorter than the current footnote to allow for more space on the label, stating:
  *The percent daily value (%DV) tells you how much a nutrient in a serving of food contributes to a daily diet. 2000 calories a day is used for general nutrition advice. For children 1 − 3 years of age:
  *The percent daily value (%DV) tells you how much a nutrient in a serving of food contributes to a daily diet. 1000 calories a day is used for general nutrition advice.
- *A Refreshed Design*

Labeling requirements for export to the Canada

The Canadian Food Inspection Agency (CFIA) is the major federal agency that has jurisdiction over foods offered for sale in Canada, whether produced domestically or imported. All foods in Canada need to have labels that comply with the Consumer Packaging and Labeling Act (CPLA), the Food and Drug Act (FDA) and the Food and Drug Regulations (FDR). The official languages of Canada, English and French, must be present on all labels except where exemptions are permitted by law. The information in French may appear on a separate panel than the information in English or may be on the same panel but must meet the minimum height requirement for the text (type) printed on the label.

As with the labels on packaged foods destined for consumption in the US, certain information is mandatory, including the name of the product or the "common name". Both French and English common names should be on the principal display panel and should be no less than 1.6 mm based on the lower case letter "o". The English common name should appear in boldface type. Consistency in type height is encouraged for both languages. Certain voluntary information, if included on the labels or advertisement on packages, is subjected to other requirements and must be presented bilingually although voluntary information is not generally subject to bilingual requirements at the federal level. Some of the information required to be present bilingually are:

- Organic Claims
- Nutrient Content Claims
- Additional Nutritional Information
- Health Claims

Another mandatory element that should be present on all labels is the *net quantity declarations* which also must be bilingual. When words are used instead of the SI symbols in net quantity declarations, they must appear in both official languages. A list of ingredients should be presented on all pre-packaged foods with more than one ingredient and the ingredients should be declared in descending order of their proportion by weight as determined before they are combined to make the food.

Allergen labeling for Canada

If any allergens are present in the pre-packaged food, they must be declared. In Canada, the list of allergens of concern/ingredients causing adverse health reactions (collectively referred to as "allergens" in the Canadian scenario) is longer than in the US and includes:

- tree nuts (almonds, Brazil nuts, cashews, hazelnuts, macadamia nuts, pecans, pine nuts, pistachios or walnuts)
- peanuts
- sesame seeds
- wheat (gluten)
- eggs
- milk
- soybeans
- crustaceans
- shellfish
- fish
- mustard seeds
- sulfites

It is mandatory that all food allergens must be declared in the list of ingredients or in a "contains" statement. When using the ingredients list format, the prescribed source name of the food allergen should be shown in parentheses, as follows:

Ingredient List: flour (wheat), liquid albumin (egg), vegetable oil, sugar, flavor.

Other labeling requirements

The Food and Drug Regulations (FDR) requires that food additives are declared in the list of ingredients of pre-packaged foods when added to a food, whether alone or with natural flavoring agents. The word "artificial" or "imitation" must be included as an integral

TABLE 7.10   List of month abbreviations for the Canadian labels.

Month	Abbreviation	Month	Abbreviation
January	JA	July	JL
February	FE	August	AU
March	MR	September	SE
April	AL	October	OC
May	MA	November	NO
June	JN	December	DE

part of the flavoring preparation name and the declaration should be in the same type size and style as the flavoring preparation name. The identity and principal place of business of the manufacturer or the business for which the food has been manufactured or produced for sale or resale should be declared in both official languages and meet the type height requirements of 1.6 mm or greater. It is important to note that websites, telephone numbers, and virtual addresses do not fit the criteria for "principal place of business" declarations since they are not physical locations. Food exported for consumption in Canada should include "imported by/for" (importé par/pour) or the country of origin as part of the identity and principal place of business declaration whether they are labeled or relabeled in Canada or elsewhere, including bulk imports.

The date format applicable for products sold in Canada is different to that used elsewhere (Canadian Food Inspection Agency CFIA, 2019d). The abbreviations are shown in Table 7.10 (above). The format requires that the year should be listed first, followed by the month (bilingual or abbreviated), followed by the day of the month as seen in Fig. 7.19. Date markings/Best Before dates are required when the packaged food has a shelf life less than 90 days. In these cases, the date marking should be preceded by "Best Before". Food with shelf life greater than 90 days (e.g. cereals) are not required to be labeled with a "Best Before" date. However, a "Best Before" may be declared on foods with a shelf life greater than 90 days, depending on the purveyor. The "Best Before" date may appear anywhere on the package. If it is placed on the bottom, a clear indication of its location must be shown elsewhere on the label. It must be present in both English and French or indicated by using specified bilingual abbreviations (Canadian Food Inspection Agency CFIA, 2019e). In addition, foods requiring date marks should bear storage instructions once storage conditions differ from normal room/ambient temperature. Examples include "keep refrigerated" and "store in a cool, dry place".

Labeling of irradiated foods

Federal controls are applicable to the safety and labeling of foods that have been irradiated. There are certain irradiated foods that can be sold in Canada. Some of these are potatoes, wheat, flour, whole wheat flour, onions, whole or ground spices and dehydrated seasoning preparations. Labeling regulations for irradiated food are enforced by CFIA and are detailed in the Food and Drug Regulations (B.01.035) which require the identification

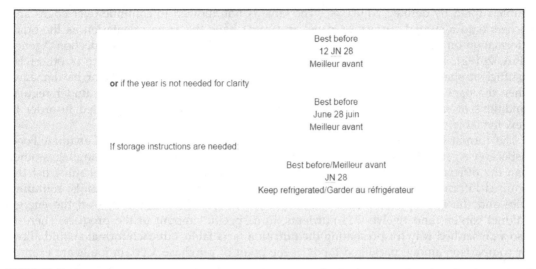

Best before
12 JN 28
Meilleur avant

or if the year is not needed for clarity

Best before
June 28 juin
Meilleur avant

If storage instructions are needed:

Best before/Meilleur avant
JN 28
Keep refrigerated/Garder au réfrigérateur

FIGURE 7.19 Examples of acceptable durability declarations for foods exported to Canada. *Source: http:// www.inspection.gc.ca/food/requirements/labeling/industry/date-markings-and-storageinstructions/eng/1328032988308/ 1328034259857?chap = 2.*

FIGURE 7.20 The symbol for irradiated products.

of wholly irradiated foods with both a written statement such as "irradiated" or "treated with radiation" or "treated by irradiation" and the international symbol shown in Fig. 7.20 on the PDP at the correct size as specified by the regulations. When there are irradiated ingredients accounting for 10% or more in the final product, they must be identified in the list of ingredients as irradiated (Canadian Food Inspection Agency CFIA, 2019f).

The labeling of irradiated food is not limited to the package but should extend to the shipping containers housing the products en route to the final destination for sale to consumers. These are also required to bear the identification of wholly irradiated foods with a written statement such as "irradiated" or "treated with radiation" or "treated by irradiation" but are not required to bear the international symbol.

Nutrition labeling

The Nutrition Facts table (NFt) must be located on one continuous surface, generally understood to be a flat surface or a slightly curved surface that is unbroken or uninterrupted by defined edges, etc. The table is not allowed to continue over edges and corners onto a second surface or panel and must have the same orientation as the other information on the label when there is sufficient space (Canadian Food Inspection Agency CFIA, 2019g). It may however be oriented in another manner when there is otherwise insufficient space available, provided that the product will not leak out or be damaged when the package is turned to view the table. The NFt should be visible under regular conditions of sale, i.e. the outer package should not have to be manipulated in order to view the table.

The format and presentation of the NFt are specifically prescribed (Canadian Food Inspection Agency CFIA, 2019h) and there is no provision for the use of languages other than the official languages (French and English) within the table. The NFt must list the required information in the correct order, using approved nomenclature, units, rounding rules and the appropriate format. The table should provide information on the energy (caloric) content and twelve (12) nutrients, in a specific amount of the product. There is also a prescribed way for presenting the nutrition facts table, but each format should allow for comparison among packaged foods at the point of purchase. Certain foods are exempt from nutrition labeling but may voluntarily display the Nutrition Facts table which should be compliant with the requirements of the regulations. Fig. 7.21 shows the mandatory

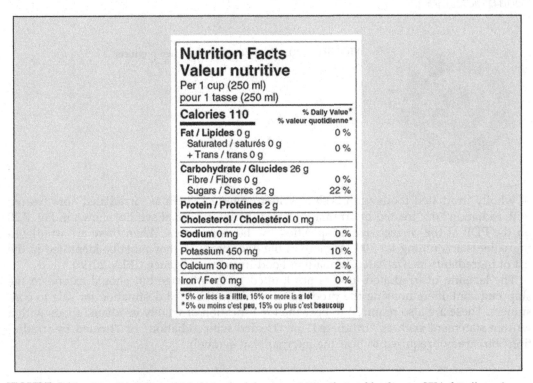

FIGURE 7.21   Canadian bilingual label standard format nutrition facts table. *Source: CFIA, http://www.inspec-tion.gc.ca/food/requirements/labeling/industry/nutritionlabeling/nutrition-facts-table/eng/1389198568400/1389198597278? chap = 1.*

TABLE 7.11 Point size measures for the wording on Canadian labels.

Type size	Leading size	Rule	Indents/spacing
6 point = 2.12 mm	9 point = 3.17 mm	0.5 point = 0.18 mm	3 point = 1.06 mm
8 point = 2.82 mm	12 point = 4.23 mm	1 point = 0.35 mm	5 point = 1.76 mm
13 point = 4.59 mm	14.5 point = 5.12 mm	2.5 point = 0.88 mm	6 point = 2.12 mm

information which must always be present in the NFt and the order in which it should appear. Generally, any voluntary information declared maybe presented within or outside of the table. Additional information may be declared and is especially required once a claim is made about the nutrient mentioned.

As required by the regulations, the characters within the nutrition facts table should not be decorative and should be presented in single standard sans serif font, an example of which is Helvetica. The characters must be displayed such that they never touch each other or the rules – the horizontal and vertical lines of the table. The requirements for font size and width are set out in the regulations for all formats of the NFt (Canadian Food Inspection Agency CFIA, 2019h). The table below indicates common point size measures in the standard formats of the NFt (Table 7.11).

There are three basic formats for the presentation of the nutrition facts table: basic, horizontal and linear. There are also specialized formats used in special cases, including:

- Simplified formats
- Dual formats for
  - Food requiring preparation
  - Different amount of food
- Aggregate formats
  - Different kinds of food (for each food in an assortment)
  - Different amounts of food

The main differences in the labels compliant with US regulations versus those compliant with Canadian regulation include but are not limited to the following.

- Even though the quantity of *trans fat* is required on both labels, a reference standard was not developed for the sum of saturated and trans fat and, as a result, no percentage daily value (DV) is required (Canadian Food Inspection Agency CFIA, 2019i).
- Although required by both countries, percent DVs for mandatory vitamins and minerals are based on Reference Daily Intakes/Reference Amounts Customarily Consumed (RACC) in the US and Recommended Daily Intakes for Canadians (Health Canada, 2006, 2016).
- The rounding rules for certain components of the foods that should be included in the nutrition table are different between countries.
- Servings per container may be expressed as "per", "serving" or "serving size" (in addition to the French equivalent). Household measures should be declared with metric values in brackets using the bilingual abbreviations (e.g. mg, g, mL).

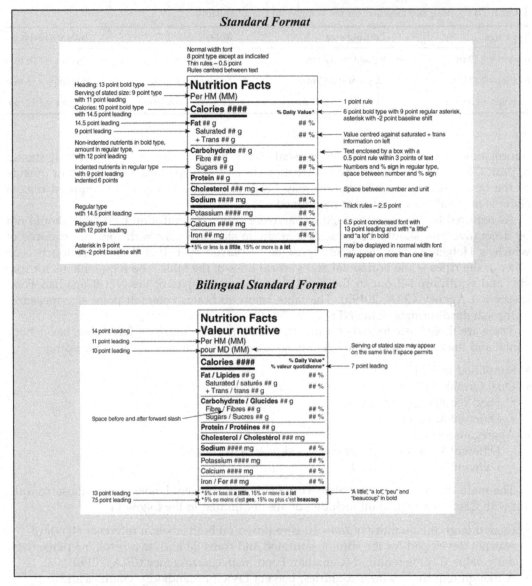

FIGURE 7.22 Standard & bilingual standard format for the nutrition facts table showing the mandatory elements and required presentation. *Source: https://www.canada.ca/en/health-canada/services/technical-documents-labeling-requirements/directory-nutrition-facts-table-formats.html.*

The Food and Drug Regulations (FDR) defines the manner in which energy content and nutrient values should be declared. It is important to note that the requirements are different for different categories of products such as simplified formats, pre-packaged foods for children under two years of age, pre-packaged foods for use in manufacturing other foods,

foods for commercial and industrial enterprises or institutions, as well as small packages ($< 100 \text{ cm}^2$). The manufacturer should ensure that they are familiar with the requirements applicable to their product. Generally, the information required for the standard and the bilingual standard format of the Nutrition Facts table are indicated in Fig. 7.22. The bilingual format uses the same specifications as for the standard format except as otherwise indicated. The order of languages may be reversed from the order shown in Fig. 7.22.

References

Black, D.G., Barach, J.T. (Eds.), 2015. Canned Foods: Principles of Thermal Process Control, Acidification and Container Closure Evaluation. GMA Science and Education Foundation.

Canadian Food Inspection Agency (CFIA), 2019a. Regulatory requirements: preventive controls. <http://www.inspection.gc.ca/food/requirements/preventive-controls-food-businesses/regulatory-requirements/eng/1524581767630/1524581834894?chap = 0/> (accessed 22.04.19).

Canadian Food Inspection Agency (CFIA), 2019b. Dairy processing systems: Retorting (canning). Container closure evaluation. <http://www.inspection.gc.ca/food/requirements/preventive-controls-food-businesses/dairy-products/dairy-processing-systems-retorting-canning-/eng/1536234798360/1536234798656?chap = 4/> (accessed 22.04.19).

Canadian Food Inspection Agency (CFIA), 2019c. Metal can defects — Identification and classification. <http://www.inspection.gc.ca/food/requirements/preventive-controls-food-businesses/metal-can-defects/eng/1510763304486/1510763304952/> (accessed 13.04.19).

Canadian Food Inspection Agency (CFIA), 2019d. Date marking and storage instructions. Manner of declaring. <http://www.inspection.gc.ca/food/requirements/labeling/industry/date-markings-and-storage-instructions/eng/1328032988308/1328034259857?chap = 2/> (accessed 13.04.19).

Canadian Food Inspection Agency (CFIA), 2019e. Interactive tool — food labeling requirements. <http://www.inspection.gc.ca/food/requirements/labeling/for-consumers/food-labeling-requirements/eng/1302802599765/1302802702946/> (accessed 22.04.19).

Canadian Food Inspection Agency (CFIA), 2019f. Irradiated foods. <http://www.inspection.gc.ca/food/requirements/labeling/industry/irradiated-foods/eng/1334594151161/1334596074872/> (accessed 13.04.19).

Canadian Food Inspection Agency (CFIA), 2019g. Graphical and technical requirements within the nutrition facts table. <http://www.inspection.gc.ca/food/requirements/labeling/industry/nutritionlabeling/presentation/eng/1387664849974/1387664998059?chap = 1/> (accessed 11.05.19).

Canadian Food Inspection Agency (CFIA), 2019h. Information within the nutrition facts table. Mandatory information. <http://www.inspection.gc.ca/food/requirements/labeling/industry/nutrition-labeling/nutrition-facts-table/eng/1389198568400/1389198597278?chap = 1/> (accessed 13.04.19).

Canadian Food Inspection Agency (CFIA), 2019i. Information within the nutrition facts table. Serving sizes and reference amounts. <http://www.inspection.gc.ca/food/requirements/labeling/industry/nutrition-labeling/nutrition-facts-table/eng/1389198568400/1389198597278?chap = 2/> (accessed 19.04.19).

Cerqueira, M.A., Pereira, R.N., Ramos, O.L., Teixeira, J.A., Vicente, A.A., 2016a. Edible Food Packaging: Materials and Processing Technologies. CRC Press, Boca Raton, Florida.

Cerqueira, M.A., Teixeira, J.A., Vicente, A.A., 2016b. Edible packaging today. In: Cerqueira, M.A., Pereira, R.N., Ramos, O.L., Teixeira, J.A., Vicente, A.A. (Eds.), Edible Food Packaging: Materials and Processing Technologies. CRC Press, Boca Raton, Florida, pp. 1—8.

Covington, L., 2014. Not just for bourbon: 8 barrel aged foods and how to buy them. Food Republic. <https://www.foodrepublic.com/2014/01/07/not-just-for-bourbon-8-barrel-aged-food-products-and-how-to-buy-them/> (accessed 22.04.19).

Day, B.P., 2008. Active packaging of food. In: Kerry, J., Butler, P. (Eds.), Smart Packaging Technologies for Fast Moving Consumer Goods. John Wiley & Sons Inc, New York, pp. 1—18.

Department of Health, UK., 2016a. Guide to creating a front of pack (FoP) nutrition label for pre-packed products sold through retail outlets. <https://www.food.gov.uk/sites/default/files/media/document/fop-guidance_0.pdf/> (accessed 12.04.19).

Department of Health, UK., 2016b. Technical guidance on nutrition labeling. <https://assets.publishing.service. gov.uk/government/uploads/system/uploads/attachment_data/file/595961/Nutrition_Technical_Guidance. pdf/> (accessed 22.04.19).

Department of Health, UK., 2017. Technical Guidance on Nutrition Labeling. Published in pdf only on March 2017 at www.gov.uk/dh. <https://assets.publishing.service.gov.uk/government/uploads/system/uploads/ attachment_data/file/595961/Nutrition_Technical_Guidance.pdf/> (accessed 23.05.19).

Deutsches Intitut fur Normung (DIN)., 2000. DIN EN 13432. Requirements for packaging recoverable through composting and biodegradation – Test scheme and evaluation criteria for the final acceptance of packaging.

Elopak and Sabic, 2019. Accessed on January 22 from <http://www.foodbev.com/news/elopak-and-sabic-collab-orate-on-virtually-100-renewable-carton/>.

EUR-Lex, 2017. Commission Notice on the application of the principle of quantitative ingredients declaration (QUID). Document 52017XC1121(01) – EN. <https://eur-lex.europa.eu/legal-content/EN/TXT/? uri = CELEX:52017XC1121(01)/> (accessed 21.04.19).

EUR-Lex, 2019a. Regulation (EC) No. 1333/2008 of the European Parliament and of the Council of 16 December 2008 on food additives. <https://eur-lex.europa.eu/legal-content/EN/ALL/?uri = CELEX:32008R1333/> (accessed 21.04.19).

EUR-Lex, 2019b. Labeling of prepacked products. <https://eur-lex.europa.eu/legal-content/EN/TXT/? uri = LEGISSUM:l32029/> (accessed 21.04.19).

Europa EU, 2019. Regulations, directives and other acts. <https://europa.eu/european-union/eu-law/legal-act-s_en/> (accessed 14.05.19).

European Commission, 2019. Food information to consumers – legislation. <https://ec.europa.eu/food/safety/ labeling_nutrition/labeling_legislation_en/> (accessed 14.05.19).

Evergreen Packaging, 2018. 2019 Food and beverage packaging trends. 18-EVP-0057 Trend White Paper FINAL. <https://evergreenpackaging.com/wp-content/uploads/18-EVP-0057-Trend-White-Paper-FINAL.pdf/> (accessed 23.04.19).

Food and Drug Administration, 1998. Container closure evaluation. <https://www.fda.gov/iceci/inspections/ inspectionguides/ucm074999.htm/> (accessed 22.04.19).

Gordon, A. (Ed.), 2016. Food Safety and Quality Systems in Developing Countries: Volume II: Case Studies of Effective Implementation. Academic Press.

Health Canada, 2006. Daily reference intake tables. <https://www.canada.ca/en/health-canada/services/food-nutrition/healthy-eating/dietary-reference-intakes/tables.html/> (accessed 13.04.19).

Health Canada, 2016. Table of daily values. <https://www.canada.ca/en/health-canada/services/technical-documents-labeling-requirements/table-daily-values.html/> (accessed 13.04.19).

International Plant Protection Convention (IPPC), 2016. International Standard for Phytosanitary Measures (ISPM) 15: Regulation of Wood Packaging Material in International Trade. IPPC, FAO, Rome.

Ivanković, A., Zeljko, K., Talić, S., Martinović Bevanda, A., Lasić, M., 2017. Biodegradable packaging in the food industry. Archiv Lebensmittelhygiene 68, 23–52. Available from: https://doi.org/https:doi.org/10.2376/ 0003-925X-68-26.

Janjarasskul, T., Krochta, J.M., 2010. Edible packaging materials. Annu. Rev. Food Sci. Technol. 1, 415–448. Available from: https://doi.org/https:doi.org/10.1146/annurev.food.080708.100836.

Kenneth Research, 2019. Biodegradable packaging: key vendors, trends, analysis, segmentation, forecast to 2018-2026. MarketWatch May 20, 2019, American News Hour via Comtex. <https://www.marketwatch.com/ press-release/biodegradable-packaging-key-vendors-trends-analysis-segmentation-forecast-to-2018-2026-2019-05-20/print/> (accessed 25.05.19).

Labuza, T.P., 1982. Shelf-Life Dating of Foods. Food & Nutrition Press, Inc.

LaCroix, J.-P., 2017. Packaging Trends 2020 and Beyond. Shikatani LaCroix Design, Inc., Toronto, Ontario, Canada, <https://www.sld.com/blog/food-beverage/7-packaging-trends-for-2020/> (accessed 18.04.19).

McCarthy, N., 2018. Countries banning plastic bags. Statisa, June 5, 2018. <https://www.statista.com/chart/ 14120/the-countries-banning-plastic-bags/> (accessed 22.04.19).

Michalopoulos, S., 2017. 'Traffic light' food labels gain momentum across Europe. <https://www.euractiv. com/section/agriculture-food/news/traffic-light-food-labels-gain-momentum-across-europe/> (accessed 22.04.19).

News Desk (Ed.), 2015. Belgian Brewer Prints Interactive Characters on Beer Bottles. Food Bev Media 17 (August 2015), <http://www.foodbev.com/news/belgian-brewer-prints-interactive-characters-on-beer-bottles/> (accessed 14.10.18).

Ramos, Ó.L., Pereira, R.N., Cerqueira, M.A., Martins, J.R., Teixeira, J.A., Malcata, F.X., Vicente, A.A., 2018. Bio-based nanocomposites for food packaging and their effect in food quality and safety. Food Packaging and Preservation. Academic Press, pp. 271–306.

Robertson, G.L., 2012. Edible, biobased and biodegradable food packaging materials. In: Robertson, G.L. (Ed.), Food Packaging: Principles and Practice, third ed. CRC Press, Boca Raton, FL, USA.

Scroggins, J., 2019. Five food packaging trends for 2019. <https://foodchannel.com/2019/five-food-packaging-trends-for-2019/> (accessed 20.04.19).

Stier, R.F, Ahmed, M.S. & Weinstein, H. (2002). Constraints to HACCP Implementation in Developing Countries II. Food Safety Magazine, October 2002/November 2002 Issue. Available at <https://www.foodsafetymagazine.com/magazine-archive1/octobernovember-2002/constraints-to-haccp-implementation-in-developing-countries-part-ii/>

The European Parliament and the Council of the European Union, (1994). European Parliament and Council Directive 94/62/EC of 20 December 1994 on packaging and packaging waste, Official Journal L 365, 31/12/1994 P. 0010 – 0023. <https://eur-lex.europa.eu/LexUriServ/LexUriServ.do?uri = CELEX:31994L0062:EN:HTML/> (accessed 22.04.19).

The European Parliament and the Council of the European Union. (2018). Directive (EU) 2018/852 of the European parliament and of the council amending Directive 94/62/EC on packaging and packaging waste, 30 May 2018. <https://eur-lex.europa.eu/legal-content/EN/TXT/PDF/?uri = CELEX:32018L0852/> (accessed 22.04.19).

United States Department of Agriculture, (2019). EU labeling requirements. <http://www.usda-eu.org/trade-with-the-eu/eu-import-rules/eu-labeling-requirements/> (accessed 20.04.19).

Wang, L., Kerry, J.P., 2018. Edible/biodegradable packaging for food. <https://www.newfoodmagazine.com/article/215/edible-biodegradable-packaging-for-food/> (accessed 23.04.19).

Wiley, C., 2018. 2019 Food Packaging Trends: Clean Packaging. Food Industry Executive, <https://foodindzustryexecutive.com/2018/12/2019-food-packaging-trends-clean-packaging/> (accessed 20.04.19).

Further reading

Canadian Food Inspection Agency (CFIA), 2018. Serving size and reference amounts. <http://www.inspection.gc.ca/food/requirements/labeling/industry/nutrition-labeling/nutrition-facts-table/eng/1389198568400/1389198597278?chap = 2/> (accessed 13.04.19).

Directory of nutrition facts table formats, 2016. <https://www.canada.ca/en/health-canada/services/technical-documents-labelingrequirements/directory-nutrition-facts-table-formats.html/> (accessed 14.04.19).

The Packaging Manufacturers Association, 2019. <http://www.ambalaj.org.tr/en/environment-composite-packaging.html/> (accessed 23.04.19).

CHAPTER

8

Food safety and quality considerations for cassava, a major staple containing a natural toxicant

Jose Jackson[1], Linley Chiwona-Karltun[2] and André Gordon[3]

[1]Alliance for African Partnership, Michigan State University, East Lansing, MI, United States
[2]Department of Urban and Rural Development, Swedish University of Agricultural Sciences, Uppsala, Sweden [3]Chairman & CEO, Technological Solutions Limited, Kingston, Jamaica, West Indies

Introduction

Cassava (*Manihot escutenta* Crantz) is native to South America and southern and western Mexico. It was one of the first crops to be domesticated, is thought to have originated in Brazil (Phillips, 1974) and is known to have also been grown in Columbia, Guatemala,

FIGURE 8.1 Cassava peeled and prepared for further processing.

Venezuela and southern Mexico (Lokko et al., 2007). Also known as manioc, yucca, or tapioca, it is one of the most important staple food crops grown in developing countries. The native peoples of the Caribbean and northern South America were probably some of the earliest cultivators of cassava (Henry and Hershey, 2002), and many of their customs of cultivation and processing remain virtually unchanged today. Practices from ancestral times also have been conserved throughout the Amazon basin. Most tropical countries, including eastern countries such as Indonesia and Thailand produce cassava, but its cultivation is most highly concentrated in four areas: northern and eastern coastal Brazil, southern Brazil and eastern Paraguay, northwestern South America (especially the Caribbean coast of Columbia), and the Greater Antilles in the Caribbean, which includes Haiti, the Dominican Republic, Cuba (Henry and Hershey, 2002) and Jamaica.

Cassava (Fig. 8.1) spread from Central and South America to other parts of the world in the post-Columbian period, having been introduced into Western Africa and Zaire in the late 1500s, probably by slave ships. It was introduced into Madagascar and Zanzibar (East Africa) via Réunion by the end of the 1700s, and by 1800 it had reached India. By the 1850s, it was widely cultivated in Southeast Asia and Africa. Cassava is now the third most important source of calories in the tropics after rice and maize, is now grown in over 90 countries and is a staple for half a billion people in Africa, Asia, Latin America and the Caribbean (Jackson and Chiwona-Karltun, 2018). Global cassava production encompasses most of the developing world) which collectively accounts for 47% t of the world's population and 46% of its arable land.

Cassava production and consumption trends

The plant has become a staple in developing countries since its introduction to major growing areas, largely because of its adaptability to conditions that are often inimical to the growth of other crops. Cassava is adapted to growing in the zone that falls between latitudes 30 degrees north and south of the equator, at elevations of not more than 2,000 m above sea level. It grows well at temperatures ranging from 18 to 25°C, with rainfall of 50 to 5,000 mm annually, and in poor soils with a pH range from 4 to 9.0 (Ezumah and Okibo, 1980). Typically, it is grown by poor farmers, often on marginal land. Many of these farmers are women. Since it can withstand drought and because of its efficient production of food energy, year-round availability, tolerance to extreme stress conditions, and suitability to present farming and food systems in Africa and the Caribbean, it is sometimes a nutritionally strategic famine reserve crop in areas of unreliable rainfall (Ezumah and Okibo, 1980; Hahn et al., 1987). Cassava is also a source of commercial animal feed, fiber for paper and textile manufacturers, and starch for the food and pharmaceutical industries, with the world's largest importer of cassava and cassava derivatives being China (Prakash, 2018).

It is estimated that 70% of the world production of cassava comes from five countries: Nigeria, Brazil, Indonesia, Thailand and the Democratic Republic of the Congo, with Africa and Asia now producing more than Latin America and the Caribbean. Nigeria, Thailand and Indonesia with 47.4, 30.2 and 23.9 million tonnes in 2013, respectively, are the world's largest producers, with a total of 268 million metric tons produced in 2014 (International Fund for Agricultural Development (IFAD) & Food and Agricultural Organization of the United Nations (FAO), 2000; Worldatlas.com, 2019). Fifty six percent (56%) of the world's cassava is grown in Africa, thirty two percent (32%) in Asia and 12% in the Americas (Jackson and Chiwona-Karltun, 2018). Cassava production on the African continent has continued to increase with the annual average growth rate over the last 10 years outpacing population growth, with the exception of 2017 and 2018. The current outlook for cassava production in Asia is uncertain with China, the largest regional and global market for the product, cutting back imports as it works through its domestic surplus of maize (Prakash, 2018). There is much lower production in the Americas than other regions, with Brazil, Paraguay and Colombia producing a projected 20.9, 3.3 and 2.3 million metric tonnes respectively and all other Latin American producers, 4.2 million metric tonnes combined. Haiti (415,000 metric tonnes) and the Dominican Republic (500,000 metric tonnes) are the other major producers in the Caribbean (FAOSTAT, 2015). Still, most countries in the Americas, except for those in North America, continue production of cassava, which plays an important cultural and dietary role throughout the region. A traditional processing centre where cassava is collected and processed to make cassava flour, cassava bread and other products is shown in Fig. 8.2.

There have been consistent gains in productivity and yields for all cassava producing regions as the importance of the crop increases. In 2015, the five countries with the highest yields were India (300 Hg/Ha*), Cook Islands (250 Hg/Ha), Suriname (200 Hg/Ha), China (200 Hg/Ha) and Barbados (200 Hg/Ha) (FAOSTAT, 2015). As Asian countries like Thailand, Indonesia, Vietnam and Cambodia have increased their production, gains in productivity have also begun to rise, largely fueled by what had been a consistent increase in demand from China for more cassava chips and pellets (Prakash, 2018). This has paralleled an explosion in the growth of the area under cultivation as some Asian countries

* Hg/Ha means hectagram (100g) per hectare, the standard measurement used by FAO for crop yields.

FIGURE 8.2   Cassava at a traditional processing centre in Central America.

such as Lao Peoples Democratic Republic and Vietnam have seen the area under cultivation increase from almost nothing to 100,000 ha and three-fold to 600,000 ha, respectively.

While this current change in the fortunes of the crop in Asia is clearly driven by economic prospects, most farmers have traditionally grown cassava for food security and as a response to famine, hunger and drought. This is particularly so in Africa, a fact supported by surveys in selected African countries (Food and Agricultural Organization of the United Nations (FAO), 2005). Another important reason iss that cassava is a hardy crop that has significant resistance to pests and diseases, hence the reason it plays such an important role in developing country diets (Food and Agricultural Organization of the United Nations (FAO), 2005). As a result, the whole plant provides a nutritious basic staple during periods of drought or reduced rainfall and is important in the mitigation of famine and hunger, particularly in Africa (Jackson and Chiwona-Karltun, 2018). With the crop enjoying increased popularity as a result of greater domestic demand in producing countries, increased production yields and generally higher prices, the increased interest in, and access to, export markets has also helped to transition cassava from a crop planted for food security purposes to one that is rising in economic importance for developing countries (Food and Agricultural Organization of the United Nations (FAO), 2005; Prakash, 2018; Jackson and Chiwona-Karltun, 2018).

In Africa, cassava is almost exclusively used for consumption as food or used in foods (Food and Agricultural Organization of the United Nations (FAO), 2005; Prakash, 2018) with about 95% of the total cassava production, after accounting for waste, being used as food. Total cassava consumption in Africa has more than doubled, due largely to a significant increase in per capita consumption in countries such as Ghana and Nigeria where cassava was produced as a cash crop for urban consumption. It is likely that the importance of cassava in some countries, for example the Congo, is underestimated due to under-reporting, as many families eat cassava for breakfast, lunch and dinner and cassava is known to contribute over 1000 calories per person per day, i.e., 55% of the average daily caloric intake (Food and Agricultural Organization of the United Nations (FAO), 2005). In addition to the roots, cassava leaves are also widely consumed as a vegetable in several places where cassava is grown such as in the Congo and Tanzania. The availability of cassava in a convenient food form, such as *gari*, has played a major role in the increase in the per capita cassava consumption in Ghana and Nigeria. Future increases in cassava consumption in other African countries will depend on how well cassava is prepared into food forms to create an easy-to-use alternative to wheat, rice, maize and sorghum for urban consumers.

Cassava was found to be the cheapest source of calories among all food crops in each of the six countries reported in a study conducted by the Food and Agricultural Organization of the United Nations (FAO) (2005). As family incomes increased and lifestyles changed, the consumption of cassava as dried root flour declined while consumption in convenient food forms such as *gari* increased. This was despite the fact that dried cassava root flour was cheaper than *gari* because of the high cost of processing *gari* (Food and Agricultural Organization of the United Nations (FAO), 2005). Medium and high-income families were found to consume *gari* because it was cheaper and more convenient to cook than grains. Consequently, the future of cassava as a rural and urban food staple will depend on its ability to compete with wheat, rice, maize, sorghum and other grains in terms of cost, convenience and availability in urban markets. The role of convenience, labor-saving production, harvesting and processing technologies in this regard will be important.

Cassava toxicology and safety

Cassava cultivars grown on all continents contain naturally toxic compounds known as cyanogenic glycosides (Conn, 1969; Jackson-Malete et al., 2015). Cyanogenic glycosides (CGs), like other cyanogenic compounds, produce hydrogen cyanide (HCN) when the cells in which they reside become damaged or are being degraded, as during digestion (Nahrstedt, 1988; Kashala-Abotnes et al., 2019), through enzymatic hydrolysis by beta-glucosidase. While the cyanogenic glycosides themselves are relatively non-toxic in their native form, their ability to produce HCN upon maceration and its high level of toxicity make them very dangerous. Clinical signs of acute cyanide intoxication include a rapid pulse, rapid respiration, stomach pain, vomiting, diarrhea, a drop in blood pressure, dizziness, headache, mental confusion, twitching and convulsions. When the cyanide levels exceed the acute lethal dose of 0.5–3.5 mg per kilogram of body weight, exceeding the bodies ability to detoxify it, death due to cyanide poisoning can occur (Kwok, 2008; Kashala-Abotnes et al., 2019). Consequently, care must be taken in the handling, preparation and consumption of this important staple to avoid undesirable health outcomes due to HCN poisoning.

There are two varieties of cassava: "sweet" cassava, which has a low toxin content, and "bitter" cassava with a characteristically high toxin content. The toxins of concern, cyanogenic glycosides (CGs), are a group of chemicals that occur naturally in over 2,000 plants which are highly toxic when consumed (Kwok, 2008). CGs are found in cassava in the form of linamarin and lotaustralin or a combination thereof, as free hydrocyanic acid (HCN) and cyanohydrin (Chikezie and Ojiako, 2013). All cassava varieties can be classified into these two groups, with the roots of "bitter" cassava cultivars containing higher levels of CGs than those of "sweet" cassava cultivars (Bolhuis, 1954). Generally, the sweet varieties of cassava, because of their lower content of cyanogenic glycosides (Carmody, 1900), can be eaten without pre-preparation (i.e. "raw"). The bitter varieties require pre-processing to make the final product safe for consumption because of the much higher toxin levels (von Hagen, 1949; Chiwona-Karltun et al., 2004).

The cyanide released by the breakdown of the cyanogenic glycosides during preparation, post-ingestion or even after absorption into the body (Brimer, 2000), is toxic and can cause severe acute poisoning. Typically, however, it is more frequent that chronic poisoning is observed. This may be characterized by slower growth and development, and neurological symptoms resulting from central nervous system (CNS) tissue damage (Tylleskär et al., 1992). Ruminants, in which the fore-stomach flora contributes to the hydrolysis of cyanogenic glycosides, are considered to be more vulnerable to such compounds than monogastric animals (Brimer, 2001) and this, therefore, has implications for the use of the crop in the making of animal feed.

The main cyanogenic glycoside in cassava is linamarin (Dunstan et al., 1906) and there are smaller amounts of lotaustralin (Brimer, 2000) also typically present. They occur in the proportions of about 10:1. Linamarin and lotaustralin are synthesized from the amino-acids valine and isolucine. These cyanogenic glycosides (Seigler, 1991) can be found throughout the whole cassava plant with the highest concentrations in the young leaf shoots (Joachim and Pandittesekere, 1944), followed by the petioles, stems and roots (de Bruijn, 1973). Synthesis of the glycosides mainly occurs in the leaves and they are transported to the roots, although there is some synthesis in the root as well (Bokanga 1995). Within the root, the peel contains more cyanogenic glycosides than the edible flesh (Bokanga, 1994). Genetic and environmental factors concurrently determine the glycoside levels in the roots (Mahungu, 1994). Nowhere in the world has a cassava cultivar without any cyanogenic glycosides been identified , despite anecdotal statements to the contrary.

All cyanogenic glycoside-containing plants also contain enzymes for the decomposition of the glycosides. In cassava, this endogenous enzyme is called linamarase (Kereztessy et al., 1994). The glycosides and the enzymes are stored in separate compartments within the plant cell, the former being stored in the cell wall and the latter in the cell vacuoles (Mkpong et al., 1990). In order for cassava tissues to liberate hydrogen cyanide (HCN) contact has to be made between the substrate linamarin and the enzyme linamarase. This is done by bruising the tissues or by any other process that ruptures the cell structure such as grinding, grating, soaking and fermentation, freezing, drying or the addition of chemical agents (McMahon et al., 1995).

The hydrolysis of the glycosides yields glucose and cyanohydrins that remain stable at a pH <5 and at low temperatures. An increase in temperature or pH results in spontaneous breakdown of the cyanohydrins to acetone and HCN (McMahon et al., 1995). The breakdown may also be facilitated by the enzyme hydroxynitrile lyase that is expressed in the leaves but not in the roots. This process of releasing HCN is referred to as cyanogenesis (Conn, 1969). The glycosides, cyanohydrins and HCN are collectively referred to as cyanogens because they are all capable of forming the cyanide ion CN⁻ that is the toxin. There are other factors that affect the levels of cyanogenic glycosides such as the age of the plant, growing conditions and genetic factors.

Cassava safety

There are a number of studies from different parts of Africa that show the consequences of very small changes in processing on the amounts and types of cyanogens that remain in the final product and hence on the chemical food safety of these products. A study from Zaire was one of the first studies to reveal the consequences of shortcuts in the processing of cassava flour (Banea et al., 1992). This was later followed up with a comparative study from Tanzania of two processes named makopa and chinyanya, both resulting in flour (Mlingi et al., 1995). Similarly, a study in Nigeria showed the effects of two different methods of dewatering, continuous dewatering during fermentation versus dewatering at the end of the fermentation, on the residual level of cyanohydrins in *gari* (Onabolu et al., 2002a).

In the study from Nigeria, the authors observed a very different development of the pH of the product during the processing and of the final product pH when comparing the two different methods of dewatering. The difference may very well account, at least in part, for the observed significant difference in the level of residual cyanohydrin in the *gari*, or may stem from differences in the flora of microorganisms produced by the two dewatering processes as concluded by Onabolu et al. (2002b). There are studies that have focused on the fermentation process and how different strains of *Lactobacillus plantarum* influence the final residual total cyanogens (Lei et al., 1999). The processing of all cassava roots and leaves increases the shelf-life of these products.

Pinto-Zevallios et al., (2016) detailed the role that the CGs found in cassava play in defending the plant from pests, a side effect of which is the risk to human health from consumption of improperly processed bitter cassava. Falade and Akingbala (2010) reviewed the approaches required for the consistent safe use of cassava as a food, discussing the importance of processing such as grinding and fermentation, as well as storage and packaging. The review discussed traditional African foods such as fermented products, including cassava bread, fermented cassava flour, fermented starch, fufu, lafun, akyeke (or attieke), agbelima, and gari, and unfermented products including tapioca, cassava chips and pellets, unfermented cassava flour and starch. They also evaluated new uses of cassava as a food, including as flour in gluten-free or gluten-reduced products (e.g., bread, biscuits, etc.). Frediansyah (2017) described the making of a range of traditional Indonesian products including peyeum, tape singkong, gaplek, tiwul, gathot, timus, getuk, gemblong, opak/keripik singkong, tepung singkong/cassava flour, tepung tapioca/tapioca and mocaf from cassava. Of these, peyeum, tape singkong and mocaf are fermented, with mocaf, a fermented gluten-free flour being the most recent trend in the industry. Dórea (2004) reported that despite what may be regarded as high concentrations of naturally occurring neurotoxins (the CG linamarin) in the cassava which was typically consumed daily with fish, the consumption of the staple in large amounts over the course of a lifetime posed no health risk for native Amazonians in South America. In 2015 when MacDonald's in Venezuela could not get enough potato to offer traditional fries, they put *cassava fries* on their menu, extending the product into a non-traditional area as a food (Harkup, 2017).

Cassava has been shown to be at risk of aflatoxin (AFT) contamination should its processing, handling and storage not conform with best practices. Aflatoxins contribute to liver cancer, affect maternal and infant health, and cause stunting in young children.

Consequently, care must be taken to preclude its formation during the processing and handling of cassava and cassava products. Vasconcelos et al. (1990) discussed the detoxification of cassava during the making of *gari*. Fermentation for 48 hours significantly reduced the cyanide content of *gari* ($P < 0.05$) but did not eliminate it. It also promoted elevated levels of aflatoxin (Chikezie and Ojiako, 2013), indicating that while AFT is not seen as a major risk in cassava products, care should be taken to prepare, handle and store them under conditions which prevent the growth of AFT-producing fungi such as *Aspergillus* spp. Muzanila et al. (2000), in a comparison of wet and solid state fermentation and sun drying of bitter and sweet cassava, respectively, in Tanzania found no mycotoxin (AFT) production at the end of the period of processing. This is in keeping with the findings of Gnonlonfin et al. (2012) for *cassava chips* in Benin vs maize samples.

Manjula et al. (2009) found that while there was negligible presence of fumonisins in *dried cassava* and maize, relatively high levels of AFT (up to 13.0ppb) were found in *cassava chips and flour* samples collected from villages in Tanzania and the Congo and stored for four (4) months. Maize kernels had much higher levels of both toxins and very high (up to 39.5% and 70.5%) presence of *Aspergillus* and *Fusarium* spp., respectively on the kernels, indicating a potential food safety risk. In a study of *lafun*, a traditional fermented and dried product from Nigeria, as well as the water and fermentation broth, Lateef and Ojo (2016) found a range of potentially pathogenic organisms, including *Staphylococcus aureus*, *Escherichia coli*, *Salmonella typhimurium*, *Bacillus cereus*, *Klebsiella oxytoca*, *Aspergillus fumigatus*, *Aspergillus flavus* and *Aspergillus niger*, among others. The aspergilli were capable of producing aflatoxins up to a level of 1,600 ppb.

Effects of cassava-mediated cyanide exposure

The human toxicity of cyanides from cassava is well established and these have been identified as the cause of the tropical myeloneuropathies tropical ataxic neuropathy (TAN) and Konzo (Kashala-Abotnes et al., 2019). First described in Nigeria, TAN is a progressive myeloneuropathy that is characterized by the progressive loss of control of bodily movements (ataxia) while Konzo is a permanent upper motor neurone disease that is clinically distinct and was first described in the Democratic Republic of Congo (DRC) at the end of the 1800s (Kashala-Abotnes et al., 2019). Studies from Zaire and Mozambique showed a strong association between Konzo and dietary cyanide exposure (Tylleskär et al., 1992; Cliff et al., 2011). Konzo outbreaks have been reported in other African countries including Tanzania, Cameroon, Angola, Central African Republic (CAR) and Zambia (Chabwine et al., 2011; Cliff et al., 2011; Mlingi et al., 2011; Kashala-Abotnes et al., 2019). Outbreaks are usually occasioned by disasters such as famine, war or drought during which people resort to the consumption of inadequately processed "bitter" cassava.

Experimental studies show that cyanide can pass the placental barrier of rodents, but no information is available on humans or species with a similar placental structure. Developmental neurotoxicity is often non-specific, and the adverse effects may depend on the timing of exposure. While the etiology of Konzo has not yet been fully clarified, a toxic/nutritional etiology is strongly suggested (Tylleskär et al., 1992). Kashala-Abotnes et al. (2019) have noted that evidence suggests that the disruption of thiol-redox and protein-folding mechanisms may be the cause of the illness, potentiated by poor nutrition, dietary cyanogen exposure, poverty and pre-disposition due to genetics. Evidence from other neurotoxic substances suggests that exposures during the third trimester are most likely to cause functional deficits, e.g. decreased IQ scores (Grandjean and Landrigan, 2014). Unless they are severe, the deficits may be inconspicuous during infancy but later become apparent when the child has difficulties in school. This pattern has been widely documented in regard to such neuro-toxicants as lead, methylmercury, and polychlorinated biphenyls. Such deficits may hamper educational achievement and economic success (Grandjean and Landrigan, 2014). A useful review of the konzo and its related neuropathies is presented in Kashala-Abotnes et al. (2019).

Processing and utilization of cassava tubers (roots) and leaves

Cassava, unlike other tuberous crops, is highly perishable, being comprised of 70% moisture and requires processing as soon as the roots are harvested to increase the shelf-life. If not processed within three to four days, cassava roots will begin to discolor and rot. Consequently, quick processing of cassava roots is imperative to avoid spoilage but also because it improves the palatability of the prepared dishes (Lancaster et al., 1982; Hahn et al., 1987; Onabolu et al, 2002a). It is suggested that cassava processing in Africa has been adapted from processing methods used for other crops that have traditionally been grown and utilized in Africa (Nweke et al., 2001). To get an understanding of the development of cassava processing in Africa, one needs to look at practices related to indigenous African food crops such as yams, sorghum and millet and review Jones (1959) and Berry and Petty (1992). It has also been suggested that the processing methods for cassava in West Africa may stem from the techniques of processing toxic yams combined with the knowledge of the returnee slaves.

In order to render cassava safe for consumption for populations that depend on or consume cassava, cassava toxicity can be minimized in two ways: a conscious effort to select cassava varieties that have low cyanogenic potential or the application of scientific knowledge in the processing and removal of the cynogenic glycosides from the plant. For processing, the objective is reduction of the cyanogens to safe levels for consumption. This is the preferred approach taken by native Amazonias who have been eating cassava for centuries (Dufour, 1994) and appear to have opted for processing as a sure way of eliminating the risk of toxicity from cassava consumption. However, processing of cassava roots and leaves is not only conducted for detoxification purposes but also to prepare the cassava products desired. Processing also improves shelf-life, reduces bulk during transportation and improves palatability of the prepared dishes.

Processing of the leaves

A summary will be presented here on the processing of cassava leaves as a vegetable. A more detailed treatment of the processing of cassava roots will be covered later on and is captured in excellent discussions by Lancaster and Brooks (1983), Lancaster et al. (1982) and Nweke and Bokanga (1994).

Cassava leaves have been imperfectly understood as a vegetable especially by the scientific community. Due to a lack of data, we cannot at this point state just how much cassava leaves are consumed or how much of the cassava that is grown is actually grown for the leaves. There is a need for such studies to ascertain the production and consumption of cassava leaves in Africa. In 1959, according to Jones (1959), the absence of such studies had been attributed to the fact that cassava was viewed as an inferior crop in relation to cereal crops, particularly maize that has received much attention politically in the African context (McCann, 2001). For example, in the report from the Harvard Africa Expedition 1926—1927, it was reported that the leaves of portaluca and sweet potato were very important leaves as vegetable greens. However, although there was the observation of high cassava root consumption, the report made no mention of cassava leaf consumption. Jones (1959) postulates that this could be due to two factors: one, that there is an abundance of edible green vegetables in Africa and thus cassava leaves are not that important, or two, that cassava leaf consumption is discouragedso people are reluctant to admit or report their consumption due to the negative labeling and connotation of being economically backwards if reporting the consumption of cassava leaves.

Interestingly, green leaves are an integral and accepted part of the diet of many African societies. This is because leaves comprise part of the production system of root crops, cereals, legumes as well as other plants. In French West Africa cassava leaves have formed part of the diet for quite some time. The most preferred variety of cassava leaves the "tree manioc" also referred to as "thunder manioc" because it is perceived to have magical properties (Jones, 1959). The roots from the "tree manioc" are not consumed as they contain a lot of water and fibers. The stalks of the cassava are burned and the ashes are used as a seasoning salt. In some cases, the ashes can be used as dye, snuff or for soap-making. Similarly cassava leaves play an important role in the diets of many Eastern and Southern African countries (Lancaster and Brooks, 1983). In a study carried out across Africa, 81% of cassava growing communities reported consuming cassava leaves (Nweke, 1994).

For the most part, cassava is primarily grown for the roots while the leaves just happen to be a by-product. However, early observations which were later confirmed revealed that there are varieties such as those mentioned above that are grown and preferred for the leaves. Bokanga (1994) argues that harvesting the leaves may have detrimental effects on the root yield but that with controlled harvesting this would not be the case. On the other hand, not much research has been done on understanding the multi-functionality of the roots and leaves or trade-offs that farmers make when utilizing cassava roots and leaves (Chiwona-Karltun et al., 2015).

Unlike the roots, cassava leaves contain very high levels of CG approximately 5—20 times more than that of the edible root (Bokanga, 1994). Fortunately, there is little cause to be wary of eating cassava leaves. Cassava leaves also contain the enzyme, linamarase, a β-glucosidase that can breakdown the glycosides linamarin and lotaustralin. In order for the two to meet, there has to be a breakdown of the leaves' cell walls as they are stored in separate sections. This can happen through grinding, pounding or chopping. This reaction leads to a breakdown of the cyanohydrins to hydrogen cyanide.

Studies have shown that pounding the leaves for about 15 minutes, followed by boiling them in twice their weight of water for 15—20 minutes, even with levels of cyanogen potential of over 1,000 mg HCN equivalents kg^{-1}, can reduce these levels between 4 and

11 mg HCN equivalents kg^{-1}. As much as 30% of the cyanogenic glycosides are reduced by pounding alone and after boiling for less than 15 minutes, they are reduced to less than 1% of the initial levels (Bokanga, 1994). This is the reason why consumption of cassava leaves, if done properly, does not lead to cyanide poisoning. The high enzyme content and activity in the leaves, coupled with the pounding that brings together the linamarase and the linamarin induces hydrolysis of the cyanogenic glycoside, which is then easily removed by traditional methods.

More recently, there have been more studies on cassava leaves, mapping preparation and potential use in canning. The highest per capita consumption of cassava leaves is in what was formerly Zaire (now the Democratic Republic of the Congo) with as much as 500 g/day (Lancaster and Brooks, 1983). A study in Central Africa describes how "saka-saka" or "mpondou" is prepared with cassava leaves, palm oil, chillies, salt, onions and baking soda. Sometimes, dried fish is added to the dish along with other vegetables, depending on availability. Another cross-country study from six African countries confirmed the importance of cassava leaves as a vegetable in the African diet (Ufuan Achidi et al., 2005). The younger the leaves, the more preferred they are as a vegetable and young leaves have been reported to have a higher protein concentration and less fibers.

There does not seem to be much variation in the processing and preparation of the leaves apart from the time taken to complete the process. These differences would appear to be due to cultural practices, the age of the leaves, variety, genetics, and environmental or geographical location (El-Sharkawy, 2003). In East and Southern Africa, cassava leaves are more often cooked together with bicarbonate soda, mixed with ground peanut flour or with coconut milk (Chiwona-Karltun et al., 1998).

While processing safely reduces the amounts of cyanogenic glycosides, there is the concern that the nutritional quality is compromised with regard to the protein, vitamins and sulfur-containing essential amino acids. Studies by Bradbury and Denton (2014) and Bradbury et al. (2011) showed that slightly modifying the process and pounding the leaves for at least 10 minutes, followed by washing them with twice their weight of water at ambient temperature and changing water between washing reduces the cyanogens by 99%. Repeated washing reduces these levels even more and retains nearly all the protein, the amino acids and vitamins of the original leaves.

Agricultural studies have emphasized the role that women play as food producers and providers particularly in Africa, with as high as 80% of all food for household consumption being produced by women. Apart from providing the macro-calories in the form of staple crops, women are also largely responsible for ensuring dietary variation as they use their knowledge to gather wild crops, process and prepare the food. This is particularly the case in rural areas where socio-cultural norms and traditions has food production and preparationclearly divided between males and females. Already in the 1970s Boserup (1970) described the different roles and domains that women and men played in agricultural societies and referred to them as female and male farming systems.

With the well-documented transformation of cassava in West Africa and Southern Africa, changes in the way that men and women participate in the various phases of production are evident. Early studies in Zaire described cassava farming within a shifting cultivation system where it was necessary to clear and make available new fields every seven to ten years (Fresco, 1986). The clearingof new fields required the felling of big trees and rough bush, such activities requiring the engagement of men. In such a system, women would be more responsible for planting, harvesting and processing. The situation is clearly illustrated by Ohadike (1981), during a pandemic influenza attack when men were incapacitated or had died, and women were left to do all the chores by themselves. There was a shift from yam production to crops like cassava as women could easily grow cassava (Ohadike, 1981). Similarly, in the Barotse floods bordering Zaire and Zambia, cassava was readily adopted as there was a huge male migration to work in the mines and millet production required intensive labor (Vickery, 1989).

As technologies, both in terms of improved planting materials and processing technologies have become readily available, income from cassava has been realized by both men and women. According to Nweke et al. (2001), while cassava has been termed in some literature as a woman's crop, this is no longer the case as men participate in the production, processing and marketing. While small-scale machinery has made it possible in West Africa for women to also use machinery for processing, such is not the case in East and Southern Africa. The International Institute of Tropical Agriculture (IITA) has been promoting the use of these technologies (IITA, 1990), but the cost remains relatively high for most rural households to acquire them. Ultimately, what one observes are women continuing to be engaged at the lower end of the cassava value chain of processing, such as peeling, grating, pounding, frying, roasting and drying. Because cassava is vegetatively propagated, much of this work is done by hand with no machinery involved. In countries where the sweet varieties are readily sold in urban areas, this production is largely controlled by men, from production to marketing, as has been observed in Malawi (Moyo et al., 1998; Akoroda and Mwabumba, 2000). On the other hand, women and destitute households prefer to grow bitter cassava varieties because the processing required prior to consumption confers protection from theft (Chiwona-Karltun et al., 1998). Based on evidence from Congo, Ghana and Nigeria, as cassava becomes commercialized, men begin to participate at all levels of the production, processing and marketing making cassava predominately a male crop (Nweke et al., 2001). This has implications for women and if not carefully studied may lead to some undesirable nutritional outcomes.

Processing of the roots

Cassava has an abundance of digestible carbohydrates in the form of starch that are a major source of food energy. Raw cassava roots have more carbohydrate than potatoes but less carbohydrate than wheat, rice, yellow corn, and sorghum on a 100 g basis (Afoakwa et al., 2012). The carbohydrate content in cassava roots is reported to range from 32% to 35% on a fresh weight (FW) basis, and from 80% to 90% on a dry matter (DM) basis (Montagnac et al., 2009). Eighty percent of the carbohydrates in cassava roots is starch (Buitrago et al., 2002), with 83% in the form of amylopectin and 17% being amylose. Cassava roots contain small quantities of sucrose, glucose, fructose, and maltose (Tewe and Lutaladio, 2004). These sugars are only present in minute quantities, ranging between 1.57 and 2.89% on a dry weight basis (Aryee et al., 2006). However, it has been reported that in sweet varieties, the sucrose content can be as high as 17% and also that the fiber content is dependent on the age, variety and environment (Charles et al., 2005). In terms of vitamins, only vitamin C is present in relatively high amounts of 15 − 45 mg/100 g.

With regards to minerals, there have been reports of zinc ranging from 3 to 140 ppm and iron levels of 8–24 mg/kg (Burns et al., 2012). Cassava roots are often referred to as being mostly white in color, but it should not be forgotten that there is a spectrum of colors ranging from white to yellow. Yellow tubers contain much more β-carotene than white tubers and cassava is a good source of pro-vitamin A carotenoids compared with other root crops.

Depending on the variety, the protein content of cassava roots ranges from less than 1% and to 5% on a dry weight basis. The protein has low levels of the essential amino acids lysine and leucine as well as of the sulfur-containing methionine and cysteine (Yeoh and Chew, 1976; Gomez and Noma, 1986; Diasolua Ngudi et al., 2002). In contrast, the total protein content in leaves may reach up to 35% of the dry weight. The leaves are low in lysine and leucine as well as in histidine as compared to human nutritional needs (Lancaster and Brooks, 1983; Nassar and Marques, 2006). Currently, improving the protein content of cassava in the tubers by genetic modification is discussed as a possibility in the scientific literature, although not yet implemented at scale (Stupak et al., 2006).

The varieties of roots that are preferred for the making of starch-based foods are those with high levels of mealiness that appear to be associated with high dry matter and starch content (Safo-Kantanka and Owusu-Nipah, 1992). Therefore, knowing the proximate composition of the roots before cooking is important in the preference and selection of the tuber for cooking (Chiwona-Karltun et al., 2015). Studies by Charoenkul et al. (2006) of several varieties, showed that flours contain almost the same components as are present in the raw materials, except the moisture content. The gelatization qualities of cassava starch are very different from the flour in terms of viscosity levels as well as other properties. This has been attributed to the varying activities of amylase activity in the flours (Charoenkul et al., 2011). Low-cyanogen varieties ("sweet" cassava) are readily eaten fresh, boiled or roasted and so these cassava roots are regarded as a vegetable. According to Jones (1959) in the first book published on cassava, fresh cassava roots of the "sweet varieties" were consumed as vegetables. Most processing techniques for cassava roots entail a sequence of peeling, washing, grating, fermenting, drying, frying or baking. While roots from sweet varieties are processed to achieve the desired product, mostly flour, the bitter varieties must always be thoroughly processed. The main methods for processing bitter cassava roots are: soaking without peeling, peeling then soaking and fermenting, grating and fermenting, grating and sun drying, sun-drying, particularly the sweet or low cyanogen varieties, heap fermentation, leaching, roasting and steaming (Lancaster et al., 1982: Essers et al., 1995).

Cassava bread: traditional production in Belize, Central America

As indicated in the introduction, cassava is made into various traditional products that form the base of the diets in many developing countries. Countries in the Central American region of the Americas are no exception, with the Garifuna people of Belize, Central America, as well as other peoples throughout the Americas consuming cassava as a staple. Cassava Products Limited of Dangriga, Belize is one of the firms in that country that has sought to capitalize on this traditional food and further develop the industry and the sector by upgrading their operations to expand production and extend the range of products they can supply. They have sought to further capitalize on the extensive research

FIGURE 8.3   Steps in the making of cassava flour in Belize: (A) peeled cassava (B) inspection of the cassava flour. *Source: André Gordon (2015).*

financed by international agencies that has been done on the cassava tuber and its commercialization, particularly in Colombia in South America and in Nigeria.

In Belize, the traditional use of cassava is in the production of cassava bread. The cassava bread has an extended shelf life because of the way it is traditionally made. Cassava tuber can also be converted into flour, which can be used alone or mixed with other types of flour to be used in a variety of applications, particularly in the baking industry in developing countries. There is an attractive market for cassava products, as well as combination flour products in Belize and other Central American countries, Southern America, Jamaica in the Caribbean where a major bakery has launched a range of cassava-based bakery products, and also in the United States of America.

The traditional way of making cassava flour and cassava bread by the Garifuna people in Belize is shown in Figs. 8.3 and 8.4 respectively. The tuber is peeled (Fig. 8.3A) and then grated to produce ground cassava. The ground cassava is then gathered together, soaked to remove the toxin, formed into a "cake" by initial pressing, and then pressed to extract the liquid (Fig. 8.4A), mainly cassava starch and water, which contains the remaining cyanogenic glycosides. The dried cassava cakes which result (Fig. 8.4B) are then separated, spread and dried to form cassava flour (Fig. 8.3B) or formed into cassava bread which is then dusted with cassava flour (Fig. 8.4C) to prevent sticking/adhesion, and then stacked in preparation for packaging (Fig. 8.4D).

Nutritional value of cassava

The tubers (roots)

Cassava consumption is a significant part of diets in the Americas and Sub-Saharan Africa where consumption was estimated to be 20 and 80 kg/capita, respectively (Aerni, 2006). Being one of the most efficient converters of solar energy into soluble carbohydrate per unit area with 1 kg of moisture-free cassava meal yielding up to approximately 3,750 kcal

FIGURE 8.4   The making of cassava bread in Belize. *Source: André Gordon (2015).*

(Okezie and Kosikowski, 1982), the cassava tuber is an excellent source of energy. The tuber (root) is an excellent source of starch, the second most energy dense food of those commonly consumed in developing countries (Table 8.2) and provides an average of 286 calories/person/day in Sub-Saharan Africa (Jackson and Chiwona-Karltun, 2018). However, because the tuber contains very little protein (up to 2%) by comparison to maize, sorghum, rice and all the other major staple foods (Table 8.1), it has been identified as a cause of protein-energy malnutrition in Sub-Saharan Africa where the poor consume high quantities of this energy-dense but nutrient-poor staple. Despite this concern, cassava roots contain other nutrients such as thiamine, riboflavin, niacin and ascorbic acid, all at similar levels to those found in rice. Nevertheless, its nutrient composition requires it to be combined with other more protein and nutrient rich foods if long term nutrient deficiencies are to be avoided.

The leaves

Farmers that grow and consume cassava consider it a versatile crop as the roots provide energy while the leaves provide the basis for making a stew that can augment the

TABLE 8.1   Calories and protein provided from major staple crops consumed in Sub-Saharan Africa.

Crop	Calories/person/day	Protein/person/day (g)
Maize	337	8.6
Cassava Root	286	2.0
Sorghum	202	6.0
Rice milled	175	3.6
Millet	137	3.3
Wheat	121	3.6
Yams	78	1.2
Plantains	61	0.5
Sweet potato	33	0.4
Potatoes	10	0.2

Source: FAO statistical database 2015 (adapted from Jackson and Chiwona-Karltun, 2018).

TABLE 8.2   Mineral and vitamin content (per 100 g) of cassava and spinach leaf, soybean and yellow maize.

	Ca (mg)	Fe (mg)	Vitamin A (mg)	Thiamine (mg)	Riboflavin (mg)	Niacin (mg)	Vitamin C (mg)
Cassava leaves	300	7.6	3000	0.25	0.60	2.4	310
Amaranth leaves	410	8.9	2300	0.05	0.42	1.2	50
Soyabean	185	6.1	28	0.71	0.25	2.0	0
Maize (yellow)	13	4.9	125	0.32	0.12	1.7	4

Source: West et al. (1988) (adapted from Jackson and Chiwona-Karltun, 2018).

consumption of the roots when cooked together. Cassava leaves are consumed in the Congo, Tanzania and 60% of the countries in Sub-Saharan Africa, but are not consumed in West Africa, with the exception of Sierra Leone, and, compared to other vegetables, are a very good source of major nutrients. Unlike the tubers, cassava leaves contain more Calcium, Vitamin A, Riboflavin, Niacin and Vitamin C than other major staples such as *Amaranthus* leaves, soybean and maize (Table 8.2). The major carotenoids in the leaves are the non-vitamin A carotenoid lutein 86–290 mg/kg fresh weight (FW) and the pro-vitamin A carotenoid β-carotene at 13–78 mg/kg FW (Adewusi and Bradbury, 1993). The leaves contain between 8% and 9% crude protein, inclusive of the essential amino acid lysine, although they tend to be deficient in methionine and tryptophan. It has been suggested that combining the leaves with a source of protein such as cod or other fish could counter-balance any deficiency in the protein quality that may exist and the leaves also contain minerals including ferric oxide and calcium as shown in Table 8.2 which help to improve their overall contribution to dietary needs.

A major negative for the leaves is that they contain antinutritional factors, including tannins, polyphenols, oxalates, nitrates, saponins and phytates. Like the tubers, they also contain cyanogenic glycosides that should be removed by cooking for at least 10 minutes prior to consumption. It has also been shown that processing of the leaves by pounding or grinding, and processing of the tubers by oven drying and fermentation can reduce the presence of these antinutrients by 50–85%, respectively (Achidi et al., 2008; Montagnac et al., 2009). Nevertheless, these considerations, as well as the negative view of the consumption of cassava leaves in parts of Uganda and other parts of Africa, have depressed their use as a food and also resulted in underreporting of their consumption where they are routinely used (Jackson and Chiwona-Karltun, 2018).

Cassava's critical role in global food security

As is evident from the growing trend towards using tapioca, cassava flour, cassava itself directly and other forms of the tuber in foods and various food applications, the crop will continue to play an important role in helping to assure food security of the world's most at risk in Sub-Saharan Africa, Asia and the Americas. In Africa, about 85% of all cassava utilization is for consumption, with roughly 80% prepared as *gari* and other processed foods. With the fast rate of population growth and urbanization in West Africa, demand for cassava as food remains very high especially as a relatively cheaper source of energy. The remaining 20% is used for fresh household consumption. Haggblade et al. (2012) traces five definitive steps that have led to a cassava transformation in Southern Africa. They use the case of cassava in Malawi, Mozambique and Zambia, including improved yields and drought tolerance, its promotion as a famine reserve crop by the colonial masters, and now the rapidly increasing demand for cassava as it transitions to multiple uses in different parts of the world. This has resulted in renewed focus on the crop by private, government and non-governmental organizations (Haggblade and Hazell, 2010). Challenges with maize in Africa resulted in governments, NGOs and other development agencies more aggressively promoting cassava as a food security crop as well as for its commercial benefits.

In Southern Africa, the production of sweet cassava for chewing as a snack, chips or boiling is equally as important as the production of bitter varieties that are used for making flour. The transformation is, however, slower than in Western Africa. Nevertheless, the high productivity of modern cassava varieties has resulted in lower production costs per kilogram of carbohydrate, thereby opening up profitable commercial opportunities for cassava-based foods, starches and feeds (Jackson and Chiwona-Karltun, 2018). From a food security perspective, cassava's high productivity, coupled with drought tolerance, low input costs and a flexible harvesting calendar, enable even households of modest means to ensure food supplies seasonally and over extended periods of time (Alene et al., 2013). Farmers have responded to productivity gains and growing markets by increasing cassava production and sales (Haggblade et al., 2012).

While most of the cassava produced in Asia and Latin America is for non-human consumption, in Africa cassava still remains a major food crop, second only to maize. According to Nweke (2004), cassava has been Africa's best kept secret and that is why the transformation of cassava in Africa has taken a long time. Nweke (2004) shows five stages of cassava transformation in Nigeria: 1) cassava as a food staple 1910–1945; 2) cassava as a cash crop 1946–1977; 3) the mealybug disease invasion 1978–1983; 4) the cassava surge changing policy incentives 1984–1994 and 5) new markets and new challenges 1995 to present. To accelerate the production and utilization of cassava for food security as well as to capture markets and industry, Nigeria implemented a presidential cassava initiative. This enabled persons working with the government and the private sector to travel to Asia and Latin America to learn about how to add value to cassava. As an incentive, former President Obsanjo in Nigeria mandated a 10% mixture of cassava with wheat flour. There was also major investment in improved varieties, increasing yields with the same amounts of labor. This meant that provision of food calories became much cheaper and low-income households could benefit nutritionally. The involvement of the private sector in promoting production, processing and diversifying the utilization of cassava has enabled Nigeria to be a pace-setter in Africa. As the importance of the crop to global food security, particularly in food insecure regions continues to increase, more attention will need to be paid to ensuring its safe preparation and use. Food safety concerns will therefore remain a major focus of private, national and multilateral efforts to expand the use and consumption of cassava.

The impact of urbanization on the role of cassava in developing countries

Urbanization creates demand for cheaper sources of calories. More and more research on cassava processing and fortification for use in confectionery, bread and other industries are being explored in some developing countries, with examples being Mozambique (Tivana, 2012), Indonesia (Frediansyah, 2017) and Jamaica where cassava is being prized for its role in producing gluten-free baked products (The Gleaner, 2019), as well as its expanding role as a replacement for hops in the making of Red Stripe Beer (Serju, 2018). In Indonesia as well, the role of cassava in the expanded production of gluten-free flour is of great significance for the future growth of domestic consumption. Since 2011/2012, South African Breweries (SAB) has been brewing and bottling cassava-based Impala Beer in South Africa and Mozambique while targeting the lower income group (Maritz, 2013; France Presse, 2017). Cassava is also being used in Nigeria for making beer. It has been used in parts of the Americas for this purpose for centuries, native peoples in South America such as the Jivaro, the Yuida and the Tupinamba producing a beer-like drink called masato, caouin, and nihamanchi (depending on the region) from boiled cassava (manioc) root for thousands of years, for both daily and sacramental use (Buhner, 1998).

As urban centers in the developing world transform, so do diets. People are looking not only for cheaper sources of energy, but also foods that are easy and quick to prepare. Much of the cassava value chain studies have been about producing high quality cassava flour that can be integrated into confectionery, noodles and sweet substitutes. There is increasing demand, however, for easy-to-prepare, convenient, tasty and authentic cassava products, particularly in the Americas where an expanding diaspora in North America is looking for familiar foods in a convenient form. This has fueled demand for traditional products such as the cassava breads shown in Fig. 8.4, bammies, a Jamaican/Caribbean tortilla-like cassava product and even *attieke* and *gari* among the increasing numbers of African diaspora in the US and Canada. Studies have also shown that it is cheaper, further up the food chain, to substitute cereal grains like maize with cassava in the formulation of feed for livestock and poultry (Tewe and Lutaladio, 2004). These trends are slowly resulting in the transformation of cassava from solely a food reserve into a commercial crop, the most impactful example of which is its application in brewing, as discussed above.

References

Achidi, A.U., Ajayi, O.A., Maziya-Dixon, B.U.S.S.I.E., Bokanga, M., 2008. The effect of processing on the nutrient content of cassava (Manihot esculenta Crantz) leaves. Food. Pres. 32 (3), 486–502.

Adewusi, S.R., Bradbury, J.H., 1993. Carotenoids in cassava: Comparison of open-column and HPLC methods of analysis. J. Sci. Food Agri. 62 (4), 375–383.

Aerni, P., 2006. Mobilizing science and technology for development: The case of the Cassava Biotechnology Network (CBN). AgBioForum 9 (1), 1–14.

Afoakwa, E.O., Budu, A.S., Asiedu, C., Chiwona-Karltun, L., Nyirenda, D.B., 2012. Viscoelastic properties and physicfunctional characterization of six high yielding cassava mosaic disease-resistant cassava (Manihot esculenta Crantz) genotypes. J. Nutri. Food Sci. 2 (2), 129.

Akoroda M.O., Mwabumba M.L., 2000. Sweet success: cassava in Lilongwe East RDP, SARRNET, Lilongwe.

Alene, A., Khataza, R., Chibwana, C., Ntawuruhunga, P., Moyo, C., 2013. Economic impacts of cassava research and extension in Malawi and Zambia, 2013. J. Agric. Econ. 5 (11), 457–469.

Aryee, F.N.A., Oduro, I., Ellis, W.O., Afuakwa, J.J., 2006. The physicochemical properties of flour samples from the roots of 31 varieties of cassava. Food Control. 17 (11), 916–922.

Banea, M., Poulter, N.H., Rosling, H., 1992. Shortcuts in cassava processing and risk of dietary cyanide exposure in Zaire. Food Nutr. Bull. 14 (2), 1–7.

Banea, J.P., Bradbury, J.H., Mandombi, C., Nahimana, D., Denton, I.C., Kuwa, N.L., et al., 2014. Effectiveness of wetting method for control of konzo and reduction of cyanide poisoning by removal of cyanogens from cassava flour. Food Nutr. Bull. 35 (1), 28–32.

Berry, V., Petty, C., 1992. The Nyasaland Survey Papers 1938-1943: Agriculture. Food and Health. Academy Books, London.

Bokanga, M., 1994. Distribution of cyanogenic potential in cassava germplasm. Int. Workshop Cassava Saf. 375, 117–124.

Bokanga, M., 1995. Biotechnology and cassava processing in Africa: food biotechnology applications in developing countries. Food Technol. 49 (1), 86–90.

Bolhuis, G.G., 1954. The toxicity of cassava roots. Neth. J. Agric. Sci. 2, 176–186.

Boserup, E., 1970. Present and potential food production in developing countries. Geography and a crowding world. A symposium on population pressures upon physical and social resources in the developing lands. Oxford University Press, New York/ London/Toronto.

Bradbury, J.H., Denton, I.C., 2014. Mild method for removal of cyanogens from cassava leaves with retention of vitamins and protein. Food Chem. 158, 417–420.

Bradbury, J.H., Cliff, J., Denton, I.C., 2011. Uptake of wetting method in Africa to reduce cyanide poisoning and konzo from cassava. Food Chem. Toxicol. 49 (3), 539–542.

Brimer, L., 2001. Chemical hazards and their control: endogenous compounds. In: Adams, M.R., Nout, M.J.R. (Eds.), Fermentation and Food Safety. Aspen Publishing, Gaithersburg, pp. 71–98.

Buhner, S.H., 1998. Sacred and Herbal Healing Beers: The Secrets of Ancient Fermentations. Siris Books/Brewers Publications, Boulder, CO.

Buitrago Arbeláez, J., Gil Llanos, J.L., Ospina Patiño, B., 2002. Cassava in poultry nutrition.

Burns, A.E., Gleadow, R.M., Zacarias, A.M., Cuambe, C.E., Miller, R.E., Cavagnaro, T.R., 2012. Variations in the chemical composition of cassava (Manihot esculenta Crantz) leaves and roots as affected by genotypic and environmental variation. J. Agri. Food Chem. 60 (19), 4946–4956.

Carmody, A., 1900. Prussic acid in sweet cassava. Lancet. 156, 736–737.

Chabwine, J.N., Masheka, C., Balol'ebwami, Z., Maheshe, B., Balegamire, S., Rutega, B., et al., 2011. Appearance of konzo in South Kivu, a wartorn area in the Democratic Republic of Congo. Food Chem. Toxicol. 40, 644–649.

Charles, A.L., Sriroth, K., Huang, T.C., 2005. Proximate composition, mineral contents, hydrogen cyanide and phytic acid of 5 cassava genotypes. Food chem. 92 (4), 615–620.

Charoenkul, N., Uttapap, D., Pathipanawat, W., Takeda, Y., 2006. Molecular structure of starches from cassava varieties having different cooked root textures. Starch-Stärke 58 (9), 443–452.

Charoenkul, N., Uttapap, D., Pathipanawat, W., Takeda, Y., 2011. Physicochemical characteristics of starches and flours from cassava varieties having different cooked root textures. LWT-Food Sci. Technol. 44 (8), 1774–1781.

Chikezie, P.C., Ojiako, O.A., 2013. Cyanide and aflatoxin loads of processed cassava (Manihot esculenta) tubers (Garri) in Njaba, Imo State, Nigeria. Toxicol. Int. 20 (3), 261–267. Available from: https://doi.org/10.4103/0971-6580.121679.

Chiwona-Karltun, L., Mkumbira, J., Saka, J., Bovin, M., Mahungu, N.M., Rosling, H., 1998. The importance of being bitter – a qualitative study on cassava cultivar preference in Malawi. Ecol. Food Nutr. 37, 219–245.

Chiwona-Karltun, L., Brimer, L., Kalenga Saka, J.D., Mhone, A.R., Mkumbira, J., Johansson, L., et al., 2004. Bitter taste in cassava roots correlates with cyanogenic glucoside levels. J. Sci. Food Agriculture 84 (6), 581–590.

Chiwona-Karltun, L., Nyirenda, D., Mwansa, C.N., Kongor, J.E., Brimer, L., Haggblade, S., et al., 2015. Farmer preference, utilization, and biochemical composition of improved cassava (Manihot esculenta Crantz) varieties in southeastern Africa. Econ. Bot. 69 (1), 42–56.

Cliff, J., Muquingue, H., Nhassico, D., Nzwalo, H., Bradbury, J.H., 2011. Konzo and continuing cyanide intoxication from cassava in Mozambique. Food Chem. Toxicol. 49 (3), 631–635.

Conn, E.E., 1969. Cyanogenic glycosides. J. Agri. Food Chem. 17 (3), 519–526.

De Bruijn, G.H., 1973. Cyanogenic character of cassava (Manihot esculenta). Chronic cassava toxicity. IDRC, Ottawa, ON, CA.

Diasolua Ngudi, D., Kuo, Y.H., Lambein, F., 2002. Food safety and amino acid balance in processed cassava cossettes. J. Agri. Food Chem. 50 (10), 3042–3049.

Dórea, J.G., 2004. Cassava cyanogens and fish mercury are high but safely consumed in the diet of native Amazonians. Ecotoxicol. Environ. Saf. 57 (3), 248–256.

Dufour, D.L., 1994. Cassava in Amazonia: lessons in utilization and safety from native peoples. Int. Workshop Cassava Saf. 375, 175–182.

Dunstan, W.R., Henry, T.A., Auld, S.J.M., 1906. Cyanogenesis in plants. Part V.—The occurrence of phaseolunatin in cassava (Manihot Aipi and Manihot utilissima). Proceedings of the Royal Society of London. Series B, Containing Papers of a Biological Character, 78(523), 152-158.

El-Sharkawy, M.A., 2003. Cassava biology and physiology. Plant molecular biology, 53(5), Kouamé, A. K., Djéni, T. N., N'guessan, F. K., & Dje, M. K. (2013). Post processing microflora of commercial attieke (a fermented cassava product) produced in the south of Côte d'I voire. Lett. Appl. Microbiol. 56 (1), 44–50.

Essers, A.A., Ebong, C., van der Grift, R.M., Nout, M.R., Otim-Nape, W., Rosling, H., 1995. Reducing cassava toxicity by heap-fermentation in Uganda. Int. J. Food Sci. Nutr. 46 (2), 125–136.

Ezumah, H.C., Okibo, B.N., 1980. Cassava planting systems in Africa. In: Cassava Cultural Practices: Proceedings of a Workshop Held in Salvador, Bahia, Brazil, 18-21 March 1980. IDRC, Ottawa, ON, CA.

Falade, K.O., Akingbala, J.O., 2010. Utilization of cassava for food. Food Rev. Int. 27 (1), 51–83.

FAO (Food and Agriculture Organization of the United Nations), 2015. FAOSTAT database.

Food and Agricultural Organization (FAO), 2005. FAO PRODUCTION YEAR BOOK 2005, FAOSTAT. Statistics Division of the food and FAO Rome Italy data http://faostat.fao.org/faosto.

France Presse, A., Beyond Barley: Cassava Beer Creating a Buzz in the Market In Food and Drink, 13 July 2017, http://www.ndtv.com. Accessed on 27 July 2019 from https://food.ndtv.com/food-drinks/beyond-barley-cassava-beer-creating-a-buzz-in-the-market-1239017.

Frediansyah, A., 2017. Microbial fermentation as means of improving cassava production in Indonesia. Cassava. Intech Open. Available from: http://doi.org.10.5772/intechopen.71966.

Fresco, L.O. 1986. Cassava in shifting cultivation: a systems approach to agricultural technology development in Africa, Fresco.

Gnonlonfin, G.J.B., Adjovi, C.S.Y., Katerere, D.R., Shephard, G.S., Sanni, A., Brimer, L., 2012. Mycoflora and absence of aflatoxin contamination of commercialized cassava chips in Benin, West Africa. Food Control. 23 (2), 333–337.

Gomez, G., Noma, A.T., 1986. The amino acid composition of cassava leaves, foliage, root tissues and whole-root chips. Nutr. Rep. Int. 33 (4), 595–601.

Grandjean, P., Landrigan, P.J., 2014. Neurobehavioural effects of developmental toxicity. Lancet Neurol. 13 (3), 330–338.

Haggblade, S., Hazell P.B., 2010. Successes in African agriculture: Lessons for the future. Intl Food Policy Res Inst.

Haggblade, S., Djurfeldt, A.A., Nyirenda, D.B., Lodin, J.B., Brimer, L., Chiona, M., et al., 2012. Cassava commercialization in southeastern Africa. J. Agribusiness in Developing and Emerging Economies, May 25.

Hahn, S.K., Mahungu, N.M., Otoo, J.A., Msabaha, M.A.M., Lutaladio, N.B., Dahniya, M.T., 1987. Cassava and the African food crisis. In Tropical Root Crops at the African Food Crisis". Proceedings of the 3rd Triennial Symposium of the International Society for Tropical Root Crops. African Branch. Owerri, Nigeria.(Editors: Terry, ER, Akoroda, MO and Arene, OB). IDRC–258e Publ. Canada.,24-29.

Harkup, K., 2017. Cassava crisis: the deadly food the doubles as a vital Venezuelan crop, 22 June 2017, The Guardian. Accessed on November 2019 from https://www.theguardian.com/science/blog/2017/jun/22/cassava-deadly-food-venezuela.

Henry, G., Hershey, C., 2002. Cassava in South America and the Caribbean. Cassava: Biology, Prod. utilization 17–40.

IITA (International Institute of Tropical Agriculture), 1990. Cassava in tropical Africa: a reference manual. Balding and Mansell International, Wesbech, U.K..

International Fund for Agricultural Development (IFAD) & Food and Agricultural Organization of the United Nations (FAO), 2000. The World Cassava Economy. <http://www.fao.org/3/x4007e/X4007E04.htm/> (accessed 29.0719).

Jackson, J., Chiwona-Karltun, L., 2018. Cassava production, processing and nutrition. In: Handbook of Vegetables and Vegetable Processing, pp. 609–632.

Jackson-Malete, J.C., Blake, O., Gordon, A., 2015. Natural Toxins in Fruits and Vegetables: *Blighia sapida* and hypoglycin. In: Gordon, A. (Ed.), Food Safety and Quality Systems in Developing Countries: Volume One: Export Challenges and Implementation Strategies. Academic Press, London, UK, p. 4.

Joachim, A.W., Pandittesekere, D.G., 1944. Investigations of the hydrocyanic acid content of maniac (Manihot utilissima). Trap. Agric.(Ceylon) 100, 156–163.

Kashala-Abotnes, E., Okitundu, D., Mumba, D., Boivin, M.J., Tylleskär, T., Tshala-Katumbay, D., 2019. Konzo: a distinct neurological disease associated with food (cassava) cyanogenic poisoning. Brain Res. Bull. 145, 87–91.

Keresztessy, Z., Kiss, L., Hughes, M.A., 1994. Investigation of the Active Site of the Cyanogenic β-D-Glucosidase (Linamarase) from Manihot esculenta Crantz (Cassava).: I. Evidence for an Essential Carboxylate and a Reactive Histidine Residue in a Single Catalytic Center. Arch. Biochem. biophysics 314 (1), 142–152.

Kwok, J., 2008. Cyanide poisoning and cassava. Food Safety Focus, 19th Issue, February 2008 – Incident in Focus. Risk Communication Section, Centre for Food Safety, The Government of the Hong Kong Special Administrative Region, <https://www.cfs.gov.hk/english/multimedia/multimedia_pub/multimedia_pub_fsf_19_01.html/> (accessed 28.07.19).

Lancaster, P.A., Brooks, J.E., 1983. Cassava leaves as human food. Econ. Bot. 37 (3), 331–348.

Lancaster, P.A., Ingram, J.S., Lim, M.Y., Coursey, D.G., 1982. Traditional cassava-based foods: survey of processing techniques. Econ. Bot. 36 (1), 12–45.

Lateef, A., Ojo, M.O., 2016. Public health issues in the processing of cassava (Manihot esculenta) for the production of lafun and the application of hazard analysis control measures. Qual. Assur. Saf. Crop. Foods 8 (1), 165–177.

Lei, V., Amoa-Awua, W.K., Brimer, L., 1999. Degradation of cyanogenic glycosides by Lactobacillus plantarum strains from spontaneous cassava fermentation and other microorganisms. International. J. food microbiology 53 (23), 169–184.

Lokko, Y., Okogbenin, E., Mba, C., Dixon, A., Raji, A., Fregene, M., 2007. Cassava. Pulses, Sugar and Tuber Crops. Springer, Berlin, Heidelberg, pp. 249–269.

Mahungu, N., 1994. Relationships between cyanogenic potential of cassava and other agronomic traits. Int. Workshop Cassava Saf. 375, 125–130.

Manjula, K., Hell, K., Fandohan, P., Abass, A., Bandyopadhyay, R., 2009. Aflatoxin and fumonisin contamination of cassava products and maize grain from markets in Tanzania and republic of the Congo. Toxin Rev. 28 (2-3), 63–69.

Maritz, J., 2013. Multinational brewers turn to cassava for low-cost beer. In: How We Made it in Africa, Africa Business Insight, 4 November 2013. <https://www.howwemadeitinafrica.com/multinational-brewers-turn-to-cassava-for-low-cost-beer/32070/> (accessed 28.07.19.).

McMahon, J.M., White, W.L., Sayre, R.T., 1995. Cyanogenesis in cassava (Manihot esculenta Crantz). J. Exp. Bot. 46 (7), 731−741.

Mkpong, O.E., Yan, H., Chism, G., Sayre, R.T., 1990. Purification, characterization, and localization of linamarase in cassava. Plant. Physiol. 93 (1), 176−181.

Mlingi, N.L., Bainbridge, Z.A., Poulter, N.H., Rosling, H., 1995. Critical stages in cyanogen removal during cassava processing in southern Tanzania. Food Chem. 53 (1), 29−33.

Montagnac, J.A., Davis, C.R., Tanumihardjo, S.A., 2009. Nutritional value of cassava for use as a staple food and recent advances for improvement. Compr. Rev. Food Sci. Food Saf. 8, 181−194.

Moyo, C.C., Benesi, I.R.M., Sandifolo, V.S., Teri, J.M., 1998. Current status of cassava and sweet potato production and utilization in Malawi. South. Afr. Root Crop. Res. Netw. (SARRNET) 51−67.

Muzanila, Y.C., Brennan, J.G., King, R.D., 2000. Residual cyanogens, chemical composition and aflatoxins in cassava flour from Tanzanian villages. Food Chem. 70 (1), 45−49.

Nahrstedt, A., 1988. Cyanogenesis and the role of cyanogenic compounds in insects. Ciba Found. Symp. 140, 131−150.

Nassar, N.M., Marques, A.O., 2006. Cassava leaves as a source of protein. J. Food Agri. Env. 4 (1), 187.

Nweke, F.I., Bokanga, M., 1994. Importance of cassava processing for production in sub-Saharan Africa. Int. Workshop Cassava Saf. 375, 401−412.

Nweke, F., Spencer, I., Dunstan, S., Lynam, J.K., 2001. The Cassava Transformation: Africa's best kept secret. Michigan State University Press, East Lansing.

Ohadike, D.C., 1981. The influenza pandemic of 1918−19 and the spread of cassava cultivation on the Lower Niger: A study in historical linkages. J. Afr. History 22 (3), 379−391.

Okezie, B.O., Kosikowski, F.V., 1982. Cassava as a food. Crit. Rev. Food Sci. Nutr. 17 (3), 259−275.

Onabolu, A.O., Oluwole, O.S., Rosling, H., Bokanga, M., 2002a. Processing factors affecting the level of residual cyanohydrins in gari. J. Sci. Food Agri. 82 (9), 966−969.

Onabolu, A.O., Oluwole, O.S., Bokanga, M., 2002b. Loss of residual cyanogens in a cassava food during short-term storage. Int. J. food Sci. Nutr. 53 (4), 343−349.

Phillips, Truman P., 1974. Cassava utilization and potential markets. International Research and Development Centre (IDRC), Ottawa, Canada.

Pinto-Zevallos, D.M., Pareja, M., Ambrogi, B.G., 2016. Current knowledge and future research perspectives on cassava (Manihot esculenta Crantz) chemical defenses: An agroecological view. Phytochemistry 130, 10−21.

Prakash, A., 2018. Cassava Market Development and Outlook. In: Food Outlook - Biannual Report on Global Food Markets − November 2018. Rome. 104 pp. Licence: CC BY-NC-SA 3.0 IGO., pp 13−23.

Safo-Kantanka, O., Owusu-Nipah, J., 1992. Cassava varietal screening for cooking quality: relationship between dry matter, starch content, mealiness and certain microscopic observations of the raw and cooked tuber. J. Sci. Food Agri. 60 (1), 99−104.

Seigler, D.S., 1991. Cyanide and cyanogenic glycosides. In Rosenthal, G.S. and Berenbaum, M.R., (eds): Herbivores: Their interaction with secondary plant metabolites, Volume I: The Chemical Participants, 35-77.

Serju, C., 2018. Red Stripe Taking Cassava to New Heights. In: The Gleaner, 7 May 2018. <http://jamaica-gleaner.com/article/news/20180507/red-stripe-taking-cassava-new-heights/> (accessed 28.07.19.).

Stupak, M., Vanderschuren, H., Gruissem, W., Zhang, P., 2006. Biotechnological approaches to cassava protein improvement. Trends food Sci. & Technol. 17 (12), 634−641.

Tewe, O.O., Lutaladio, N., 2004. The global cassava development strategy. Cassava for Livestock Feed in Sub-Saharan Africa. FAO, Rome, Italy, pp. 1−64.

The Gleaner, Shaw Lauds First Brand Cassava Bread 2019 In The Gleaner, 1 April 2019. <http://jamaica-gleaner.com/article/news/20190401/shaw-lauds-f1rst-brand-cassava-bread/> (accessed 30.07.19).

Tivana, L., 2012. Cassava processing: safety and protein fortification. Lund University.

Tylleskär, T., Banea, M., Bikangi, N., Fresco, L., Persson, L.A., Rosling, H., 1991. Epidemiological evidence from Zaire for a dietary etiology of konzo, an upper motor neuron disease. Bull. World Health Organ. 69 (5), 581.

Ufuan Achidi, A., Ajayi, O.A., Bokanga, M., Maziya-Dixon, B., 2005. The use of cassava leaves as food in Africa. Ecol. Food Nutr. 44 (6), 423−435.

Vasconcelos, A.T., Twiddy, D.R., Westby, A., Reilly, P.J.A., 1990. Detoxification of cassava during gari preparation. Int. J. Food Sci. Technol. 25 (2), 198−203.

Vickery, K.P., 1989. The Second World War revival of forced labor in the Rhodesias. Int. J. Afr. Historical Stud. 22 (3), 423−437.

von Hagen, W.V., 1949. The bitter cassava eaters. Nat. Hist. 58, 120−124.

Worldatlas.com, 2019. Top Cassava Producing Countries in the World, updated 25 April 2017. <https://www. worldatlas.com/articles/top-cassava-producing-countries-in-the-world.html/> (accessed 29.07.19.).

Yeoh, H.H., Chew, M.Y., 1976. Protein content and amino acid composition of cassava leaf. Phytochemistry 15 (11), 1597−1599.

Further reading

Adjovi, Y.C., Bailly, S., Gnonlonfin, B.J., Tadrist, S., Querin, A., Sanni, A., et al., 2014. Analysis of the contrast between natural occurrence of toxigenic Aspergilli of the Flavi section and aflatoxin B1 in cassava. Food Microbiol. 38, 151−159.

Alitubeera, P.H., Eyu, P., Kwesiga, B., Ario, A.R., Zhu, B.P., 2019. Outbreak of cyanide poisoning caused by consumption of cassava flour—Kasese District, Uganda, September 2017. Morbidity Mortal. Wkly. Rep. 68 (13), 308.

Andersson, K., Bergman Lodin, J., Chiwona-Karltun, L., 2016. Gender dynamics in cassava leaves value chains: the case of Tanzania. J. Gender, Agric. Food Security (Agri-Gender) 1 (302-2016-4753), 84−109.

Bangyekan, C., Aht-Ong, D., Srikulkit, K., 2006. Preparation and properties evaluation of chitosan-coated cassava starch films. Carbohydr. Polym. 63 (1), 61−71.

Bellotti, A.C., Smith, L., Lapointe, S.L., 1999. Recent advances in cassava pest management. Annu. Rev. Entomol. 44 (1), 343−370.

Brimer, L., 2000. Cyanogenic glycosides: occurrence, analysis and removal from food and feed: comparison to other classes of toxic and antinutritional glycosides: Technology and biotechnology for the removal of plant toxins, KVL.

Cardoso, A.P., Mirione, E., Ernesto, M., Massaza, F., Cliff, J., Haque, M.R., et al., 2005. Processing of cassava roots to remove cyanogens. J. Food Comp. Anal. 18 (5), 451−460.

Coursey, D.G., 1973. Cassava as food: toxicity and technology. Chronic Cassava Toxicity. IDRC, Ottawa, ON, CA.

Ernesto, M., Cardoso, A.P., Nicala, D., Mirione, E., Massaza, F., Cliff, J., Haque, M.R., Bradbury, J.H., et al., 2002. Persistent konzo and cyanogen toxicity from cassava in northern Mozambique. Acta Tropica 82 (3), 357−362.

Essono, G., Ayodele, M., Akoa, A., Foko, J., Filtenborg, O., Olembo, S., 2009. Aflatoxin-producing Aspergillus spp. and aflatoxin levels in stored cassava chips as affected by processing practices. Food Control. 20 (7), 648−654.

Gacheru, P.K., Abong, G.O., Okoth, M.W., Lamuka, P.O., Shibairo, S.A., Katama, C.K.M., 2016. Microbiological safety and quality of dried cassava chips and flour sold in the Nairobi and coastal regions of Kenya. Afr. Crop. Sci. J. 24 (1), 137−143.

Gomes, B.C., Franco, B.D.G.D.M., De Martinis, E.C.P., 2013. Microbiological food safety issues in Brazil: bacterial pathogens. Foodborne Pathog. Dis. 10 (3), 197−205.

Hillocks, R.J., Thresh, J.M., Bellotti, A. (Eds.), 2002. Cassava: Biology, Production and Utilization. CABI.

Ingenbleek, L., Sulyok, M., Adegboye, A., Hossou, S.E., Koné, A.Z., Oyedele, A.D., et al., 2019. Regional Sub-Saharan Africa Total Diet Study in Benin, Cameroon, Mali, and Nigeria Reveals the Presence of 164 Mycotoxins and Other Secondary Metabolites in Foods. Toxins (Basel) 2019 Jan; 11(1): 54. Published online2019 Jan 17. doi: 10.3390/toxins11010054.

Jansson, C., Westerbergh, A., Zhang, J., Hu, X., Sun, C., 2009. Cassava, a potential biofuel crop in (the) People's Republic of China. Appl. Energy 86, S95−S99.

Lebot, V., 2009. *Tropical Root and Tuber Crops: Cassava, Sweet Potato, Yams and Aroids* (No. 17). Cabi.

Legg, J.P., Fauquet, C.M., 2004. Cassava mosaic geminiviruses in Africa. Plant. Mol. Biol. 56 (4), 585−599.

Okafor, P.N., Okorowkwo, C.O., Maduagwu, E.N., 2002. Occupational and dietary exposures of humans to cyanide poisoning from large-scale cassava processing and ingestion of cassava foods. Food Chem. Toxicol. 40 (7), 1001−1005.

Olsen, K.M., Schaal, B.A., 1999. Evidence on the origin of cassava: phylogeography of Manihot esculenta. Proc. Natl Acad. Sci. 96 (10), 5586−5591.

Omafuvbe, B.O., Adigun, A.R., Ogunsuyi, J.L., Asunmo, A.L., 2007. Microbial Diversity in Ready-to-eat Fufu and La fun-Fermented Cassava Products Sold in Ile-Ife, Nigeria. Res. J. Microbiol. 2 (11), 831−837.

Pinto-Zevallos, D.M., Pareja, M., Ambrogi, B.G., 2016. Current knowledge and future research perspectives on cassava (Manihot esculenta Crantz) chemical defenses: An agroecological view. Phytochemistry 130, 10−21.

Prochnik, S., Marri, P.R., Desany, B., Rabinowicz, P.D., Kodira, C., Mohiuddin, M., et al., 2012. The cassava genome: current progress, future directions. Tropical Plant. Biol. 5 (1), 88–94.

Teles, F.F.F., 2002. Chronic poisoning by hydrogen cyanide in cassava and its prevention in Africa and Latin America. Food Nutr. Bull. 23 (4), 407–412.

Tshala-Katumbay, D.D., Ngombe, N.N., Okitundu, D., David, L., Westaway, S.K., Boivin, M.J., et al., 2016. Cyanide and the human brain: perspectives from a model of food (cassava) poisoning. Ann. N. Y. Acad. Sci. 1378 (1), 50.

World Health Organization, 2015. *WHO Estimates of the Global Burden of Foodborne Diseases: Foodborne Disease Burden Epidemiology Reference Group 2007-2015* (No. 9789241565165). World Health Organization.

CHAPTER

9

Market & technical considerations for spices: Nutmeg & Mace case study

André Gordon

Chairman & CEO, Technological Solutions Limited, Kingston, Jamaica, West Indies

OUTLINE

Food Safety and Quality Systems in Developing Countries
DOI: https://doi.org/10.1016/B978-0-12-814272-1.00009-7

367

Introduction to the global spice industry

The global industry for spices is a diverse mixture of subsectors and different applications of herbs and spices including culinary, medicinal and therapeutic applications, uses in the chemicals industry and personal care, wellness and cosmetics (PCWC) applications. Collectively, these consume just under US$10 billion worth of the spices shown in Fig. 9.1 (ITC Trade Map, 2016). The industry continues the expansion that has been ongoing for several years driven by social changes. In the culinary arena, this has included the diversification of global cuisine, the desire for new flavors, the increasing importance of "ethnic" food and the increasing importance of making processed food more natural, which requires the use of more condiments and aromatic herbs. Estimates indicate that the global market for culinary uses of seasonings, spices and herbs exceeded US$6.5 billion per year by 2018 (The ITC Trade Map, 2019). Consistent growth has taken place in global use of peppers, cinnamon and vanilla, while turmeric has also shown consistent growth in demand (Fig. 9.1).

The major spice producers are in the developing and least developed countries in the tropics, with the industry being comprised of between 40 and 50 herbs and spices of global economic and culinary importance. In terms of value, the most important spices are pepper (of the genii *Pimienta* and *Capsicum*), nutmeg and mace, cardamom, allspice/pimento, vanilla, cloves, ginger, cinnamon, cassia and turmeric (Fig. 9.1). Coriander, cumin, mustard, and sesame seeds and the herbs sage, oregano, thyme, bay and the mints are the most important spice crops that are grown in non-tropical environments. In the category ginger, turmeric, saffron, curry, thyme, bay leaves and other spices, ginger is by far the most important (in value) and while the demand for most of the other spices has continued to grow, demand for curry, thyme and bay leaves has stagnated or declined (Fig. 9.1).

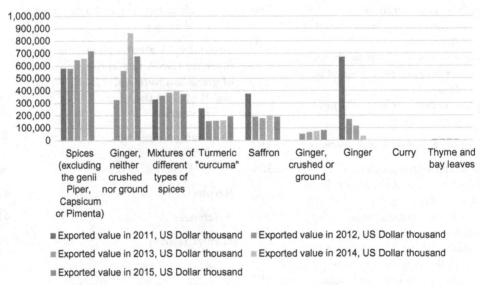

FIGURE 9.1   Export value for selected herb and spice products, 2011–2015.

Depending on the use to which the herbs and spices are to be put, they are sold in their post-harvest (unprocessed) form, dried, or as extracts (e.g. essential oils), the latter giving them a higher value per unit weight.

Of interest in these data are the absence of exports for curry, a spice in high demand but for which there are several challenges. These include issues with contamination with *Salmonella* spp. that have lead to the European Food Safety Authority (EUFSA) imposing strict scrutiny on curry imports into the EU and the US Food & Drug Administration (US FDA) doing likewise (Gurtler and Keller, 2019). In fact, the US FDA has imposed an import alert on all curry powder and related spices coming into the US (US Food & Drug Administration, 2019a), a situation that requires specific approaches to successfully address[1]. This has been further exacerbated because of recent recalls of curry powder because of the finding of lead in the product (Mackin, 2018; US Food & Drug Administration, 2019b). An important observation in this regard is that the US finds 1.9 times more pathogen positive spices than for any other category of food being imported into the country (Gurtler and Keller, 2019). This has resulted in efforts to standardize the approach by which spices and other low water activity products are examined for pathogens, a process being led by the US FDA, in collaboration with other laboratories (Gurtler et al., 2019).

This issue with curry powder and other spices being imported to developed country markets from developing countries brings into focus the importance of food safety and quality systems (FSQS) implementation in ensuring the safety of these products and protecting market opportunities that have arisen. Developing countries have a significant opportunity to benefit from the ongoing increase in demand for spices if the right approach is taken towards the development of the sector, inclusive of a focus on food safety and quality (FSQ) issues. These products could become a much more significant and profitable source of culinary, personal care, wellness, cosmetics, medicinal and related exports for these countries, with the benefits extending to all facets of the value chain, particularly primary producers. This, however, will require a much more focussed approach to the management of FSQ along the entire value chain for the production of spices which will draw on many of the issues discussed earlier in this volume. It is in this context that this chapter focussing on spices, the quintessential exports from developing countries that were responsible for much of the development of the existing trade with Europe and the rest of the developed world, examines the role of FSQS and the market considerations involved.

The nature of the global trade in spices

The trade in spices represents a centuries-old practice which paved the way for much of the rest of the global trade in food. It was the search for, and wealth derived from, the much sought-after spices that led European maritime nations into the Far East seeking a range of precious commodities in the mid-to-late 1400s AD, the trade having been developed much earlier by Middle Eastern and Eastern peoples in antiquity. It was this quest that not only led to the development of many of the trading relationships and markets

---

[1] The author's firm, Technological Solutions Limited (TSL) has worked with several exporters to deal effectively with import alerts. Most recently, they have assisted a firm from Trinidad and Tobago to successfully petition the US FDA to allow unhindered imports of their curry powder into the US.

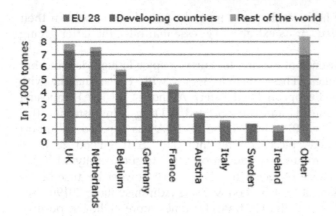

FIGURE 9.2   Import from EU and global exporters and trade of spices in the EU. *Source: Eurostat.*

that are in existence today but was also a precursor for the transnational sourcing of food and food ingredients that are now the norm. Spices have therefore played a special role in the development of the global food industry and remain an important component of the industry. This is no less so when issues of specific market considerations and the attendant food safety and quality issues are discussed.

The largest importers and users of spices globally are in the European Union (EU), with Germany being the leading importer, followed by the United Kingdom (UK), where the market for herbs was estimated to be about US$100−150 million in 2016[2]. Much of these imports come from developing countries (Fig. 9.2), particularly in the case of dried, crushed and ground peppers, a major import that exceeded €500 million (US$550 million) in 2014 (Table 9.1), although the major source of spices for most EU countries is other EU exporters (Figs. 9.2 and 9.3). Other major spices imported into the EU are curry (mainly by the UK), mixtures of whole spices and crushed and/or ground spices (Fig. 9.3). In the EU countries, 60−70% of the total herbs and spices are used for industrial consumption, 15−20% by the retail sector and 10−15% by the catering sector (Fig. 9.4). The high industrial sector use reflects the growing popularity of ready-to-use spice mixtures. Another reason is the increasing consumption of processed foods and ready to eat dishes, which often rely on herbs and spices to retain and enhance food flavor. The leading spices consumed in the EU are pepper, paprika and allspice (pimento), while the leading herbs include parsley, thyme and oregano.

Despite the large demand for spices and the strict requirements for quality, food safety and traceability, very few spices are grown in the EU although there is significant manufacturing and re-export of imported raw materials or further processed products. While not a major source of spices, countries within the EU do grow and provide spices to other importing EU countries that have heavy demand for these products. The most commonly grown herbs in the EU are: basil, bay leaves, celery leaves, chives, coriander, dill tips, chervil, juniper, marjoram, oregano, parsley, rosemary, sage, savory, tarragon, thyme and watercress. France, Italy and Greece are important producers of dried herbs, although

[2] Source: Produce Business UK: http://www.producebusinessuk.com/.

TABLE 9.1 EU pepper imports from exporting countries in 2014.

EU imports (value & quantity) for pepper (dried, crushed or ground)

Indicators	Value imported to the EU/MS (EURO)	Qty imported to the EU/MS (Kg)
EU Member State(s)	EU28	EU28
Years	2014	2014
Partners		
Total-EU28	**537,933,807**	**141,022,000**
Vietnam	179,665,777	29,842,000
China, People's Republic of	90,815,697	49,532,000
Brazil	72,568,118	13,275,000
India	61,980,413	16,425,000
Indonesia	60,592,525	7,894,000
Peru	10,358,263	5,604,000
Sri Lanka	9,183,477	1,386,000
Mexico	7,848,056	2,757,000
Malaysia	7,726,870	1,103,000
Serbia	4,559,459	1,385,000
Israel	3,978,619	1,665,000
United States	3,439,701	1,017,000
Turkey	3,415,836	1,121,000
South Africa	2,965,597	542,000
Thailand	2,526,935	3,253,000
Singapore	2,175,907	262,000
Madagascar	2,004,269	560,000
Guatemala	1,530,191	550,000

they, themselves have a low consumption of dried herbs, most of the production being destined for intra-EU export (Fig. 9.2). The UK, The Netherlands, Belgium and Germany are among the major importers and re-exporters, with Germany being a major consumer market (Figs. 9.2 and 9.3). Total EU production of spices amounted to 120,000 tonnes in 2008, of which 63% consisted of paprika, chilies and allspice. The production of spice seeds accounted for 33% and the remaining 67% was other spices.

While the EU is the most important market for spices, the US and Japan are the two largest single-country importers of spices, with the US being the largest importer of spices for the 4 year period 2012–2015, with imports of US$1.67 billion in 2015. Among the

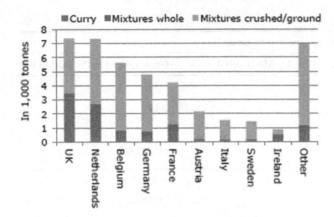

FIGURE 9.3   Most important EU importing countries of spice mixtures, 2010−2014. *Source: Eurostat.*

spices, the US imported 55,000 tonnes of peppers in 2014, the same as in 2010, after importing a high of 65,000 tonnes in 2012; they also imported just under 59,000 tonnes of ginger in 2014 up from 42,000 tonnes in 2010 (Fig. 9.4). The US is also a major importer of cloves, importing 1480 tonnes in 2014 with Madagascar, Indonesia and Brazil being their major suppliers (Fig. 9.5). Japan imported US$371 million in herbs and spices in 2015. It would be important, therefore for exporters in this sector to also examine opportunities in these markets, bearing in mind the considerations such as demand, pricing, quality and safety standards and appropriate routes to market that would impact on the returns derived from the exports.

The major exporters vary by the herb or spice and by the format of the finished product when it comes to exports of H&S products in their various forms. In developing countries, much of the spices are used to make sauces which are then exported. The Dominican Republic, South Africa, Jamaica, Trinidad and Tobago and Kenya are the top five African Caribbean and Pacific (ACP) exporters of sauces, accounting for almost all of the US$240 million in ACP exports (Fig. 9.6). However, globally, the US is the largest exporter, followed by Germany, China, Thailand, Italy and The Netherlands, with none of the ACP exporters (except Thailand) figuring in the top 20 (Fig. 9.7). For cinnamon, Sri Lanka, Indonesia and China dominate global exports (US$472 million in 2015). Allspice (or pimento as it is called in the Caribbean) is primarily exported to the USA, with Vietnam, India and Indonesia being the major exporters, although Jamaican pimento, prized as it is for its flavor, has carved out a niche in the global marketplace, being exported primarily to the United States. For peppers, the primary exporters are Vietnam, India and Indonesia, in that order, with Brazil and China rounding out the top five in 2015, with total pepper and pepper product exports exceeding US$3.8 billion in each year since 2014 (Gordon, 2012b). Other spices of interest such as mace, ginger, cinnamon (Fig. 9.8), allspice and turmeric are exported by a range of countries with Indonesia, India, Vietnam, Mexico, and other developing countries, as well as China, being among the major sources of these herbs and spices. The major exporting companies and countries (Table 9.2) are also among the largest importers of spices.

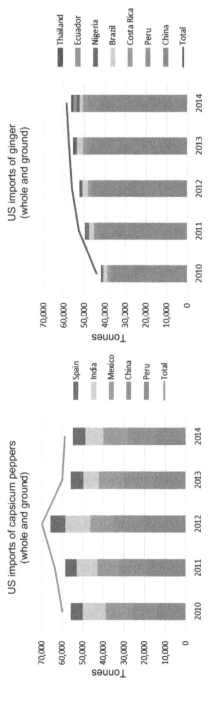

FIGURE 9.4 US imports and sources of peppers and ginger 2010–2014. *Source: ITC, Market Insider, 2015.*

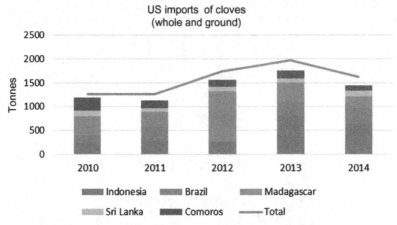

FIGURE 9.5 US imports and sources of cloves 2010–2014. *Source: ITC, Market Insider, 2015.*

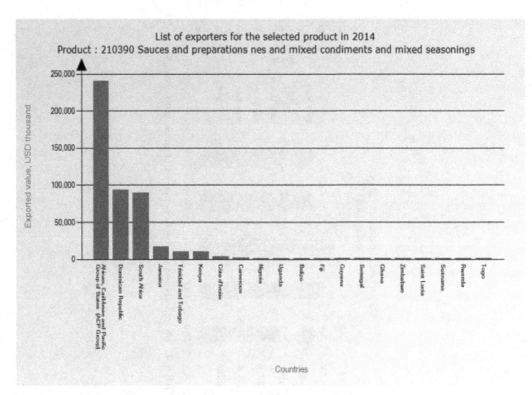

FIGURE 9.6 ACP exports of herb & spice-based sauces, condiments & seasonings, 2014. *Source: Trade Map, International Trade Centre, https://www.trademap.org.*

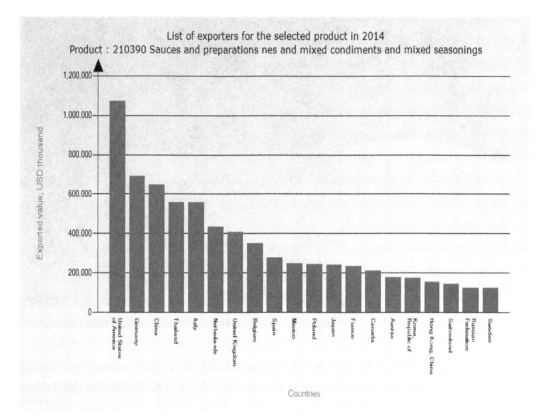

FIGURE 9.7   Global exports of herb & spice-based sauces, condiments & seasonings, 2014. *Source: Trade Map, International Trade Centre, https://www.trademap.org*

Food safety, quality & other technical considerations

The importance of spices to the food industry globally is evident in not only the value of dry spices, spice blends (Table 9.2), traditional spice products (Fig. 9.8) and herb and spice-based sauces traded (Fig. 9.7), but also in the extensive array of products in which they are used. This ranges from artisanal sauces (Fig. 9.9) and seasonings, traditional teas (Fig. 9.10) and drinks, to a wide range of other beverages, baked goods, savory dishes, vegetable-based food items to a variety of meat products and prepared meals. They have applications in the industrial, food service and retail segment of the food industry (Fig. 9.11) and are important ingredients in many of the foods that consumers around the world eat every day. This wide range of applications, combined with the variation in type, origin and sources of spices, as well as their harvesting, storage, handling, processing and transportation significantly influence and often complicate the food safety, quality and technical considerations involved in this sector.

As has already been noted, the majority of globally traded spices originate in developing countries, although much of their processing to finished retail products and application in

FIGURE 9.8   Traditional cinnamon sticks made from cinnamon bark in Dominica. *Source: André Gordon (2018).*

further processing is done in developed countries. Much of the growing, harvesting, handling and other aspects of the production, precedent to final processing is done under traditional conditions, although this has been changing with the increasing involvement and backwards linkages of major buyers into primary production. The market entry and buyer requirements for compliance with traceability, wholesomeness and other food safety and quality systems and standards has also significantly impacted the sector, as has requirements for authenticity, product differentiation and product diversification. In addition to these drivers, there has also been an increasing amount of research on the constituents of major spices and their functionality in a range of applications in the food industry (Wijesekera, 1977; Takikawa et al., 2002; Paranagama et al., 2010; Rodianawati et al., 2015; Balasubramanian et al., 2016; Cardoso-Ugarte et al., 2016), as well as in medicinal products (Mang et al., 2006; Nichol, 2014; Jham et al., 2005), themselves becoming an increasingly important sub-category of the global food product offerings. Research into the presence of natural toxicants, mycotoxins, heavy metals and other naturally-derived potential contaminants of spices has also been critical in establishing the parameters for best practices for important spice sub-categories.

In examining the opportunities for the spices sector of the global food trade, specific note must be taken of the ever more stringent regulations governing the international trade in herbs, spices and their derivatives for culinary and other uses which will mean that spice producers who ignore these requirements will eventually lose their markets to producers that fully comply. For example, EU importers have increasingly been insisting that aflatoxin limits must be adhered to for them to accept shipments of nutmegs coming from Indonesia and Grenada, a move that has caused losses for producers from these

TABLE 9.2 Global imports of dry spices & spice blends in 2015.

Importers	Trade indicators							
	Value imported in 2015 (USD thousand)	Trade balance in 2015 (USD thousand)	Quantity imported in 2015	Quantity Unit	Unit value (USD/ unit)	Annual growth in value between 2011 and 2015 (%)	Annual growth in quantity between 2011 and 2015 (%)	Annual growth in value between 2014 and 2015 (%)
World	416,833	−43,805	105,782	Tons	3940	3	7	−6
Netherlands	41,800	32,812	13,938	Tons	2999	−4	6	−23
United Kingdom	39,987	−30,425	14,508	Tons	2756	8	14	19
Saudi Arabia	31,157	−30,804	6103	Tons	5105	4	−6	−11
United States of America	27,970	−10,418	5285	Tons	5292	15	15	11
Germany	27.910	10,567	5072	Tons	5503	1	−3	−1
Belgium	24,274	−15,967	6214	Tons	3906	5	11	−10
France	22,722	−1263	4397	Tons	5168	3	13	−9
Singapore	17,286	−14,162	5707	Tons	3029	16	23	2
United Arab Emirates	15,366	−10,381	4054	Tons	3790	20	14	25
Canada	13,697	−12,946	2945	Tons	4651	17	3	6

Unit: US Dollar thousand.

countries and is driving change in the industry. Likewise, stringent assessment at US ports of entry of incoming container of curries, other spices and pepper sauces for the presence of pathogens, heavy metals, or adulteration with banned colorants has led to import alerts and the prohibition of imports (Gordon, 2012b; US Food & Drug Administration, 2019a,b). This is both a threat and a great opportunity, the latter being to carve out a niche in the global marketplace for safe, high quality herb and spice products.

Some producers have taken another route to remain competitive. Countries such as India are moving into the value-added segment of the spice market, producing more of their herbs and spices in the form of essential oils, oleoresins, specialty extracts and blends, and powders to take advantage of growing and higher-priced markets for these products. In addition, India has established spice Agri-Export Zones and they are actively developing capabilities in quality management and food safety, as well as improved packaging and technology innovation and advances in production and processing. Balasubramanian et al. (2016) examined a range of issues regarding the effect of where the spices are grown on the differential characteristics observed, as well as the application of spices from multiple countries in high end product development. They also assessed and reported on the cultivation, post-harvest handling, chemical composition, uses, health, and medicinal benefits of the selected spices. Among the spices covered were black pepper,

FIGURE 9.9   Examples of spice-based sauces from developing countries. *Source: Gordon (2016).*

FIGURE 9.10   Traditional herbal teas made from herbs & spices. *Source: André Gordon (2018).*

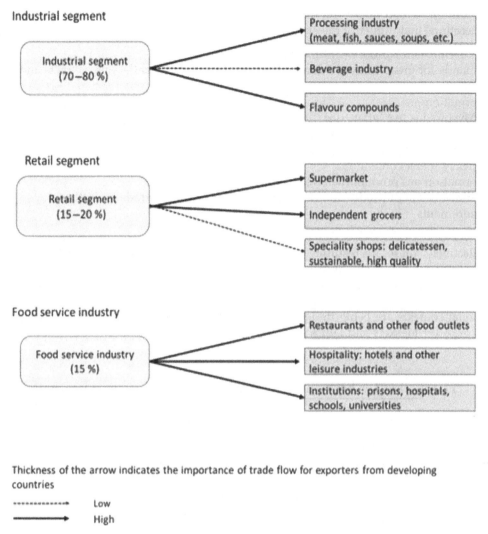

Industrial segment

Industrial segment
(70–80 %)

Processing industry
(meat, fish, sauces, soups, etc.)

Beverage industry

Flavour compounds

Retail segment

Retail segment
(15–20 %)

Supermarket

Independent grocers

Speciality shops: delicatessen,
sustainable, high quality

Food service industry

Food service industry
(15 %)

Restaurants and other food outlets

Hospitality: hotels and other
leisure industries

Institutions: prisons, hospitals,
schools, universities

Thickness of the arrow indicates the importance of trade flow for exporters from developing
countries

············▶ Low

──────▶ High

FIGURE 9.11 Structure of the herb and spices sub-markets in the EU. *Source: Adopted from CBI Market Intelligence (2015a). Market Channels and Segments: Spices and Herbs.*

coriander, cinnamon, fenugreek, turmeric, and technological advances in processing of spices viz., super critical fluid extraction, cryogenic grinding, and microencapsulation. These approaches to moving spice production up the value chain and the application of better technologies to production and packaging can deliver greater returns to producers and exporters. They will, however, require more detailed understanding of the science behind the specific spice and its production, including the consideration of appropriate food safety and quality systems and application of more sophisticated science-based controls.

In addition to food safety and quality systems certification requirements, other technical considerations that influence the spice sub-sector include:

1. Requirements for proof of "authenticity"
2. Demands for greater information on the composition of essential oils, extracts and tinctures
3. Information on
   a. Mycotoxins (e.g. aflatoxin)
   b. Natural constituents/potential toxicants such as myristicin in nutmeg and mace and coumarin in cinnamon
   c. heavy metal content
4. Information on/proof of the effectiveness of herbal products
5. Documentation of production practices compliant with Fairtrade and other requirements
6. Proof of "green" production and extraction practices

Collectively, these are resulting in an enhancement of the technical capabilities of the spices sector as exporting countries and exporters strive to meet the needs of their customers in the destination market for their product. Consequently, much more research, analysis, systems implementation and adaptation of international best practices to local and traditional contexts are being undertaken in producer countries, facilitating better quality, safer spices for markets and tables globally.

The market for herb and spice products globally — considerations

Considerations for future growth of the spices sector

Exporters will therefore need to specifically understand the markets into which they are seeking to expand exports and undertake the necessary and often times non-traditional approaches to getting to the best segments of the market for their products through the most suitable market channel available. They need to understand the differences within the market in various countries and continents and seek to differentiate their product offerings between different target markets. This includes the traditional ethnic market and the new foothold gained in the multiples as well as what may be required by the high value boutique product markets. They will have to specifically understand and ensure that they meet the requirements to access new marketing channels that are available to them. In the developed country market, these include convenience stores, boutique retail outlets and stores such as Whole Foods and Kroger's in the US and numerous boutique brands in Europe that are focussing on authenticity, variety, quality and safety of herb and spice products. In the Caribbean, this will require an understanding of why some countries such as the Bahamas and Barbados have been successful in breaking into the potentially lucrative but difficult to access cruise ships market while others (e.g. Jamaica) have not. What follows in the situational assessment and analysis presented encompasses these considerations and looks at the situation in each CARICOM territory, each major market, as well as changes taking place in both the culinary herb and spices market as well as that for therapeutic, medicinal, personal care, wellness and cosmetic (TMPCWC) products. As is evident from the quotation from the CBI (below), the Caribbean has an opportunity, if the right things are done, to capitalize on a growing market for herb and spice products. This body of work seeks to lay the basis for this, with this section of the report providing targeted information on markets.

Some producers in the Far East, Central America, and Latin America and the Caribbean, the main exporting regions, already traditionally use many of the herbs and spices required in the TMPCWC segment of the EU market. What is required is for there to be concerted, focused efforts to increase direct supply to high value markets in the EU. These trends in both the culinary and TMPCWC segments of the herbs and spices market in the EU make it a market on which significant efforts should be focused as price is not typically the main driver in buying decisions (CBI Market Intelligence, 2015a,b). Also, because the EU already imports almost all of its herbs and spices from developing countries (Fig. 9.2), with India, Pakistan and Sri Lanka being the major suppliers, there are already well-established trading systems with this market. While the Netherlands, the UK, Germany, Belgium and France are the main destinations within the EU for spice blends (Table 9.2), the UK is the largest importer from developing countries, importing mainly curry, with only France importing significant quantities of other spices directly from developing countries.

Making further inroads in the global markets for herb and spice products will require a detailed understanding of the existing raw material supply and production capabilities, the detailed structure and operations of the markets being targeted and the market requirements for each product category in the specific markets. Successfully capitalizing on the opportunities indicated will necessitate a different approach than what has traditionally been the case. In order to do this, developing country herbs, spices and sauce producers and exporters will need to identify specifically what their best product offerings are, successfully develop and manage the value chain for these products, identify what their best market opportunities are and plan and execute successful market access and development strategies, specific to product/market combinations. This will also require them to identify what are the specific challenges that must be overcome, including a detailed knowledge of the technical requirements of the country and market being targeted and any required technical or scientific inputs into production or supply chain management process. Against this background, this chapter has sought to highlight key information required to provide the guidance to stakeholders in the sector as they seek to develop or exploit opportunities for expanding exports from herb and spice-producing countries.

Specific considerations for the market for herb and spice products in the European Union

With over five hundred (500) million consumers, the EU is an important market for herbs and spices. The EU remains an attractive region for suppliers of all sizes and types of products and the best opportunities for small and medium-sized enterprises in developing countries are to be found in the high end of the market and in the marketing of value-added products. The important trends influencing the EU market for spices and herbs are:

- Increased demand for convenience food requires the food industry to add the required flavors to various prepared foods and this has led to an extensive range of ready-to-use spice mixes.
- The trend towards internationalization and increasing consumption of ethnic foods has created growing interest in spices, as well as a variety herbal products. As a result, a range of spices can be found in most consumer kitchens.
- European consumers have a strongly increased interest in a healthy lifestyle and, consequently, in the consumption of healthy food (e.g. herbal teas, Fig. 9.10). For example, specific spices and herbs are replacing sugars and salts, as well as artificial additives in prepared food, restaurant dishes and in home-prepared meals. Herbal teas are also becoming increasingly popular.
- The market for organic food as well as FairTrade food is increasing. However, the mainstream retail market for these certified herbs and spices is likely to remain relatively small until supermarket chains offer a full range of organic and FairTrade herbs and spices.
- The growing sense that raw materials in this market are scarce makes EU importers interested in developing long-term partnerships directly with suppliers that can meet their needs for a controlled supply chain. This creates an opportunity for suitably equipped suppliers to set up direct, and more lucrative, relationships with EU end users.
- Large actors in the EU herb and spice industry are working more closely with farmers and exporters in country of origin. They are facilitating contract packing for their retail operation (e.g. Tesco in the UK) or setting up/expanding their own packing facilities (e.g. Olam and Nedspice from the Netherlands). This creates opportunities for going into joint ventures with retailers or other buyers (CBI Market Intelligence, 2015b).
- The sophisticated EU market continues to be the main user of herbs and spices for the therapeutic, medicinal, personal care, wellness and cosmetics (TMPCWC) industries, with demand continuing to increase. This creates significant opportunities for suppliers who take the time to understand and make the correct connections with buyers and users in this sector.

Identification of opportunities for future growth in the EU market requires an understanding of the current market structure. The EU market is currently structured such that 15% of imports go to the food service sector, about 15% go to retail and 70–80% is sold to the industrial sector (Fig. 9.11, above). The spices are used for food and meal preparation in their current form, as delivered, or further processed to make derivatives which are then added to food. The value chain in the EU (i.e. the value of product at each stage of the chain of handling) is such that over 50% of the overall value is derived at the retail end of the chain in the EU (Fig. 9.12). What this means is that producers who can sell branded or contract-prepared finished product in the retail marketing channel can harvest significantly higher returns for their product than were they to sell raw materials into the EU market. Unfortunately, this is where most of the developing country herbs and spice for culinary use is now sold, except for selected exporters (e.g. from India, Trinidad and Tobago and Thailand) who have been able to sell packaged mixtures, other spices and

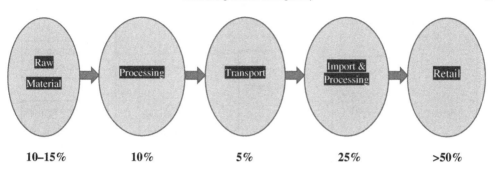

FIGURE 9.12 Relative value along the value chain in the EU culinary H&S market. *Adapted from CBI Product Fact Sheet: Spices & Herb Mixtures in Europe (CBI Market Intelligence, 2015b).*

sauces into the retail end of the EU and US markets, which translates to the higher prices/kg that they have been able to get for their product . While it may be unlikely that they can easily supply directly to large retailers like Tesco or Carrefour because of their requirements for just-in-time (JIT) delivery, volumes and other logistics, producers need to take advantage of the current trends and seek to supply as close as possible to end users, particularly in the EU market. In this regard, collective efforts with selected EU-based buyers can yield better returns to developing country suppliers. With both global and EU demand for herbs and spices set to continue to grow at or around 4—5% per annum (CBI Market Intelligence, 2015a), the time to invest in expansion and upgrading of developing country spice businesses is now.

Market channels in the EU for culinary spices

The market channels for culinary herbs and spices in the EU are standard and are captured in Figs. 9.11 and 9.13. Products either directly produced and delivered by the farmers or their representatives, or a processor who buys from them and processes in developing countries, are typically sold to an importer/wholesaler or to a broker who then on-sells to a processor or packer (Fig. 9.13). McCormick Europe, headquartered in Aylesbury in the UK, and Raps GmbH & Co. KG based in Kulmbach, Germany are among the largest traders and processors, respectively. In some cases, the processor or packer buys directly from the processor/exporter in the developing country (Fig. 9.13). In any case, this intermediary then sells to the industrial, retail or food service sector, as does the importer/wholesaler. Current trends and the growing importance of sustainability (e.g. organic, Fairtrade, Fairwild or Rainforest Alliance-certified product) suggest that this may be an area of opportunity for those Caribbean suppliers that can meet the requirements and/or get certified (Fig. 9.14).

The nutmeg & mace trade globally

For nutmegs, Indonesia, is by far the major exporter globally. Other major exporters are India, the Netherlands (a non-producing country), Sri Lanka and Vietnam, in that order,

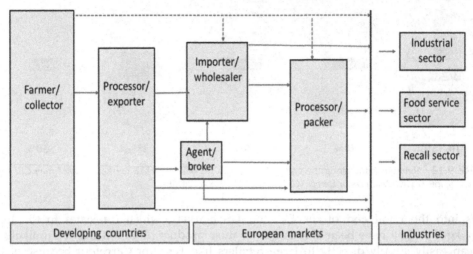

FIGURE 9.13 Overview of the marketing channel for nutmeg, Mace and other culinary spices in the EU Market. *Source: CBI Market Intelligence (2015a). Market Channels and Segments: Spices and Herbs (CBI Market Intelligence, 2015a).*

with Grenada now lying sixth in total exports, post-Hurricane Ivan (Fig. 9.9). The Caribbean nutmeg value chain, while typically getting better prices than the world market average or the average in the EU as shown for Grenadian nutmeg (Table 9.3), still got much lower prices than are available from some other buyers (see China at €27.26 vs. €11.13 per tonne for Grenada nutmeg in Table 9.3). The situation was even less favorable for Grenadian mace in 2014 with the product attracting €9.73 as against the €10.58 and €16.48 paid to Indonesian and Sri Lankan suppliers, respectively (Table 9.4). EU trade figures also indicate substantial opportunities for getting better prices from markets to which it re-exports herbs and spices, both within Europe and beyond (Tables 9.3–9.5). The same applies to crushed and ground mace (Table 9.5). A common consideration, however, should be the strict EU requirements for reliability and on-time delivery as well food safety, quality and compliance with regulatory limits (CBI Market Intelligence, 2015a,b), failure to meet which have cost processors significantly in the past (Fries et al., 2013; Gordon, 2016).

Case study: nutmeg & mace

Background to the nutmeg industry

The nutmeg (*Myristica fragrans* Houtt) is one of the more important spices that has found application in a wide range of culinary, food and beverage applications and also in the medicinal products industry. The plant is grown for two spices which are derived from the fruit: nutmeg which comes from the seed and mace which is the derived from the seed covering (Figs. 9.15 and 9.16). The seed is also a source of nutmeg essential oil

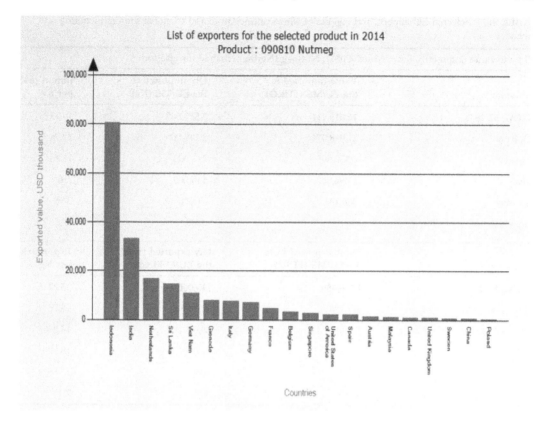

FIGURE 9.14 Global exports of nutmegs in 2014. *Source: Trade Map, International Trade Centre, https://www. trademap.org.*

and nutmeg butter, both of which have found applications in the food industry as well as personal care and the medicinal products industry.

What is generally known as nutmeg is the whole dried kernel of the seed or the powder derived from the ground dried kernel of "fragrant nutmeg" (*M. fragrans*). While the whole fruit is known as "nutmeg", the pericarp (the fruit covering − Fig. 9.15) is used mainly in the countries where the fruit is grown for culinary purposes, while the kernels ("the nutmeg") and the mace are mainly processed and packed for export. Once the fruit is mature, the fruit is harvested, the pericarp removed and the mace which covers the seed is also removed. The seed is then dried, shrinking the nutmeg kernel within the seed until they rattle when shaken. The shell is then cracked to harvest the kernel, this being done a variety of ways, including in the traditional way in Indonesia with a club. The nutmeg kernel and the mace can then be packaged for retail sale (Fig. 9.17), packaged in bulk or distributed for local use. Both nutmeg and mace have distinctive fragrances and unique flavors, making them the spice of choice in traditional recipes in their home countries and regions, as well as in cuisines around the world.

TABLE 9.3 Selected EU imports and exports of whole nutmeg from and to global sources including Grenada.

EU trade value & quantity for product 090811: Nutmeg (Neither crushed nor ground)

Indicators	Value imported to the EU/MS (EURO)	Qty imported to the EU/MS (Kg)	import price per Kg
TOTAL-EU28	26,015,911	2,569,000	10.13
Indonesia	21,908,074	2,119,000	10.34
Grenada	2,359,508	212,000	11.13
India	1,094,967	148,000	7.40
Sri Lanka	380,141	71,000	5.35
People's Republic of China	81,789	3000	27.26

Indicators	Value exported from the EU/MS (EURO)	Qty exported from the EU/MS (Kg)	Export price per Kg
TOTAL-EU28	1,756,359	115,000	15.27
Switzerland	511,178	40,000	12.78
People's Republic of China	500,066	28,000	17.86
United States	396,977	21,000	18.90
Norway	107,327	2000	53.66
United Arab Emirates	35,959	5000	7.19

The nutmeg tree

Nutmeg trees are tall spreading diecious evergreen trees (Fig. 9.15) which were endemic to the islands of the Moluccas and New Guinea in Indonesia and grew traditionally in the rainforest regions (Marcelle, 1995). The plant is also grown in India, Sri Lanka, Malaysia, North-Eastern Australia, Taiwan and the Pacific, including the Solomon Islands, Fiji and Samoa (Purseglove et al., 1981), as well as several islands in the Caribbean, with Grenada being the main producer in this region. The variety cultivated is mainly the Banda nutmeg, *Myristica fragrans* Houtt., although in Indonesia, there is some cultivation of other varieties, including the Papuan nutmeg, *Myristica argentea* Warb. In Grenada, the second largest producer, the Banda variety is the one that has almost exclusively been grown there for many years although in recent years, some Malayan plants have been imported and introduced to improve yields and shorten time to harvest (Gordon, 2012a,b).

Nutmeg trees are propagated both sexually (from seeds) and from grafts They typically take between seven (7) to nine (9) years after planting before they begin to bear and may take up to twenty (20) years before they reach full maturity and full productivity in terms of yield per tree. Among the issues that affect the agronomy of nutmeg is the inability to determine the sex of the tree before it flowers, resulting in propagation yields which are often 50%

TABLE 9.4 Selected EU imports and exports of whole Mace from and to global sources including the Caribbean.

EU trade value & quantity for product 090821: mace (Neither crushed nor ground)

Indicators	Value imported to the EU/MS (EURO)	Qty imported to the EU/MS (Kg)	Import price per Kg
TOTAL-EU28	6,439,355	597,000	10.79
Indonesia	5,520,741	522,000	10.58
Sri Lanka	642,845	39,000	16.48
Grenada	175,129	18,000	9.73
India	82,103	18,000	4.56
Hong Kong	6868	–	–

Indicators	Value exported from the EU/MS (EURO)	Qty exported from the EU/MS (kg)	Export price per kg
TOTAL-EU28	132,304	9000	14.70
Switzerland	97,269	7000	13.90
Indonesia	14,112	1000	14.11
Canada	8418	–	–
Kenya	5120	1000	5.12
Australia	2693	–	–

male (the non-productive sex) when planted from seeds. This had resulted in grafting being the preferred means of propagation as it is much more efficient and can be used to ensure that the tree at flowering is the female which will produce the nutmeg fruit. Addressing this issue can make a significant difference to the productivity and yields for traditional farmers and other value chain participants as it would significantly increase the availability of particularly the highly sought after varieties of delicately flavored nutmeg and mace.

Culinary uses of the fruit

The fruit covering (pericarp) is used to make jams and jellies in countries in the Far East and in the Caribbean where the fruit is grown. In Indonesia, it is sliced and made into sweets known as *manisan* (a fragrant fruit candy) or desert known as *maisan pala* while in Penang, Malaysia a similarly sliced, dried and sugar-coated product is used as a topping on Penand *ais kacang*. The fresh pericarp (rind) is also made into juices in Indonesia, Malaysia, the Caribbean, and India, among other countries, being used directly as harvested or boiled, with sugar added, depending on the type of beverage that is being made. It is also used for chutney and pickles in India.

Ground mutmeg's distinctive flavor, taste and olfactory stimulation makes it ideal to flavor a wide range of products. These include confectionary, meats and meat products,

TABLE 9.5   Selected EU imports and exports of crushed or ground Mace from and to global sources.

EU trade value & quantity for product 090822: mace (crushed nor ground)

Indicators	Value imported to the EU/MS (EURO)	Qty imported to the EU/MS (kg)	Import price per kg
TOTAL-EU28	4,139,338	368,000	11.25
Indonesia	2,996,031	260,000	11.52
Vietnam	882,714	90,000	9.81
Sri Lanka	171,995	7000	24.57
India	76,265	8000	9.53
Hong Kong	11,084	3000	3.69

Indicators	Value exported from the EU/MS (EURO)	Qty exported from the EU/MS (kg)	Export price per kg
TOTAL-EU28	684,839	47,000	14.57
United states	374,285	36,000	10.40
Norway	89,851	5000	17.97
Switzerland	59,037	3000	19.68
China, People's Republic of	46,072	1000	46.07
Chile	24,002	-	-

FIGURE 9.15   Nutmeg trees and the nutmeg fruit with the pericarp (top arrow) and Mace (bottom arrow) surrounding the nutmeg Seed. *Source: A. Gordon (2016).*

including sausages, sauces, spice blends, beverages, including egg nog, cakes and other baked goods, among other foods. While having similar sensory properties to ground nutmeg, mace has a more delicate flavor and is often used for the unique coloring that is provides to dishes. In European cuisine, mace is used in rice pudding and, along with

FIGURE 9.16 Mace covering the outer shell of the nutmeg Seed. *Source: A. Gordon (2016).*

FIGURE 9.17 Adding value to nutmeg and Mace by packaging for consumer use.

nutmeg, in meat and potato dishes and meat products, baked items, soups and sauces. The Italians use nutmeg in tortellini, meat-filled dumplings, meatloaf and pumpkin pie while the Dutch add it to vegetables, mulled wine and, of course, egg nog. The traditional Scottish meat dish, haggis, benefits from the use of both nutmeg and mace while Caribbean people incorporate grated nutmeg into egg nog, cakes and baked goods and a range of alcoholic beverages, particularly those made with rum. In the East, Indians use nutmeg in meat preparations and other savory dishes, desserts, in *garam masala* in Kerala, Mughlai and other regional cuisines while soups such as *soto*, *oxtail soup*, *basko* and *sup kambing* and the meat dishes *bistik lidah*, *bistik* and *rolade* are among the dishes made with nutmeg in Indonesian cuisine.

Impact on health

Nutmeg has been reported to be an effective antifungal agent (Pande, 2010; Rodianawati et al., 2015) and also to have cytotoxic properties against specific cell lines of cancer cells (Martins et al., 2014), with the essential oil extract and myristicin being involved in mediating these effects (Piras et al, 2012: Thuong et al. 2014). Nutmeg has also been shown by Takikawa et al. (2002) to be effective in reducing the populations of some types of pathogenic *E. coli*. These include enterohaemorrhagic *E. coli O157* and enteroptahogenic *E. coli O111*, with β-pinene identified as a major active ingredient (Takikawa et al., 2002). Nutmeg was however found not to be effective against other *E. coli*, notably enteroinvasive *E. coli O29* and *O124* and enterotoxigenic *E. coli O6* and *O148*. Nutmegs have also been found to have analgesic properties (Hayfaa et al. 2013), to have the potential to play a role in reducing obesity through AMP-activated protein kinase (Nguyen et al., 2010) and to exhibit significant and sustained increases in sexual activity in male rats (Tajuddin, et al., 2005), one of the traditional uses for which nutmegs are used in Indonesia, India and other parts of the Far East.

Despite these potentially beneficial properties, there have been various studies that have reported toxicity associated with nutmeg consumption. Nutmeg has been associated with various toxic (tachycardia, nausea, vomiting and agitation) and psychoactive effects if abused. The latter include hallucinations, delusions and feelings of euphoria typically associated with the use of some narcotic drugs (El-Alfy et al., 2009; Quin et al, 1998), with myristicin, a major component of nutmeg being identified as the major active ingredient. The effects have been found to vary, depending on the quantum of the myristicin consumed in the seed, powder or extracted oil. It has been reported that overconsumption of nutmegs has resulted in at least two fatalities (Stein et al., 2001; Demetriades et al., 2005), although one of these included the consumption of a toxic dose of flunitrazepam (Stein et al, 2001), making it difficult to attribute the fatality to nutmeg consumption. Consequently, the association between nutmeg consumption and potential fatality is inconclusive, even when taken as part of a cocktail of psychoactive agents and narcotic drugs. This observation is supported by studies, some involving extensive case reviews that have shown that even at elevated levels of consumption, life-threatening situations have not been observed (Castairs & Cantrell, 2011; Ehrenpreis et al., 2014; Stein et al., 2001).

Grenada nutmeg & mace industry development case study

The Grenadian nutmeg industry

The nutmeg plant was introduced to Grenada in the 1843 from Banda in Indonesia, with the first commercial plantations being established in the 1850s and Grenadian nutmeg first being available on the world market in significant quantities 1865 (Marcelle, 1995). The variety cultivated in Grenada is mainly the Banda nutmeg, *Myristica fragrans*

Houtt, although in recent years, some Malayan plants have been imported and introduced to improve yields and shorten time to harvest. In Indonesia, there is some cultivation of other varieties, including the Papuan nutmeg, *Myristica argentea* Warb.

The nutmeg industry in Grenada had, for many years, been the second major supplier of nutmeg to the world market after Indonesia (Fries et al., 2013) up until Hurricane Ivan in 2004, after which it is now the sixth largest exporter, as is evident from the data from 2015 (Fig. 9.14). Grenada is also the only significant supplier of West Indian nutmeg to the global market although St. Vincent & the Grenadines, St. Lucia and Jamaica also grow some nutmegs. While volumes have fluctuated dependent on the vagaries of the industry and significant weather-mediated events, Grenada has remained a major supplier to the EU, US and, to a lesser extent, Canada (Fries et al., 2013), with the EU taking the majority of the products exported. The industry is run by a cooperative, the Grenada Cooperative Nutmeg Association (GCNA) which, by law, provides technical support, marketing, processing, export services and industry development services to nutmeg farmers across the country. GCNA works with the Ministry of Agriculture and the Grenada Bureau of Standards (GBS) to ensure that Grenada produces and exports high quality, safe nutmeg and mace that meet the requirements of their market. Nutmeg (Fig. 9.18) is grown around the island and is collected at Receiving Stations which are located in major producing centres. The nutmegs are then taken to Processing Stations, some of which also double as Receiving Stations, at which the mace is removed graded using a pictorial chart (see Figure 4.8, Chapter 4), processed and packaged as shown in Fig. 9.19, below. The nutmeg seeds are dried, processed and packed, as described for Grenada Unsorted Nutmegs (GUNS), below (Fig. 9.20).

FIGURE 9.18 The nutmeg fruit showing the pericarp and the Mace. *Source: A. Gordon (2016).*

FIGURE 9.19   Production of high-quality number 1 Mace in Grenada (A and B) Mace on nutmeg, (C) removing, sorting, cleaning & grading Mace, (D) placement in Jute bags for Transportation, (E) sifting to remove foreign matter, (F) packaged # 1 Mace.

Process Flow Diagram for Grenada Unassorted Nutmegs (GUNS)

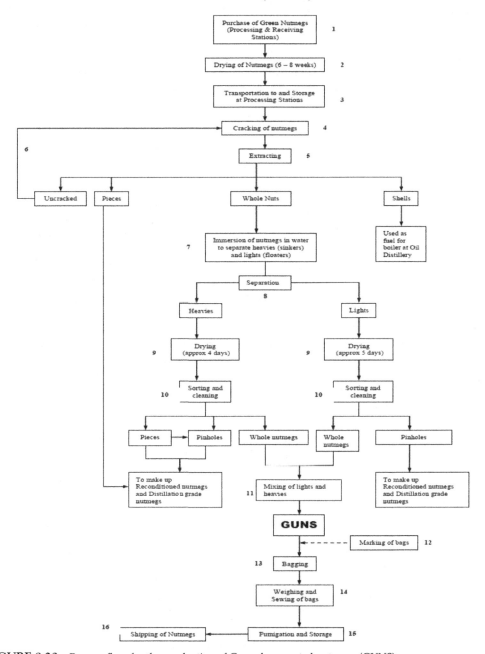

FIGURE 9.20　Process flow for the production of Grenada unsorted nutmegs (GUNS).

The processing of nutmeg and mace involves both products undergoing quality assurance, prior to shipment, a process which has been significantly upgraded as part of the interventions under a series of programmes to improve the viability of the industry. This has further augmented the reputation of the Grenadian products, about which it has been said *"As for quality and food safety testing, even without sophisticated laboratory infrastructure and services, European buyers report confidence in the consistent quality of Grenada's nutmeg ... based on years of successful trade relationships, as well as the fulfillment of additional testing requirements for European standards at the port of destination"* (Fries et al., 2013). For nutmeg exports to the United States, aflatoxin testing is required with samples traditionally being sent to the US for testing, a situation being changed by upgrading of capabilities locally.

Description of the production of Grenada #1 and #2 mace

Grenadian mace is sold on the international market as Grenada #1 and Grenada #2 Mace, based on its quality, as determined by industry-accepted criteria. The categorization is largely based on the integrity and color of the aril which covers the seed (i.e. the mace itself). Number one (#1) mace is a bright red and has arils, the great percentage of which are largely whole while Number two (#2) mace has a less bright color and/or arils which have a higher percentage of broken arils, even though they may have the characteristic color. Where the color of the mace has faded to a yellow-red hue, this mace is still usable as long as it has not begun to deteriorate or become moldy. Such mace is categorized and sold as #3 Mace. Characterization and grading is done in Grenada by way of experience and pictorial charts.

Mace is produced as shown in Fig. 9.19 which is a mixture of traditional production and the more commercial scale production that characterizes the industry. The aril (mace) is harvested from the nutmeg seed, cleaned and sorted by hand (Fig. 9.19 C). Hand cleaning is done as shown traditionally at homes or on tables in Receiving Station and Processing Plants. The mace is then spread out to dry for between 10 and 14 days, after which it is placed in bags (Fig. 9.19D). It then undergoes a final examination, sifting and selection (Fig. 9.19E) after which it is packaged, either in bulk or in retail packages (Fig. 9.19F). In Grenada, mace was typically harvested and dried in communities and then transported to the Receiving Stations for grading and sale. Much of this practice has changed with time, as described below.

Description of the production of Grenada nutmegs

Nutmegs are harvested from trees or picked up (if they had fallen), removed from the seed covering and placed in jute bags or sacks (typically on the farms) and delivered to the Receiving/Processing Stations where they are air dried on racks for 6–8 weeks (Fig. 9.20 and 9.21A and B). They are then taken to the Processing Stations (or Processing Area), and passed through a machine which cracks open the nuts (Fig. 9.22A), after which they are removed from the shells by hand (Fig. 9.22B) and sorted into four categories as shown in Fig. 9.20. The shells and shrunken kernels (whole nuts) are shown in Fig. 9.23. The whole nuts are then separated by immersion in water which differentiates the "heavies" from the "lights", both of which are then dried and further sorted to separate the whole nutmegs from those which are broken or have pin holes. The heavy and light nutmegs are then mixed if Grenada Unsorted Nutmegs (GUNS) are being produced

FIGURE 9.21 (A) Purchasing and (B) drying of nutmegs at a receiving station in Grenada. *Source: André Gordon (2016).*

FIGURE 9.22 (A) Cracking and (B) extracting (sorting) of nutmegs in Grenada. *Source: André Gordon (2016).*

FIGURE 9.23 Nutmeg shells and kernels (the Nut) after sorting. *Source: André Gordon (2016).*

(Fig. 9.19) or kept separate if being delivered to markets that specifically want only light or heavy nutmegs. Whatever the grade, the nutmegs are then placed in jute bags (Figure 24A) and fumigated to eliminate any pests which may be present among the nutmegs (Figure 24B). The product is the delivered a central warehouse for further quality assurance checks, warehousing and eventual delivery to market (Fig. 9.19).

Technical interventions in Grenada

There have been many interventions in the Grenadian nutmeg industry over the years to improve yields, quality, husbandry, food safety and also to diversify the range of products derived from the fruit. This led to the introduction of new varieties of nutmeg (the Malayan), the establishment of a central nursery for nutmeg trees, an oil extraction facility at Marli and diversification of product streams, as well as differentiation of nutmeg products into the grades currently offered to the international market (Marcelle, 1995; Fries et al., 2013; Gordon, 2012a). All of these have resulted in a much more diversified and differentiated industry today, even with a significantly reduced volume of production as the sector is still recovering from hurricane damage in 2004 and 2005, more than 15 years ago. The author has been involved in several of these interventions[3], including introducing the sector to FSQS in 1997 and subsequently working with it on systems implementation and upgrading of regulations, legislation, market identification and development, as well as product differentiation. This case study focuses on the more recent series of interventions and work that has helped to build on the foundation for the industry laid through earlier interventions. This can prove instructive for other areas of the spices sector as they seek to adapt a traditional industry to meet the current realities of the global marketplace where FSQS, inclusive of full traceability back to each field, product knowledge, product differentiation and diversification of income stream are critical for viability.

A series of interventions took place over the period 2012–2014 as a part of the Grenada Nutmeg Industry Assistance Programme funded by the Centre for Development of Enterprise in Brussels and Compete Caribbean and the Government of Grenada's programme for rehabilitation and modernization of the nutmeg industry. Under these programmes, specific interventions were undertaken to address the following technical aspects of the industry, among other market, legislation and investment-related considerations:

1. *Food safety assessments and development of upgrading programmes* for selected export-oriented entities, including GCNA and other value chain participants, including a private oil extraction facility and manufacturers of food and medicinal products made from nutmeg
2. *Building the technical capacity* to support analytical work and research in Grenada in order to improve the capability of the industry to comply with export requirements and market trends

---

[3] These were done through Technological Solutions Limited (TSL), the authors' firm that provides a range of technical, research and analytical support services to the productive sectors in developing countries.

3. Use biotechnology to assist the Grenadian nutmeg industry in being able to *characterize and differentiate its nutmegs and nutmeg products* from its competitors
4. Intervening along the value chain, including with farmers to *improve their understanding of HACCP[4]-based food safety, of Good Agricultural Practices (GAPs)*, international requirements and the industry as a whole.

These interventions continued the process of building on and consolidating the gains in FSQS and practices in the industry as a whole and sought to lay the foundation for attaining sustainable competitiveness through a series of carefully implemented, FSQS-based reforms and transformations. The approach to the overall project and outcome of the several interventions is described below as well as the impact they have had on the industry.

Value chain interventions to improve good agricultural practices (GAPs) & compliance with export requirements

The approach to the series of interventions involved first working with members of the industry value chain, including the public sector stakeholders, farmers, exporters, private manufacturers and the GCNA to define the expected outcomes. Discussions were held and detailed technical interventions done with all of the critical stakeholders in the sector and all worked collectively towards agreed goals. These discussions were also important as they helped to have all beneficiaries understand the technical work that would was required to be done as well as the changes that would be required of them as the interventions progressed. The activities involved were:

1. Identification of the opportunities for gaining better leverages and positioning in the marketplace through differentiation of Grenadian nutmeg and mace. This was tied to the need for a focussed marketing and branding programme, supported by technical information to facilitate the differentiation and branding;
2. Gathering detailed technical information on the existing variation among local nutmegs and specific nutmeg varieties with favorable horticultural and commercial traits; these were then related to the genetic profiles of each cultivar through the use of DNA fingerprinting technology;
3. A review of the existing capacity to do ongoing research on various nutmeg varieties and their characteristics with recommendations as to how this could be organized;
4. A review of existing capacity for the testing of nutmeg and mace for export and for monitoring quality, including the capacity for sensorial and physical testing. This included a review of the laboratory capacity at the GCNA, the Grenada Bureau of Standards (GBS), and others;
5. Undertaking a programme of gap assessments at different levels of the nutmeg value chain to identify the needs of value chain participants in ensuring compliance with market entry requirements and, where relevant and possible, certification to meet the stipulation of buyers;

[4] HACCP – Hazard Analysis Critical Control Point.

6. Identifying the most relevant food safety standard for market access for nutmeg in the EU market. The British Retail Consortium (BRC) Food Standards Version 6 (at the time[5]) was recommended;

7. Group training/sensitization and consensus-building for farmers gave them technical information regarding food safety standards, advised them on the status of the industry and the international market, and sought to gain their support for focus on production for value-added exports. This aspect of the project also involved training trainers who will be able to further provide training to farmers in Good Agricultural Practices (GAPs). They were also provided with training tools to assist with this process.

8. Training for processors in Good Manufacturing Practices, HACCP and EU and US regulatory requirements to access their markets, as well as the requirements of compliance with Global Food Safety Initiative (GFSI) Standards.

All of these activities were encapsulated under three broad sub-programmes:

- Food Safety Assessments & Guidance to Selected Export-oriented Entities (#5, #6, #7 and #8, above).
- Strengthening the Value Chain and Building the Technical Capacity to Support Analytical Work and Research (activities #3 and #4 above)
- Characterization and Differentiation of Nutmeg (activities #1 and #2 above)

The outcome of these and their impact on the Grenadian nutmeg industry follows as an example of how approaches such as these can be used to transform spice food industry sectors.

Food safety & quality system assessments & guidance to selected export-oriented entities

Programmes assessed

Exporters of Grenada nutmegs (GN), Grenada mace (GM) and value-added products need to comply with food safety requirements to access various European Union (EU) and other markets, as well as to secure remunerative business with buyers in these markets as far up the value chain as possible. As such, a series of assessments were done against specific globally accepted food safety and quality standards depending on the entity being assessed and what was best suited for their existing capabilities, products and preferred markets. Assessments were done of five (5) selected Receiving and Processing Stations owned and operated by the GCNA. The facilities at Grenville, Gouyave, Union, Hermitage and Marli were the ones that were assessed, the facilities at Gouyave and Grenville being both Receiving and Processing Stations and the one at Marli being where the GCNA-run oil extraction facility was located. Assessments were also done for the value-added producers West India Spices (WIS), the developers of the technology and owners of a patent for extracting the essential oil from nutmegs (Webbe et al., 2003) and De LaGrenade Industries, makers of a range of nutmeg-based sauces, spices, jams and jellies. Two firms involved in

---

[5] This standard was chosen based on advice of the participants in the sector who had been so advised by their buyers. At this current time, other standards such as SQF and FSSC 22000 may be considered also as viable for the value chain, along with the current BRC version.

the medicinal products sub-sector, Belzeb and Noelville Limited were also assessed. The assessments were against GMPs, general HACCP-based food safety systems requirements and, where relevant, a specific, identified GFSI standard (in this case BRC, version 6). A brief assessment against the requirements of the United States Food and Drug Administration (FDA) Food Safety Modernization Act (FSMA) was also done. The assessments resulted in programmes of development for each operation to facilitate compliance, implemented with varying degrees of success by the different entities.

Based on the nature of the businesses and their markets at the time, it was recommended the entities seek to get British Retail Consortium (BRC) certification for their operations. BRC is among the globally accepted Global Food Safety Initiative (GFSI) systems and appeared to be best suited for most of the Grenadian operations. Although many firms were operating in general compliance with GMPs, none of them had pre-existing ISO 9000 based-systems in place, nor any other *certified* food safety or quality system. For the GCNA facilities, it was determined that the facility at Gouyave and Marli were best placed to first seek Good Manufacturing Practices (GMP) certification, and then full GFSI (BRC) certification. The detailed findings for each entity was provided to the GCNA for future action. The overall findings are summarized below.

Overview of findings

General observations GCNA facilities

While the operations had systems in place, including elements of a HACCP programme for some entities, at the time of the assessment none had documentation and operations that would fully meet the requirements of the FSMA, HACCP or BRC. Production-related records, including traceability, were in place at some facilities but the documentation was insufficient to facilitate the implementation or operation of a fully compliant food safety system. The records available at these facilities were mainly commercial records.

The facilities had some of the required prerequisite programmes in place in terms of the practices but with insufficient or deficient documentation that did not meet the requirements. Screening, signage and security of the premises were deficient as were document control systems at some of the facilities. The employees performed their duties as regards the handling of nutmeg with acceptable proficiency, but prior to the intervention, many practices that breached basic GMPs and food safety were observed. Housekeeping at several facilities needed improvement, both internally and externally.

Areas requiring attention and actions to be taken were identified for each of the facilities, including the need for improved documentation in the following areas:

Good Manufacturing Practices
Sanitation and Hygiene
Standard Operating Procedures
Pest Control
Preventive Maintenance
Traceability and Recall
Coding and Labeling
Customer Complaints Management
Internal Auditing

Materials Safety Data Sheets (MSDS)

Sampling Programmes (in support of the Quality Assurance & Food Safety programmes)

Several programmes were identified as needing further development and more comprehensive documentation including their:

Quality Assurance Programme
Training Programme
Food Defence/Food Security Programme
Food Safety Programme
Allergen Management Programme
HACCP Programme

Some of the other areas requiring attention were the development of schematic diagrams of the facilities as well as flow diagrams to show the movement of material and personnel through the facilities. Waste management (processing dry waste), staff training and record keeping were among other areas identified for improvement.

Industry-wide practices

The nutmeg industry globally has remained traditional in its practices and Grenada at the time of the interventions was no different. Nutmeg and mace were being handled outside of the processing facilities in the manner they had always been, with several practices that would not comply with food safety best practices, such as the traditional sorting and drying of mace on a tarpaulin covering the floor of a building (Fig. 9.19C). Other traditional practices requiring change included putting bags of product directly on the floor (Fig. 9.21A and 9.24A) and the attire of staff in contact with, or in the vicinity of product prior to packaging (Fig. 9.19E). The transportation of nutmeg and mace to the facilities in open vehicles, handling practices in general and storage practices that could predispose affected product to mold growth were among

FIGURE 9.24   (A) Bagging of nutmeg and (B) packing in fumigation chamber. *Source: André Gordon, 2016.*

other sub-optimal industry practices identified. Many of these practices are traditional in countries across the world where nutmeg and similar tree crops (e.g. cashews, Chapter 4) are grown and handled. The objective, therefore, was to identify and seek to change these practices.

Traceability

The process of reviewing the system of traceability of the fruit throughout the industry from farms to export was undertaken during the series of interventions. Traceability back to the farm and eventually to the cultivar was not possible at key processing facilities at the time of the initial intervention as all data were being collected manually and did not include sufficient detail to trace back to the location on the farm and specific trees from which the nutmeg came. This was of particular importance as some buyers (e.g. a German retailer) wanted product from farms and specific trees on those farms that had desired characteristics for the finished product being offered for sale in the destination market. Systems had been established at selected facilities to make this possible but had not been adopted elsewhere.

The requirement for traceability back to specific farms, quadrants on farms and, in some cases, specific trees (with GPS locational markers) is important not only for compliance with FSQS requirements, including GAPs, but also for commercial reasons. This process was identified as being required to facilitate specific buyers wanting products with specific characteristics (as in the case of the German firm, discussed above) and also to facilitate certification of authenticity. It was recommended that the industry management system include upgrading of systems to implement better external and internal traceability and augment this by the use of technology. Each receiving station was recommended to be fully computerized as part of a network linked to the head office of the GCNA a process which had started by the time the series of interventions concluded. In addition to this, specific training of receiving clerks, labeling and pre-planned allocation of storage, racking, drying and holding areas were also identified as necessary to facilitate full traceability. The Good Agricultural Practices (GAPs) training component of the project then sought to lay the groundwork for supporting the changes required in the way that nutmegs (including mace) are handled and tracked such that effective traceability could be established.

The upgrade of the value chain partner firms

The companies assessed during the series of interventions made differing levels of improvements towards compliance with import regulations and being able to attain certification. WIS was the firm that was the most aggressive in implementing the changes recommended throughout its operations. They implemented the documentation and systems improvements recommended, significantly enhanced their physical environment and put themselves in a position to be fully FDA and FSMA compliant. By the middle of 2013, WIS was able to attain GMP certification of its systems, practices and facilities by a European-based, global certification body, setting the stage for other processors to follow.

Specific interventions to strengthen the value chain

Compliance with export requirements: GAPs, GMPs, HACCP & capacity building

Improving long term sustainable market outcomes required the transformation of a traditional industry to adapt and use technology in positive ways compatible with the culture. As has been discussed in detail by Yiannas (2008), culture is a critical determinant of success in the sustainable implementation of food safety and quality systems. It was therefore of importance to identify key aspects of this culture and seek to transform it across the industry to a more positive food safety culture through targeted interventions along the value chain. In this regard, success is dependent not only on the technical competence and knowledge of the change leaders, but also in their ability to engage stakeholders in a traditional agri-food based industry such that they agree to and lead the change themselves. This required a series of carefully managed sessions which included information sharing, discussions and collaboration building initiatives, involving multiple stakeholders and the GCNA, the backbone of the industry.

The initial focus was on building consensus as to the outcomes desired that would be in everybody's interest, including the farmers. This was followed by active engagement around Global Market Trends and Opportunities (such as some of those discussed in this chapter), followed by a series of sessions on Nutmeg and Mace Quality Considerations for Global Markets, Good Manufacturing Practices, Good Agricultural Practices, Food Safety & Quality Systems, and basic EU and US regulatory requirements. These were used to sensitize the farmers to GAPs and the current realities in the industry globally. They were done through very hands-on, lively interaction sessions with single farmers or groups of farmers in different fora, as well as staff members, managers and board directors for the firms and organizations involved, as required. This included the Ministry of Agriculture (MOA), GCNA, the GBS, the Spice Research Project (SRP) team and all of the firms involved. While the objective was to engender a willingness to make the changes required to assure compliance with regulatory and market requirements, this never appeared as the major focus of the discussions. The interest and benefit to the participants was highlighted. The series of meetings and discussion were very useful in advising the approach that was taken as well as updating the farmers and stakeholders and getting their buy-in for the approach and the objectives. The success of the approach was evident in the transformation that was brought to overall practices in the industry: today, you will not see nutmeg or mace being handled as it was when the process started and Grenadian nutmeg and mace are among the most compliant with industry requirements.

Building the technical capacity to support the industry

In order for GN and GM to be able to consistently meet all industry requirements and command the kind of returns required to improve the viability of the industry, it was necessary to ensure that effective and adequate scientific and technical (S&T) support was available to support the industry value chain. As part of the project, an assessment was done of the S&T requirements of the industry and gaps identified. These were then used to direct a programme of capacity building in support of the industry, ensuring the focussing of resources (avoidance of unnecessary duplication in a developing country with limited resources) and also building of collaboration to ensure the effective use of S&T capacity, once built.

Capabilities were identified as residing within the MOA, SRP, GCNA, the GBS and, among the industry stakeholders, at West India Spices (WIS). The nature of the specific capabilities and skill sets were documented and decisions made about best use of these skills, following which the parties were engaged and the resources deployed to achieve the best possible outcomes. It was identified that technical skills were required in the identification and production of preferred cultivars as planting and propagation material and that appropriately trained agronomists were needed to support the farmers in the sector. These were already available and were met through the SRP team, as indicated earlier. The GBS and WIS had the equipment (Fig. 9.25) and personnel who could undertake the analyses required for aflatoxin, microbiological contamination and several of the key components of nutmeg and mace, with the appropriate training. It was also recognized that international laboratory accreditation to ISO 17025:2005 would be required for whichever laboratory would be doing the analytical work. The GBS was therefore supported to embark on a programme to attain certification, which continued to be supported by the GOG and which was successfully completed in 2016*. The outcome was support for the sector based on the capabilities that exist at the GBS which included the ability to do the key analysis required for determining the sensory and chemical characteristics of the nutmeg and mace, as well as undertaking routine analysis for aflatoxin. This has put the country in a position to get critical aflatoxin results prior to shipment and to verify those provided by buyers for product arriving at port of entry.

Another important component of the building of the sector was to develop the capability to implement a certification programme that could help to differentiate GN and GM in the marketplace. This programme would hinge on identifying specific characteristics of GN and GM that distinguish them from others. Among the interventions, therefore, was a component that sought to provide this information based on DNA fingerprinting of and detailed physicochemical and phenotypic characterization of GN (described below). The detailed characterization of Grenada nutmeg was done and can be used to form the basis for a certification

FIGURE 9.25   High performance liquid chromatograph for analyzes of nutmeg quality.

system for authentic Grenadian nutmeg which can be registered and protected internationally. This programme can be administered by the GBS which already has the capabilities to implement such a programme on behalf of the industry. A mechanism for getting this to be able to work for the benefit of the sector remains to be developed.

Improved collaboration along the value chain

An important objective of the interventions was to get deeper collaboration among players in the sector who represented key nodes of the value chain. This was to ensure that the industry as a whole could supply its markets with the optimal amount of high quality, safe and fully compliant product. The project sought to and was successful in creating a structured, working relationship among value chain participants that didn't exist before. It also sought to leverage existing strengths and capabilities for the benefit of the entire industry. Technical support in the area of research was provided by the Ministry of Agriculture mainly through the Spice Research Project (SRP). The approach was also successful in facilitating a collaboration between Baron Foods Limited, a major local producer and exporter of dried spices and the GCNA. This resulted in Barons using more Grenadian grown nutmeg in its dry blended products. The partnership approach extended to brokering an agreement between WIS and GCNA for the supply of nutmeg to facilitate the making of nutmeg oil, including a contract packing component. This ensured that Grenadian nutmeg oil was more readily available for Noelville and Belzeb, other manufacturers in the value chain, to be able to use the oil in their products to the benefit of the industry. The potential impact of this on the viability of the industry is addressed by the following:

> "The markets for essential oils, oil resins and extracts are growing in importance in the industrial seasonings sector, but also in cosmetics, toiletries and pharmaceuticals. This could provide opportunities for suppliers of raw materials, spices and herbs. Another option is to invest in extraction facilities to enter this market." (CBI Market Intelligence, 2015a)

Chemical, phenotypic and genetic characterization and differentiation of nutmeg

The objective of this study was to assess of the characteristics of Grenadian nutmeg and relate them to the genetic profiles of specific cultivars to advise the characteristics that are unique to Grenadian nutmeg. This was intended to assist the Grenadian nutmeg industry to be able to specifically differentiate its products from its competitors based on its genetic and phenotypic characteristics. The approach to this aspect of the project involved collecting a range of nutmeg, mace and leaf samples, which specimen were then evaluated in terms of their phenotypic characteristics, their chemical composition and their genetic profiles to tie, as best as possible, specific characteristics to specific nutmegs from specific trees. Further study could also assist in identifying specific characteristics that confer special attributes (e.g. yield, thickness of mace covering, size of kernel, etc.) to the nutmeg. The overall results of these studies are presented below and were to form the basis on which further differentiation and selection of Grenadian nutmeg is done for certification/branding purposes and also to assure better agronomic results.

Technical information on varieties

Detailed technical information on the existing variation among local nutmegs and specific nutmeg varieties with favorable horticultural and commercial traits was gathered by the technical consultants working closely with the Spice Research Project (SRP) team who were responsible for technical and agronomic assistance to the farms and propagation of the plants for grafting. These trained agronomists had good knowledge of the nutmeg fruit and the industry and were instrumental in helping to decide the characteristics on which to focus. They have the capability to provide the necessary technical support to the industry for propagation and varietal selection, and have done considerable work in this area. Samples of nutmeg (leaves, nutmeg and mace) were collected from farms all over the island for determination of their DNA profile (DNA fingerprinting). This information was used to identify differences between nutmeg trees that produce fruit with different favorable characteristics, and generated data that could be used in a national marketing programme. This process involved identification of specific primers that were able to differentiate between nutmeg varieties and also developing approaches that would allow for the identification of related trees as a result of the DNA fingerprinting that was done. DNA fingerprinting produces unique bar code-like patterns that allow the distinguishing of one individual (plant) from another (Selvakumari et al., 2017). The outcome of the many studies and analyses of chemical and physical characteristics of the nutmeg varieties selected was the identification of unique characteristics for the nutmeg trees which can be characterized as "typically Grenadian". This information in detail has been made available to the SRP who are custodians of the information for the Ministry of Agriculture, Grenada (MOAG).

Some of the specific special characteristics identified that form part of this Grenadian identity include:

- Low saffrole and myristicin levels as compared to other nutmegs (from the Far East).
- High essential oil content
- Good thick mace coverage
- Large seeds
- Good productivity
- Low aflatoxin levels (husbandry-related)

These characteristics are derived from the original (Banda) variety, information that the SRP has. The recommendations from this work were that Grenada should craft and undertake an ongoing research programme that would produce critical information be used to guide the transformation of the industry through varietal selection and selective propagation. This would continue and better target the work already being done by the Ministry of Agriculture's SRP. This process should allow the SRP to continue to refine the varieties it has targeted based on the data generated.

Relationship between cultivars

The DNA fingerprinting of nutmeg leaf samples showed co-sharing of several DNA bands (Fig. 9.26). Only primer OPW-01 was able to identify unique DNA bands in samples

M 1 2 3 4 5 M 6 7 8 9 10 M

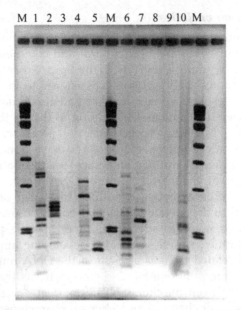

M: Marker (1 kb DNA ladder); Lane 1: E0161102; Lane 2: E016201, Lane 3: E016202, Lane 4: N0180, Lane 5: N01450, M: Marker Lane 6: N01530' Lane 7: (HMF) WSTC001M1, Lane 8: WSTC003F1, Lane 9: WSTC002F1, Lane 10: NTC, M: Marker (See Appendix 1 below).

FIGURE 9.26 Example of a RAPD profile of DNA isolated from leaves of *M. fragrans* with primer OPW-01 (5′CTCAGTGTCC3′). *Source: Adapted from Gordon (2012b).*

E016102 and E016201. These trees are between 70 to over 90 years old and represent the trees which are closest to the original Grenadian nutmeg trees and have characteristics that are "typically Grenadian". Their unique DNA fingerprint patterns should therefore form the basis of the unique differentiation of Grenadian nutmegs from those of nutmegs of other origin. While both have phenotypic differences, those genetic characteristics identified, as well as specific phenotypic characteristics can form the basis of this system of differentiation.

The following were some of the characteristics of selected varieties used for the analyses. The data presented is just for illustration purposes only, although the myristicin and saffrole contents are as determined (Table 9.6).

The data for relationships showed a strong similarity between E0002, E0001 and N0018, both varieties showing a 80–90% relationship to N0018. While E0002 and E0001 were similar, they were not of the same age and displayed differences in phenotypic characteristics that would arise from their difference in DNA profiles.

The DNA fingerprinting study was successful in identifying some specific characteristics of typical Grenadian nutmeg and in relating these to phenotypic and chemical profiles. This will allow the Grenadian industry to be able to develop a range of branding and agronomic programmes that will redound to their benefit in the future.

TABLE 9.6 Characteristics of selected Grenadian nutmeg varieties used in the study.

Descriptor	Origin	Saffrole	Myristicin	Region
N0001	Malayan	0.025	8.557	North
E0002	Original (Banda)	0.230	1.541	East
E0001	Original (Banda)	NA	NA	East
N0030	Malayan	0.097	0.043	North
N0018	Uncertain (55 yr)	0.064	0.422	North

Technical support for branding programme

The information derived from the biotechnology-based and other research that was done on various nutmeg varieties has provided detailed, specific and traceable information for the industry in Grenada. Because of how the research project was designed, all of the data is traceable to specific trees at specific locations with specific, defined and verified characteristics. It is therefore not only possible to use this additional information in breeding and propagation programmes, it will also be critical in supporting the process of potentially developing a certification mark programme. In addition to already being widely recognized for its' quality and flavor characteristics, this work has laid the basis by which Grenadian nutmeg producers can definitively differentiate their product from other nutmegs. It provided the foundation on which a science-based future branding programme can be built. This will, however, require future work to refine the findings and recommendations.

Appendix 9A Studies on genetic diversity (DNA fingerprinting) of Grenadian nutmegs — PCR and RAPD methodology

Introduction

This study was undertaken as part of a process to identify the characteristics that are typical to Grenada nutmeg for further use in the development of specific markers that can be used to differentiate these nutmegs from others being offered for sale on the international market. A fully scientific approach was followed as outlined below.

Materials and methods

DNA isolation

Plant material - leaves

Fresh leaves from adult male and female trees of nutmeg were collected from plantations in Grenada and stored on ice until reaching the laboratory. The leaves were dipped in liquid nitrogen and stored at −80 °C until extraction of DNA. The leaf DNA was extracted using the CTAB method. Below is detailed the approach to be followed during the extractions:

1. The plant material was ground in liquid nitrogen (2.5 g fresh tissue).
2. The extraction buffer was pre-heated with 0.3% mercaptoethanol and PVP (1.5%) powder added directly to the tube.
3. The ground material was transferred to 30 ml polypropylene tube; add 5 ml of the extraction buffer and mix.
4. The tubes were incubated at 65 °C for 45 min in a shaking water bath.
5. The tubes were then plunged into ice immediately and then brought to room temperature
6. An equal volume of chloroform: isoamyl alcohol (24:1) was added and mixed by inverting the tubes several times. This was kept for 10 min at room temperature with one or two intermittent inversions.
7. The tubes were then centrifuged at 12,000 g for 10 min at 25 °C, the aqueous phase carefully transferred and 2/3rd volume ice cold isopropanol added.
8. The tubes were incubated at −20 °C for 2 hr or kept overnight prior to centrifuging at 12,000 g for 10 min at 4 °C
9. The supernatant was discarded and 70% ethanol added, the pellet washed, dried and dissolved in nuclease free water.
10. The aqueous solution was transferred to 2 ml Eppendorf tubes and 10 µg/ml of RNase added to the DNA solution; it was incubated at 37 °C for 1 hr.
11. An equal volume of buffer saturated phenol: chloroform: isoamyl alcohol
12. (25:24:1) mix was added by inverting the tubes, the mixture being incubated at room temperature for 10 min.
13. The tubes were centrifuged at 12,000 g for 10 min at 4 °C, the supernatant extracted with an equal amount of chloroform: isoamyl alcohol (24:1), the aqueous phase transferred to a new Eppendorf tube and sodium acetate (3 M; 1/10 Vol) and ice cold isopropanol added.
14. This was kept at −20 °C for 30 min
15. It was then centrifuged at 12,000 rpm for 10 min at 25 °C, the supernatant discarded, 70% ethanol added and the pellet washed.
16. The final DNA pellet was air-dried and dissolved in nuclease free water.

Plant material – nutmeg & mace

Export grade seeds of *M. fragrans* and mace were collected from plantations in Grenada and used for DNA isolation as per the protocol given below:

1. Grind the nutmeg seeds or mace to a fine powder
2. Preheat extraction buffer (3% CTAB, 150 mM Tris, 30 mM EDTA, 2 M NaCl) at 60 °C with freshly prepared 0.3% β-mercaptoethanol and add PVP (1.5%) powder directly to the buffer.
3. Transfer 1 gm of ground material to a 30 ml polypropylene tube; add 5 ml of the extraction buffer and mix.
4. Incubate tubes at 65 °C for 90 min in a shaking water bath.

5. Plunge tubes in ice immediately and bring to room temperature.
6. Add an equal volume of buffer saturated phenol: chloroform: isoamylalcohol (25:24:1) and mix by inverting the tubes several times, incubate at room temperature for 15 min with one or two intermittent inversions.
7. Centrifuge the tubes at 12,000 rpm for 15 min at 25 °C.
8. Carefully transfer the aqueous phase to fresh tubes, and add an equal amount of chloroform: isoamylalcohol (24:1).
9. Mix the tubes and incubate at room temperature for 20 min with two or three intermittent inversions.
10. Centrifuge the tubes at 12,000 rpm for 15 min at 25 °C.
11. Carefully transfer the aqueous phase to fresh tubes, add 2/3rd volume of ice cold isopropanol.
12. Incubate the tubes at −20 °C for 2 hr or keep overnight.
13. Centrifuge the tubes at 12,000 rpm for 15 min at 25 °C.
14. Discard the supernatant and dissolve the pellet in nuclease free water.
15. Transfer the aqueous solution to 2 ml Eppendorf tubes and add 10 μg/ml RNase to the solution and incubate the samples at 37 °C for 1 hour.
16. Add an equal volume of buffer saturated phenol: chloroform: isoamylalcohol (25:24:1) and mix by inverting the tubes, incubate the tubes at room temperature for 10 min
17. Add an equal volume of chloroform: isoamylalcohol and mix by inverting tubes several times; incubated the tubes at room temperature for 10 min with one or two intermittent inversions.
18. Transfer the aqueous phase to a new Eppendorf tube, add 2/3rd vol of ice cold isopropanol and incubate the tubes at −20 °C for 2 hr.
19. Centrifuge the tubes at 12,000 rpm for 20 min at 25 °C.
20. Discard the supernatant; add 70% ethanol, wash the pellet.
21. Air-dry the DNA pellet and dissolve in nuclease free water.

PCR analysis

PCR parameters

The concentration of the extracted DNA was determined at 260 nm in a UV spectrophotometer (Nanodrop 2000 spectrophotometer). The purity of the DNA was determined at 260 nm/280 nm in a UV spectrophotometer (Nanodrop 2000 spectrophotometer).

RAPD analysis

Random amplified polymorphic DNA (RAPD) analysis was carried out using random decamer primers OPW-01 (5'CTCAGTGTCC3'), OPW-11 (5'CTGATGCGTG3'), OPW-12 (5'TGGGCAGAAG3') and OPW-15 (5'ACACCGGAAC3') in accordance with Williams et al. (1990) to establish the genetic diversity of *Myristica* sp. (plant) and to differentiate the population based on sex of the individuals. Different random decamer primers were screened to establish the genetic diversity of *Myristica* sp. (plant and nutmeg) and to differentiate the population based on sex of the individuals (Tables 9A.1).

TABLE 9A.1   Random decamer primers.

5′CCCAAGGTCC3′

5′GGTGCGGGAA3′

5′CCAGATGCAC3′

5′GTGACATGCC3′

5′TCAGGGAGGT3′

5′AAGACCCCTC3′

5′AGATGCAGCC3′

5′TCACCACGGT3′

5′CTTCACCCGA3′

5′CACCAGGTGA3′

5′GAGTCTCAGG3′

5′TTATCGCCCC3′

5′CCCGATTCGG3′

5′TGCGGCTGAG3′

5′ACGCACAACC3′

5′GGTGACTGTG3′

5′CTACTGCCGT3′

5′GGACTGCAGA3′

5′ACGGCGTATG3′

5′GGACCCTTAC3′

PCR parameters

Perform the RAPD reaction in a 25 µl reaction volume with 50 ng genomic DNA, 0.4 mM dNTPs, 10 picomols primer, 2.5 mM $MgCl_2$ and 0.5U Taq DNA polymerase.

Amplification parameters

Perform amplification in a programmable thermocycler with the following cycles:

Initial denaturation at 93 °C for 4 min
45 cycles of:
93 °C for 1 min
35 °C for 1 min
72 °C for 2 min

Final extension at 72 °C for 8 min

TABLE 9A.2 DNA isolation and purification of nutmeg leaves.

SAMPLE	ID	A260	A280	A320	A260/A280	CONC μg/ml	CONC μg/μl	CONC ng/μl
N01	E0161102	0.019	0.015	0.014	1.27	95.00	0.095	95.00
N02	E016201	0.030	0.038	0.028	0.79	150.00	0.150	150.00
N03	E016202	0.079	0.107	0.065	0.74	395.00	0.395	395.00
N04	N0180	0.097	0.116	0.091	0.84	485.00	0.485	485.00
N05	N01450	0.047	0.05	0.045	0.94	235.00	0.235	235.00
N06	N01530	0.029	0.035	0.039	0.83	145.00	0.145	145.00
N07	(HMF) WSTC001M1	0.075	0.07	0.075	1.71	375.00	0.375	375.00
N08	WSTC003F1	0.062	0.070	0.068	0.89	310.00	0.310	310.00
N09	WSTC002F1	0.080	0.075	0.058	1.07	400.00	0.400	400.00

Analysis of PCR products

The amplified PCR products were separated by at 5 v/cm for 2.5 hr on 1.5% agarose gel containing ethidium bromide 0.5 mg per ml and photographed under a UV transilluminator. The amplification products (bands) were scored with the presence of a band scored as 1 and its absence as 0. Based on these scores, the average similarity in the banding patterns (or amplification products) among trees within a sex and between the sexes were computed. The similarity index (SI) were computed as $SI = 2N_{ab}/(N_a + N_b)$, where, N_a and N_b refer to the total number of amplification products or bands present respectively in individuals "a" and "b" and N_{ab} refers to the total number of bands commonly shared by the two individuals "a" and "b".

Results

DNA isolation and purification of nutmeg leaves were undertaken using the CTAB method. The CTAB method was shown to give good quality and quantity of DNA. DNA yield from leaves of *Myristica* ranged from 95 ng/μl to 400 ng/μl. The quality of the DNA recovered was good, having little contamination with proteins and carbohydrates with the A260/A280 ratio ranging from 0.79 to 1.71 (Tables 9A.2).

RAPD analysis using four RAPD primers (OPW-01, OPW-11, OPW-12, OPW-15) shows a high diversity between the nine nutmeg plants that were analyzed (Fig. 9A.1-4). The four primer gave good amplification for analysis of genetic relationship among the accessions. Non-specific DNA banding was observed but was improved by adjusting the annealing temperature in the PCR. The CTAB method also proved useful for obtaining good quality DNA for further analysis such as RADP, AFLP, RFLP, among others.

M 1 2 3 4 5 M 6 7 8 9 10 M

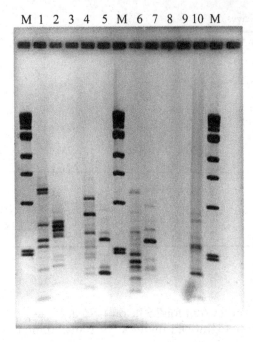

FIGURE 9A.1 Example of one of the RAPD profiles of DNA isolated from leaves of *M. fragrans* with one of the primers, OPW-01 (5'CTCAGTGTCC3'). M: marker (1 kb DNA ladder); lane 1: E0161102; lane 2: E016201, lane 3: E016202, lane 4: N0180, lane 5: N01450, M: marker lane 6: N01530' lane 7: (HMF) WSTC001M1, lane 8: WSTC003F1, lane 9: WSTC002F1, lane 10: NTC, M: marker.

References

Balasubramanian, S., Roselin, P., Singh, K.K., Zachariah, J., Saxena, S.N., 2016. Postharvest processing and benefits of black pepper, coriander, cinnamon, fenugreek, and turmeric spices. Crit. Rev. Food Sci. Nutr 56 (10), 1585–1607.

Cardoso-Ugarte, G.A., López-Malo, A., Sosa-Morales, M.E., 2016. Chapter 38 – Cinnamon (Cinnamomum zeylanicum) essential oils. In: Preedy, R. (Ed.), *Essential Oils in Food Preservation, Flavor and Safety*. Academic Press, Amsterdam, pp. 339–347.

Carstairs, S.D., Cantrell, F.L., 2011. The spice of life: an analysis of nutmeg exposures in California. Clin. Toxicol. 49 (3), 177–180.

CBI Market Intelligence, 2015a. CBI Market Channels and Segment: Spices and Herbs. CBI Ministry of Foreign Affairs, The Hague, The Netherlands, June 2015.

CBI Market Intelligence, 2015b. CBI Product Factsheet: Spices and Herb Mixtures in Europe. CBI Ministry of Foreign Affairs, The Hague, The Netherlands, October 2015.

Demetriades, A.K., Wallman, P.D., McGuiness, A., Gavalas, M.C., 2005. Low cost, high risk: accidental nutmeg intoxication. Emerg. Med. J. 22, 223–225.

Ehrenpreis, J.E., DesLauriers, C., Lank, P., Armstrong, P.K., Leikin, J.B., 2014. Nutmeg poisonings: a retrospective review of 10 years experience from the Illinois Poison Center, 2001–2011. J. Med. Toxicol. 10 (2), 148–151.

El-Alfy, A.T., Wilson, L., ElSohly, M.A., Abourashed, E.A., 2009. Towards a better understanding of the psychopharmacology of nutmeg: Activities in the mouse tetrad assay. J. Ethnopharmacol. 126 (2), 280–286.

Fries, G., Weiss, E., White, K., 2013. Agro-Logistics for Nutmeg and Cocoa Exports from Grenada – A Logistics Chain Approach. World Bank, <https://openknowledge.worldbank.org/bitstream/handle/10986/16702/826100WP0P1457720Box379867B00PUBLIC00ACS.pdf?sequence = 1/> (accessed 28.07.19).

Gordon, A., 2012a. The Grenada Nutmeg Industry: Support in the Implementation of Market Expansion. A presentation prepared for The Project Steering Committee, DFID/GOC/IADB Caribbean Competitiveness Project, Grenada Cooperative Nutmeg Association Board Room, Grenada, W.I. July 2012.

Gordon, A., 2012b. Report on biotechnology for the *Grenada Nutmeg Sector Project* under the Grenada Nutmeg Sector Support in the Implementation of Market Expansion Project, GDA/1101/R01/FO, St. Georges, Grenada, 2012.

Gordon, A. (Ed.). (2016). Food Safety and Quality Systems in Developing Countries: Volume II: Case Studies of Effective Implementation. Academic Press.

Gurtler, J.B., Keller, S.E., 2019. Microbiological safety of dried spices. Annu. Rev. Food Sci. Technol. 10, 409–427.

Gurtler, J.B., Keller, S.E., Kornacki, J.L., Annous, B.A., Jin, T., Fan, X., 2019. Challenges in recovering foodborne pathogens from low-water-activity foods. J. Food Prot. 82 (6), 988–996.

Hayfaa, A.A.S., Sahar, A.M.A.S., Awatif, M.A.S., 2013. Evaluation of analgesic activity and toxicity of alkaloids in Myristica fragrans seeds in mice. J. Pain. Res. 6, 611.

ITC. TradeMap, 2016. List of importers for the selected product 2014 to 2016. Product: 0904 – 0910. Available from: https://www.trademap.org/Country_SelProduct_TS.aspx?nvpm = 1%7c%7c%7c%7c%7c%7c0904%7c%7c%7c4%7c1%7c1%7c1%7c2%7c1%7c2%7c1%7c1.

ITC. TradeMap, 2019. List of importers for the selected product 2014 to 2018. Product: 0904 – 0910. Available from: https://www.trademap.org/Country_SelProduct_TS.aspx?nvpm = 1%7c%7c%7c%7c%7c%7c0909%7c%7c%7c4%7c1%7c1%7c1%7c2%7c1%7c2%7c1%7c1.

Jham, G.N., Dhingra, O.D., Jardim, C.M., Valente, V.M., 2005. Identification of the major fungitoxic component of cinnamon bark oil. Fitopatol. Brasil. 30 (4), 404–408.

Mang, B., Wolters, M., Schmitt, B., Kelb, K., Lichtinghagen, R., Stichtenoth, D.O., et al., 2006. Effects of a cinnamon extract on plasma glucose, HbA1c, and serum lipids in diabetes mellitus type 2. Eur. J. Clin. Invest. 36 (5), 340–344.

Marcelle, G.B., 1995. Production, Handling and Processing of Nutmeg and Mace and their Culinary Uses (No. SB298 M31). FAO Regional Office for Latin America and the Caribbean.

Mackin, K.M., 2018. Six brands of curry powder recalled because of excessive lead. 25 October 2018, Food Safety News. <https://www.foodsafetynews.com/2018/10/six-brands-of-curry-powder-recalled-because-of-excessive-lead/> (accessed 25.08.19).

Martins, C., Doran, C., Silva, I.C., Miranda, C., Rueff, J., Rodrigues, A.S., 2014. Myristicin from nutmeg induces apoptosis via the mitochondrial pathway and down regulates genes of the DNA damage response pathways in human leukaemia K562 cells. Chemico-biol. Interact. 218, 1–9.

Nguyen, P.H., Le, T.V.T., Kang, H.W., Chae, J., Kim, S.K., Kwon, K.I., et al., 2010. AMP-activated protein kinase (AMPK) activators from Myristica fragrans (nutmeg) and their anti-obesity effect. Bioorg. Med. Chem. Lett. 20 (14), 4128–4131.

Nichol, K., 2014. Provital's nutmeg-derived active offers non-invasive plumping of wrinkles <http://tinyurl.com/kgbk5su/> (accessed 25.04.14).

Pande, S., 2010. Essential oil of mystica fragrans as an effective fungitoxicant. J. Med. Aromatic Plant. Sci. 32 (4), 416–419.

Paranagama, P.A., Wimalasena, S., Jayatilake, G.S., Jayawardena, A.L., Senanayake, U.M., Mubarak, A.M., 2010. A comparison of essential oil constituents of bark, leaf, root and fruit of cinnamon (Cinnamomum zeylanicum Blum) grown in Sri Lanka. J. Natl Sci. Found. Sri Lanka 29, 3–4.

Piras, A., Rosa, A., Marongiu, B., Atzeri, A., Dessì, M.A., Falconieri, D., et al., 2012. Extraction and separation of volatile and fixed oils from seeds of Myristica fragrans by supercritical CO_2: Chemical composition and cytotoxic activity on Caco-2 cancer cells. J. Food Sci. 77 (4), C448–C453.

Purseglove, J.W., Brown, E.G., Green, C.L., Robbins, S.R.J., 1981. Spices, Vol. 2. Longman Group Ltd, pp. 447–813.

Quin, G.I., Fanning, N.F., Plunkett, P.K., 1998. Nutmeg intoxication. J. Accid. Emerg. Med. 15 (4), 287.

Rodianawati, I., Hastuti, P., Cahyanto, M.N., 2015. Nutmeg's (Myristica fragrans Houtt) oleoresin: effect of heating to chemical compositions and antifungal properties. Proc. Food Sci. 3, 244–254.

Selvakumari, E., Jenifer, J., Priyadharshini, S., Vinodhini, R., 2017. Application of DNA fingerprinting for plant identification. J. Acad. Ind. Res. (JAIR) 5 (10), 149.

Stein, U., Greyer, H., Hentschel, H., 2001. Nutmeg (myristicin) poisoning—report on a fatal case and a series of cases recorded by a poison information centre. Forensic Sci. Int. 118 (1), 87–90.

Tajuddin, S., Ahmad, S., Latif, A., Qasmi, I.A., Amin, K.M., 2005. An experimental study of sexual function improving effect of Myristica fragrans. Houtt BMC Complement Altern. Med. 5, 16.

Takikawa, A., Abe, K., Yamamoto, M., Ishimaru, S., Yasui, M., Okubo, Y., et al., 2002. Antimicrobial activity of nutmeg against Escherichia coli O157. J. Biosci. Bioeng. 94 (4), 315–320.

Thuong, P.T., Hung, T.M., Khoi, N.M., Nhung, H.T.M., Chinh, N.T., Quy, N.T., et al., 2014. Cytotoxic and anti-tumor activities of lignans from the seeds of Vietnamese nutmeg Myristica fragrans. Arch. Pharmacal Res. 37 (3), 399–403.

US Food & Drug Administration, 2019a. Import Alert 99-19. "Detention Without Physical Examination Of Food Products Due To The Presence Of Salmonella". US Dept. of Health and Human Services, 09/05/2019. <https://www.accessdata.fda.gov/cms_ia/importalert_263.html/> (accessed 07.09.19).

US Food & Drug Administration, 2019b. UBC Food Distributors Recalls Hot Curry Powder and Curry Powder Due to Lead. Recalls, Market Withdrawals, & Safety Alerts. US Dept. of Health and Human Services, 09/05/2019. <https://www.fda.gov/safety/recalls-market-withdrawals-safety-alerts/ubc-food-distributors-recalls-hot-curry-powder-and-curry-powder-due-lead/> (accessed 07.09.19).

Williams, J.G.K., Kubelik, A.R., Livak, K.J., Rafalski, J.A., Tingey, S.V., 1990. DNA polymorphisms amplified by arbitrary primers are useful as genetic markers. Nucl. Acids Res. 18, 6231–6235.

Webbe, J., Duncan, V., Forsyth, V., Bernard, L.S., 2003. Fortified nutmeg oil pain relief formulations. U.S. Patent Application No. 10/655,212, filed September 5, 2003.

Wijesekera, R.O., 1977. Historical overview of the cinnamon industry. CRC Crit. Rev. Food Sci. Nutr. 10 (1), 1–30.

Yiannas, F., 2008. Food Safety Culture: Creating a Behavior-Based Food Safety Management System. Springer Science & Business Media.

Further reading

CBI Market Information, 2013. Market insights for Indonesian spices. Available at <http://tinyurl.com/kteasuv/> (accessed 26.05.16).

CBI Market Information, 2018. Exporting nutmeg to Europe. CBI Ministry of Foreign Affairs, 13 September 2018. <https://www.cbi.eu/node/2108/pdf/> (accessed 02.08.19).

Doyle, J.J., Doyle, J.S., 1987. A rapid DNA isolation procedure for small quantities of fresh leaf tissue. Phytochem. Bull. 19, 11–15.

Gopalakrishnan, M., 1992. Chemical composition of nutmeg and mace. J. Spices Aromatic Crop. 1 (1), 49–54.

Inter-American Institute for Cooperation on Agriculture (IICA), 1997. Agriculture in Grenada: 1991-1995 and Beyond.

International Trade Centre (ITC), 2006. Marketing Manual and Web Directory for Organic Spices, Culinary Herbs and Essential Oils, second ed. International Trade Centre (ITC), Geneva, ITC, 2006. vi, 53 p. (Technical paper). Available at: <http://www.intracen.org/organics/technical-assistance-publications.htm/>.

International Trade Centre (ITC), 2015. US Spice Imports. International Trade Centre (ITC), Geneva, Switzerland.

Lele, S.S., 2011. Isolation and PCR amplification of genomic DNA from traded seeds of nutmeg (M. Fragrans). J. Biol. Agric. Healthc. 1 (2), Online.

NewHop360, 2014. Numelle SKIN ingredient backed by new sales deal. <http://newhope360.com/botanicals/numelle-skin-ingredient-backed-new-sales-deal/> (accessed 25.04.14).

Sheeja, T.E., George, K.J., Jerome, J., Varma, R.S., Syamkumar, S., Krishnamoorthy, B., et al., 2008. Optimization of DNA isolation and peR parameters in Myristica sp. and related genera for RAPD and ISSR analysis. J. Spices Aromatic Crop. 17 (2), 91–97.

Shibu, M.P., Uma Shaankar, R., Ganesshaiah, K.N., Ravishankar, K.V., Anand, L., 2001. Identification of sex-specific DNA markers in the dioecious tree, nutmeg (Myristica fragrans Houtt.). Plant. Genet. Resour. Newsl. 1, 59–61.

Smith, M., 2014. Nutmeg. Encyclopedia of Toxicology. pp. 630–631. Reference Module in BiomedicalSciences, from Encyclopedia of Toxicology (Third Edition), 2014, Pages 630-631.

UNIDO, FAO, 2005. Herbs, Spices and Essential Oils: Post-Harvest Operations in Developing Countries. United Nations Industrial Development Organization and Food and Agricultural Organization of the United Nations, Rome, Italy.

Van Gils, C., Cox, P.A., 1994. Ethnobotany of nutmeg in the Spice Islands. J. Ethnopharmacology 42 (2), 117–124.

Index

Printed in the United States
By Bookmasters